REMAKING PARTICIPATION

Changing relations between science and democracy – and controversies over issues such as climate change, energy transitions, genetically modified organisms and smart technologies – have led to a rapid rise in new forms of public participation and citizen engagement. While most existing approaches adopt fixed meanings of 'participation' and are consumed by questions of method or critiquing the possible limits of democratic engagement, this book offers new insights that rethink public engagements with science, innovation and environmental issues as diverse, emergent and in the making. Bringing together leading scholars on science and democracy, working between science and technology studies, political theory, geography, sociology and anthropology, the volume develops relational and co-productionist approaches to studying and intervening in spaces of participation. New empirical insights into the *making*, construction, circulation and effects of participation across cultures are illustrated through examples ranging from climate change and energy to nanotechnology and mundane technologies, from institutionalised deliberative processes to citizen-led innovation and activism, and from the global north to global south. This new way of seeing participation in science and democracy opens up alternative paths for reconfiguring and *remaking* participation in more experimental, reflexive, anticipatory and responsible ways.

This ground-breaking book is essential reading for scholars and students of participation across the critical social sciences and beyond, as well as those seeking to build more transformative participatory practices.

Jason Chilvers is Senior Lecturer and Chair of the Science, Society and Sustainability (3S) Research Group in the School of Environmental Sciences, University of East Anglia, UK.

Matthew Kearnes is an Associate Professor in the School of Humanities and Languages, University of New South Wales, Australia.

'The insightful chapters collected in this book show how concerns raised by technosciences provide a tremendous opportunity for remaking democracy. The editors and authors invite us to consider the so-called participatory turn neither as a masquerade nor as a mere social technology but as a global multisite construction place where new forms of collective life and government are imagined and experimented. A brilliant book that should be read by all those interested in the future of our planet.'

Michel Callon, Professor of Sociology, École des mines and Centre de sociologie de l'innovation, Paris, France

'Do not mistake the modesty advocated by this book for half-heartedness. Remaking Participation argues that we should expand our perspectives on participation, and need to get better at appreciating the incredible variety of locations, devices and genres with which participation is done in today's technological societies. This situation makes it necessary to "un-fix" our understanding of participation. In practice, participation often does not conform to the democratic ideal of participation that we know so well – it is not necessarily good, necessary, authentic. But neither would it do to declare that participation has turned into its opposite (that it has become co-opted, trivial, ineffective). Bringing together leading intellectual voices on science, technology and democracy, *Remaking Participation* shows that participation lies at the very heart of current technological, environmental and political transformations, and outlines a much needed research agenda that engages with the intensely ambivalent situations that result from this.'

Noortje Marres, Centre for Interdisciplinary Methodologies, University of Warwick, UK

'Modern societies remain hampered by myths about the relationship between science and democracy. The myths produce unwelcome practices, such as attempts to scientize political decisions or to discredit science by politicizing it. This landmark volume explodes the myths and shows how science and democracy can achieve a new relationship underpinned by the core value of public participation. It shows how and why science needs to rethink its relationship with society, and how societies can make science and democracy far more responsive to their needs and desires. The book takes readers to the cutting-edge of debates about the proper relationships between science and democracy. More than this, it also explores new territory, showing how science and democracy can be more richly infused with the practices of both. The editors and authors have together done a brilliant job of showing us what needs to change, and how. It will be a key reference for many years to come.'

Noel Castree, Professor of Geography, University of Wollongong, Australia and University of Manchester, UK

'Whether sparked by gene editing or geoengineering, fracking or food crops, arguments about the possibilities and pitfalls of advances in science and technology ripple through our societies with increasing frequency. How, and on what terms, experts, policymakers and wider publics engage in these debates is a topic of constant and fierce negotiation. In *Remaking Participation*, Jason Chilvers and Matthew Kearnes have brought together an exciting and original series of contributions from some of the leading thinkers in this field. The end result is a collection of rare quality, insight and relevance to real-world questions. It should be read by scholars, students, practitioners, policymakers, and all those who care about the future of science, technology and society.'

James Wilsdon, Professor of Science & Democracy, University of Sussex, UK and Chair, Campaign for Social Science

'"Participation" is the word that covers all sins, a term so elastic that it can be used to both challenge and legitimize any given decision-making process. *Remaking Participation* shows how to redeem this slippery concept and sharpen its critical edge. By examining in detail how citizens engage with controversial scientific and environmental issues, this book invites us to see the objects and the subjects of participation, the problems that trigger political action and the collectives that gather around them, as emergent, mutually constitutive realities. Far from being a recipe for relativism and detachment, the authors' embrace of the contingency that besets participatory democracy in the making reinvigorates the ideal of civic engagement and recasts the role of social scientists as participants in open-ended political experiments.'

Javier Lezaun, Deputy Director, Institute for Science, Innovation and Society, University of Oxford, UK

'This is the book that many have long been waiting for. It tackles head-on, some of the most important current issues at the meeting of social science and wider politics: What does participation mean? Where is it going? Transcending the usual dichotomised tropes, these essays take diverse and highly nuanced critically reflectively views – with many very practical implications. The conclusions are of enormous importance to all those academics and practitioners working in policy arenas touched by the language and practice of participation.'

Andy Stirling, Professor of Science and Technology Policy and Co-Director of the STEPS Centre, University of Sussex, UK

'Exercises of participatory technology assessment are a fascinating window onto relations of science, citizens, and state. Bringing together a rich diversity of cases and arguments, the book builds on the idea that public assessment of technology is a form of democratic experiment by analyzing the variety of ways in which this is so. In the process, we gain a useful theoretical framework for understanding the modern enterprise of "public engagement" as a co-constructive process of making publics, democratic idioms, and technoscience itself.'

David Winickoff, Director, Berkeley Program in Science & Technology Studies, USA

'This important book argues for a new approach to public participation in science and technology, one which understands participation as co-produced, relational and emergent. Written by the leading contributors in the field, and combining theoretical depth with engaging empirical material, this refreshing and timely collection is essential reading for all those concerned with science, innovation and democracy.'

Jane Calvert, Science Technology & Innovation Studies,
School of Social and Political Science, University of Edinburgh, UK

'Participatory politics are all the rage. This is especially the case when science, technology, corporate and political power shape innovation and policymaking. Such forces also manipulate opinion and even political and social outlooks. So the very act of participation could, in the wrong hands, reinforce the tools of power and influence. Jason Chilvers and Matthew Kearnes are very much alive to these dangers. They have brought together an impressive array of contributors who show that effective participation can be truly revolutionary and politically transforming. They are all on their guard that such a rewarding outcome has constantly to be fought for and reinvented through genuine partnerships and dialogue. The ultimate test is how far power is progressively shared and social justice genuinely created.'

Tim O'Riordan, Emeritus Professor, School of Environmental Sciences,
University of East Anglia, UK

'Participation is a key field within the study of international development. This book adds significantly to existing approaches to participation by adding insights from science and technology studies and theories of democracy. It should be read by students and analysts working on international development, and anyone interested in participation as a research and policy tool.'

Tim Forsyth, Department of International Development,
London School of Economics and Political Science, UK

REMAKING PARTICIPATION

Science, environment and emergent publics

Edited by Jason Chilvers
and Matthew Kearnes

Routledge
Taylor & Francis Group
LONDON AND NEW YORK

First published 2016
by Routledge
2 Park Square, Milton Park, Abingdon, Oxon OX14 4RN

and by Routledge
711 Third Avenue, New York, NY 10017

Routledge is an imprint of the Taylor & Francis Group, an informa business

© 2016 selection and editorial matter, Jason Chilvers and Matthew Kearnes; individual chapters, the contributors

The right of the editors to be identified as the authors of the editorial material, and of the authors for their individual chapters, has been asserted in accordance with sections 77 and 78 of the Copyright, Designs and Patents Act 1988.

All rights reserved. No part of this book may be reprinted or reproduced or utilised in any form or by any electronic, mechanical, or other means, now known or hereafter invented, including photocopying and recording, or in any information storage or retrieval system, without permission in writing from the publishers.

Trademark notice: Product or corporate names may be trademarks or registered trademarks, and are used only for identification and explanation without intent to infringe.

British Library Cataloguing-in-Publication Data
A catalogue record for this book is available from the British Library

Library of Congress Cataloging-in-Publication Data
Names: Chilvers, Jason, editor of compilation. | Kearnes, Matthew, editor of compilation.
Title: Remaking participation : science, environment and emergent publics / edited by Jason Chilvers and Matthew Kearnes.
Description: Abingdon, Oxon ; New York, NY : Routledge is an imprint of the Taylor & Francis Group, an Informa business, [2016] | Includes bibliographical references and index.
Identifiers: LCCN 2015020095| ISBN 9780415857390 (hardback) | ISBN 9780415857406 (pbk.) | ISBN 9780203797693 (ebook)
Subjects: LCSH: Science and state--Citizen participation. | Technology and state--Citizen participation. | Science--Decision making--Citizen participation. | Technology--Decision making--Citizen participation.
Classification: LCC Q125 .R415 2016 | DDC 338.9/26--dc23
LC record available at http://lccn.loc.gov/2015020095

ISBN: 978-0-415-85739-0 (hbk)
ISBN: 978-0-415-85740-6 (pbk)
ISBN: 978-0-203-79769-3 (ebk)

Typeset in Bembo
by GreenGate Publishing Services, Tonbridge, Kent

Printed and bound in Great Britain by
TJ International Ltd, Padstow, Cornwall

CONTENTS

List of illustrations ix
Notes on contributors x
Preface xiii

1 Science, democracy and emergent publics 1
   Jason Chilvers and Matthew Kearnes

PART I
Rethinking participation 29

2 Participation in the making: rethinking public engagement
   in co-productionist terms 31
   Jason Chilvers and Matthew Kearnes

3 Engaging in a decentred world: overflows, ambiguities
   and the governance of climate change 64
   Alan Irwin and Maja Horst

4 Engaging the mundane: complexity and speculation
   in everyday technoscience 81
   Mike Michael

5 Ghosts of the machine: publics, meanings and social science
   in a time of expert dogma and denial 99
   Brian Wynne

PART II
Making participation — 121

6 State experiments with public participation: French
   nanotechnology, Congolese deforestation and the
   search for national publics — 123
   Véra Ehrenstein and Brice Laurent

7 Technologies of participation and the making
   of technologized futures — 144
   Linda Soneryd

8 Participation as pleasure: citizenship and
   science communication — 162
   Sarah R. Davies

9 The temporal choreographies of participation: thinking
   innovation and society from a time-sensitive perspective — 178
   Ulrike Felt

PART III
Remaking participation — 199

10 An 'experiment with intensities': village hall reconfigurings
   of the world within a new participatory collective — 201
   Claire Waterton and Judith Tsouvalis

11 Against blank slate futuring: noticing obduracy in the city
   through experiential methods of public engagement — 218
   Cynthia Selin and Jathan Sadowski

12 Reflexively engaging with technologies of participation:
   constructive assessment for public participation methods — 238
   Jan-Peter Voß

13 Remaking participation: towards reflexive engagement — 261
   Jason Chilvers and Matthew Kearnes

Index — *289*

ILLUSTRATIONS

Figure

12.1	The innovation journey of citizen panels as a truncated process of 'aggregating' technical knowledge and practices of participation	243

Tables

12.1	Attendees of the Challenging Futures of Citizen Panels workshop, 26 April 2013, Berlin	248
12.2	Overview of scenarios on the future development of citizen panels	249
12.3	Agenda of the Challenging Futures of Citizen Panels workshop, 26 April 2013, Berlin	250

Box

12.1	Headings of issue descriptions for an extended innovation agenda for citizen panels	251

CONTRIBUTORS

Jason Chilvers is a Senior Lecturer, and Chair of the Science, Society and Sustainability (3S) Research Group, in the School of Environmental Sciences at the University of East Anglia. His work, situated in the disciplines of science and technology studies (STS), geography and environmental science, focuses on relations between science, technology and society, including studies of governance, appraisal and public participation relating to science and sustainability. Jason has published widely on these themes in books, journal articles and policy reports, and was director of the ESRC 'Critical Public Engagement' Seminar Series in 2009–2011.

Sarah R. Davies is Marie Curie Research Fellow in the Department of Media, Cognition and Communication at the University of Copenhagen, where her work focuses on science communication and public engagement with science. Her publications include the edited volumes *Science and its Publics* (2008) and *Understanding Nanoscience and Emerging Technologies* (2010). Her current research explores hacking and hackerspaces, science communication and scientific citizenship, and care and caring within scientific practice.

Véra Ehrenstein is a post-doctoral research fellow in Sociology at Goldsmiths, University of London. Her work builds on STS to explore globalised market interventions and political actions in relation to climate change, vaccination and agriculture.

Ulrike Felt is Professor of Science and Technology Studies and Dean of the Faculty of Social Sciences at the University of Vienna. Her research interests are centred on issues of governance and public participation in technosciences and science communication, as well as changing knowledge politics and research cultures. Her work is often comparative between national contexts and technological or scientific fields (especially biomedicine and health, life sciences, nanotechnologies and sustainability research). Between 2002 and 2007 she was editor-in-chief of the journal *Science, Technology & Human Values*.

Maja Horst is Professor of Science Communication and Head of the Department of Media, Cognition and Communication at the University of Copenhagen. Her research interests include science communication, public engagement with science and responsible research and innovation.

Alan Irwin is a Professor in the Department of Organization at Copenhagen Business School. His research focuses on science and democracy, scientific governance and the enactment of socio-technical futures.

Matthew Kearnes is an Australian Research Council Future Fellow and member of the Environmental Humanities group at the School of Humanities and Languages, University of New South Wales. His research is situated between the fields of STS, human geography, environmental sociology and contemporary social theory. His current work is focused on the social and political dimensions of technological and environmental change, and he has published widely on the ways in which the development of novel and emerging technologies are entangled with profound social, ethical and normative questions. Matthew is a co-editor of the international open-access journal *Environmental Humanities* (environmentalhumanities.org).

Brice Laurent is a researcher at the Center for the Sociology of Innovation at Mines ParisTech. His work focuses on the relationship between the making of science and the construction of democratic order.

Mike Michael is Professor of Sociology and Social Policy at the University of Sydney. Current research includes the development of speculative methodology in relation to the study of science–society dynamics and the potentialities of everyday life. His most recent book (co-authored with Marsha Rosengarten) is *Innovation and Biomedicine: Ethics, Evidence and Expectation in HIV* (Palgrave, 2013). He is a co-editor of *The Sociological Review*.

Jathan Sadowski is a PhD student in the Human and Social Dimensions of Science and Technology, Consortium for Science, Policy and Outcomes, Arizona State University. His research primarily focuses on social justice and the political economy of technology, and he's writing a dissertation on 'smart cities'.

Cynthia Selin leads the research programme on Anticipation and Deliberation at the Center for Nanotechnology in Society at Arizona State University. She is currently investigating energy foresight as a Marie Curie Fellow at the Technical University of Denmark.

Linda Soneryd is Associate Professor and Lecturer in Sociology at the Department of Sociology and Work Science, University of Gothenburg. Her research interests include methods for public participation and risk regulation in relation to science and technology and environmental decision-making.

Judith Tsouvalis is a cultural geographer currently working as a Research Associate on the Leverhulme programme 'Making Science Public' in the School of Sociology and Social Policy at the University of Nottingham in the UK.

Jan-Peter Voß is Professor of Sociology of Politics at the Technische Universität Berlin. He works on politics and science as generic modes of authority generation and collective ordering, and on their changing relations throughout history and across cultures. Empirical research foci are the governance of innovation (science and technology related with issues of energy, environment and sustainable development) and innovation in governance (development of models such as environmental markets, deliberative democracy and experimental transition management).

Claire Waterton is a Senior Lecturer within the Sociology Department at Lancaster University in the UK. Her research is concerned with the use of the theory of STS to rethink ongoing environmental issues, problems and controversies.

Brian Wynne is Emeritus Professor of Science Studies at the Centre for the Study of Environmental Change (CSEC) at Lancaster University. His research has covered technology and risk assessment, public risk perceptions and public understanding of science, focusing on the relations between expert and lay knowledge and policy decision-making. He was awarded the John Desmond Bernal Prize by the Society for Social Studies of Science in 2010.

PREFACE

The origins of this book date back to early 2008 when we proposed the idea of an international seminar series on 'critical public engagement' to the UK Economic and Social Research Council (ESRC). Our intervention was born out of a sense of concern and frustration with the ways in which burgeoning forms of public participation in issues of science, technology and the environment were being characterised in both scholarship and practice. This was by no means a new phenomenon at the time. The four preceding decades had seen a global rise in participatory practices across diverse fields – such as urban planning, natural resource management, environmental decision-making, technology assessment and international development – which were in turn rooted in longer historical trends in citizen mobilisation and engagement with science and democracy. The beginning of the twenty-first century was a time when such developments seemed to be intensifying as science and policy institutions increasingly struggled to maintain authority and credibility in the face of the radical uncertainties, complexities, inequalities and ambivalences that characterised late modern society, particularly its responses to environmental and technological risks. Indeed, we had both been involved in developing participatory approaches through earlier work at two leading research centres,[1] alongside a shared commitment to studying the emergence of participatory practices and mediators through more reflective and interpretive analyses.

Our concern was motivated by what we sensed were the relatively caricatured ways in which public participation had been depicted in scholarly reflection, policy treatment and practice. The almost breathless celebration of the transformative potential of participation, followed quickly by arguments for *more participation*, alternated with a more critical impulse to demarcate the proper limits of public involvement in governing science-related issues and efforts to distinguish lay knowledge from expert judgement. While at the same time, it seemed to us that institutional commitments to public participation, and indeed to the democratisation of science and democracy, had tended toward a kind of methodological revisionism coupled with relatively instrumental deployments of participatory methods for legitimatory ends, 'locked in' to narrow meanings of citizenship. Our sense of frustration was also prompted by a barrage of trenchant critiques, often highly dismissive and voiced by commentators quite removed from the experiments and cultures of participation in action, signalling 'the end of participation' and its replacement with something better – something *more democratic*.

Having been involved in these debates from the disciplinary perspective of science and technology studies (STS), we were acutely aware how this disciplinary field – largely through its attention to the ways in which science and technology are always 'socially' constituted and the often partial ways in which conflicts between institutionally sanctioned forms of authority and lay knowledge are resolved – had played some part in moves to democratise science and democracy. Yet a particular blind spot had formed. What had been missing from these debates was a sustained attempt to utilise the tools of symmetrical and co-productionist analysis – that had so vividly demonstrated the mutual construction of science and social order – to understand how 'participation' itself is also actively constructed and in the making. Participation was being taken for granted. Just as science is not simply concerned with the representation of truths about nature and the physical world, but is entangled with all manner of social and moral dimensions, so too it seemed to us that the use of participatory methods was not simply confined to the collation and representation of 'public truths'. It seemed that proliferating participatory practices were not only attempting to better represent publics and society but were simultaneously enacting visions of how the world ought to be, not least in terms of the versions of democracy that should make up our collective futures. Along with colleagues in STS and other critical social sciences, we saw a need to treat participation, and its construction, circulation and effects, as an object of study in itself, and through this open up new possibilities for its remaking.

In keeping with this ambition, the ESRC series – a two-year programme of five seminars between 2009 and 2011 – sought to 'consolidate a new research agenda that is more constructively critical about the potentials and pitfalls of public engagement in science and environmental risk'.[2] Rather than shaping a distant critique that perpetuated divisions between critical social science and policy practice, we sought a more constructive arrangement in the form of a collective experiment 'bringing together an interdisciplinary range of social scientists, in collaboration with scientists, participatory practitioners and policy-makers'. In a launch event at the University of Birmingham we debated the key themes and agendas to be taken forward in the series. This was followed by three workshops to explore the making and remaking of participation around emerging technologies,[3] natural hazards[4] and energy futures.[5] A final workshop at the Royal Society in London, organised in collaboration with the UK Sciencewise Expert Resource Centre, drew out the wider implications of the series for theory and reflexive practice. The series brought together over 250 critical social scientists (including from STS, geography, sociology, anthropology and political science), natural and physical scientists, participatory practitioners, science communicators, representatives from civil society organisations, and policy-makers in government and industry – mainly from the UK, but also from all corners of the globe.

This book does not represent a direct reporting or documentation of the seminar series. The main outputs of the seminar series are set out in the above-cited reports. The series was also seen as an important outcome in and of itself – a lively collective experiment where connections were made and transformations occurred within and between actors located in diverse disciplinary and institutional settings, each with their own distinct cultures, identities and purposes. As we argue towards the end of the book, this is one of many possible forms of collective experimentation necessary for remaking participation. Our intention with this particular volume as a whole, however, is to be deliberately more scholarly and academic in emphasis. What became clear through the seminar series was the crucial value and critical importance of making space for the necessary intellectual, theoretical and empirical work to rethink participation in relational and co-productionist terms. Whereas dominant perspectives assume specific pre-given models of participation and 'the public', the authors in this book develop alternative perspectives on the realities of participation as being multiple, situated, constructed and in the making. They use these perspectives to understand how participation is not a fixed or external category, but is always being made and remade through the performance of situated participatory practices and experiments, through the standardisation of participation technologies and expertise, through controversies, and in relation to political power and culture.

It is our conviction in this volume, then, that developing the necessary theoretical and interpretive-analytical resources to take forward an empirically oriented practice-based perspective on participation in the making provides the foundations and wellsprings for remaking participation in more reflexive and transformative ways. Having said this, the core ideas of the book are firmly grounded in, shaped by and reflect the spirit of the seminar series. Many of those who presented papers or delivered speeches in the series have authored chapters in this volume. The book also represents how the collective of the seminar series expanded through its lifetime and after, building solidarities with other groups working on similar agendas, enrolling additional authors and becoming more international in scope. We therefore intend that the contributions across this volume will not only be of immediate practical relevance but, we hope, also be of longer-term significance, looping back into ongoing participatory co-productions over time in unpredictable ways.

Which brings us to the cover design. The illustration on the front cover echoes and draws inspiration from what is perhaps one of the most iconic images to critique participation in the latter half of the twentieth century. The cover image references a poster produced as part of the French student-worker protests of May 1968 by an anonymous collective of art students and faculty staff under the name of Atelier Populaire. The silkscreen print poster – one of over 200 influential graphic artworks designed by the studio as part of the struggle to catalyse new levels of citizen action and resistance – featured a hand holding a paint brush writing out the radical statement: 'je participe, tu participes, il participe, nous participons, vous participez ... ils profitent' (in English: 'I participate, you participate, he participates, we participate, you participate ... they profit'). This image and text has had a powerful resonance in the participation field. Its immediate influence on participatory theory and practice was perhaps most evident as an inspiration and reference point for Arnstein's influential 'Ladder of Participation' – a typology 'arranged in a ladder pattern with each rung corresponding to the extent of citizens' power in determining the plan and/or program'.[6] A form of critique in itself, the ladder has instilled a particular vision of participatory progress, whereby climbing to the higher rungs leads to increasing devolution of power from the 'powerholding' institutions to the 'powerless' citizens. Arnstein's ladder, alongside writings on participatory and deliberative democracy, have been hugely influential in bringing forward new waves of methodological innovation to design 'better', 'more empowering', participatory processes. It stands for a perspective on participatory politics that still remains dominant today.

Our reference to the striking Atelier Populaire image on the front cover symbolises the multiple meanings of remaking participation that we promote in this book. As noted above, the authors in this volume develop an altogether different view of the reality of participation – and interpretation of the iconic poster itself – compared to Arnstein and others. In this volume we see power not as held, but as relational, tied up in the relations between all people and things, and working through the mediation of all collectives of participation. We see the multiple realities of participation not simply as externalised, fixed and pre-given models – often defined as graduating modes of citizen participation, as depicted in Arnstein's ladder[7] – but as actively and materially made and remade through the performance of situated participatory practices. For us this is more in keeping with the construction of Atelier Populaire's 'je participe ...' poster which was always seen by the group as inseparable from the situated collective struggles and ongoing processes of (re)making cultural and political action, not something to be externally validated as a final outcome of the events in May 1968.[8]

As one might expect, the cover image has many other resonances with the multiple meanings of 'Remaking Participation'. Rather than seeing the situated protest that produced the original image as a discrete event, with a particular outcome that impacted on institutional decision-making, our co-productionist and relational understanding of participation would highlight its wider significance in being shaped by – and shaping – French political culture and constitutional arrangements. For example, the poster references a shared cultural experience in which French school students were instructed in French language through the repetition of verb forms and tenses. In the context of the 1968 protests this experience was recast as a form of resistance. The political situation of the day – with increasing levels of unemployment and poverty under a conservative government and political regime – created a setting conducive to spontaneous uninvited participation that overflowed institutional framings. Yet, this specific instance of engagement formed part of wider ecologies of participation, that in turn contributed to transformations in the reconstitution of the French state itself. The civil unrest of 1968 almost brought the country and its government to a standstill, and has had a lasting impact on French history, culture and society. These instances of public engagement were, and still are, about much more than progressing up a ladder of empowerment in specific institutional 'decision moments'. Indeed, the book cover is testament to how participation and its products are never only localised ephemeral events – but are always acting and circulating beyond their sites of performance as images, imaginaries, techniques, philosophies, artefacts and so on. In this instance the Atelier Populaire image has travelled and been remade in a way that is inseparable from the situated collective of participation that led to the formation of this book.

Forming a collective publication from a wider collective experiment by way of a transdisciplinary seminar series depends on much hard work, patience, goodwill and good humour of so many people. We are eternally grateful to the authors for sharing in the vision of this book project, for responding so well to our various demands, and for their commitment to producing such exemplary chapters. We owe a debt of gratitude to colleagues at Earthscan/Routledge, particularly the commissioning editor Louisa Earls for believing in the book in the first place, and editorial assistant Helen Bell for all her help and guidance in seeing the project through. Karen Wallace, along with her copy-editing and typesetting team at GreenGate Publishing, guided the book seamlessly through the final production process. Four anonymous reviewers also provided helpful and encouraging comments on our initial book proposal.

We thank the five co-applicants on our initial proposal for the seminar series – Jacquie Burgess, Judith Petts, Jack Silgoe, Sigrid Stagl and Andy Stirling – some of whom could not stay with the series for its duration but were always there in inspiration and support. Special thanks to Andy for his guidance and inputs throughout the series and for hosting the Sussex event. Dawn Turnbull, Sylvia Sheppard, Rachel Hurren and Eileen Jackson provided invaluable administrative support to help with the day-to-day running of the seminar series. The final event at the Royal Society in London greatly benefited from the generous efforts of Diane Warburton and Lindsey Colborne who collaborated in designing and facilitating the workshop. We also thank all the speakers and participants for their energy and enthusiasm in making the seminar series such a success. The series was made possible by an ESRC grant (award number RES-451-26-0623).

Beyond the series itself it would be impossible to name all our various friends and colleagues who have shaped how the book has turned out. Jason thanks past and present members of the Science, Society and Sustainability (3S) research group at UEA for providing a vibrant and critical environment in which to develop the ideas presented in this volume, and especially Noel Longhurst, Helen Pallett, Rob Bellamy and Tom Hargreaves. Discussions with Sheila Jasanoff during Jason's visit to the Science, Technology and Society Program at the Harvard Kennedy School of Government in 2012 played a strong formative role at an important stage in the book's development, for which he is very grateful. So too did his ongoing collaborative work with a small collective of European scholars on 'technologies of participation', namely Brice Laurent, Jan-Peter Voß, Linda Soneryd and Sonja van der Arend. Jason also acknowledges his work funded under the EPSRC Realising Transition Pathways Project (award number EP/K005316/1) which helped shape ideas developed in this book project. Matthew thanks the members of the UNSW environmental humanities research group for provoking stimulating discussions around the themes of this volume. Also special thanks to Phil Macnaghten and Sarah Davies for ongoing discussion and work on the 'narrative turn' in public responses to emerging technologies, and thanks to the Australian STS community, particularly Brian Cook, Lauren Rickards, Georgia Miller, Adam Lucas, David Mercer, Nicola Marks, Declan Kuch and Darrin Durant, for at times challenging discussions of the themes of this volume. Matthew is also grateful for an ARC Future Fellowship (award number FT130101302) and his involvement in the ARC Centre of Excellence in Convergent Bio-Nano Science (award number CE140100036) which provided a venue for trailing some of the ideas contained in this volume.

Finally, a huge thank you to the most important people in our lives – Hannah, Max and Flora; and Megan, Matilda and Ceara – for their love, support and inspiration, and for putting up with the trials and tribulations of two book editors working together from either side of the world!

Jason Chilvers, Norwich, and Matthew Kearnes, Sydney

Notes

1 One of us (Chilvers) had been based at the Environment and Society Research Unit (ESRU) at University College London and the other (Kearnes) with the Centre for the Study of Environmental Change (CSEC) at Lancaster University.
2 Chilvers, J. (2009) *Critical Studies of Public Engagement in Science and the Environment: Workshop Report*. Norwich: School of Environmental Sciences, UEA, https://ueaeprints.uea.ac.uk/id/eprint/38125 (accessed 8 May 2015). Further information about the seminar series is available at: http://3sresearch.org.
3 Kearnes, M. and Chilvers, J. (2010) *Democracy, Citizenship and Anticipatory Governance of Science and Technology*. A report of the ESRC Critical Public Engagement Seminar, 15 December 2009, Durham University. Norwich: School of Environmental Sciences, UEA, https://ueaeprints.uea.ac.uk/id/eprint/38126 (accessed 8 May 2015).
4 Chilvers, J. (2010a) *Natural Hazards and Critical Public Participation*. A report of the ESRC Critical Public Engagement Seminar, 10 June 2010, University of East Anglia. Norwich: School of Environmental Sciences, UEA, https://ueaeprints.uea.ac.uk/id/eprint/38127 (accessed 8 May 2015).
5 Chilvers, J. (2010b) *Participation, Power and Sustainable Energy Futures*. A report of the ESRC Critical Public Engagement Seminar, 26 October 2010, SPRU, University of Sussex. Norwich: School of Environmental Sciences, UEA, https://ueaeprints.uea.ac.uk/id/eprint/38128 (accessed 8 May 2015).
6 Arnstein, S. (1969) 'A ladder of citizen participation', *Journal of the American Institute of Planners* 35(4): 216–224 (quote appears on p216).
7 Ibid.
8 Atelier Populaire (1969) *Posters from the Revolution: Paris, May 1968*. London: Dobson Books.

1

SCIENCE, DEMOCRACY AND EMERGENT PUBLICS

Jason Chilvers and Matthew Kearnes

A central feature of democratic systems of government wherever they have emerged – in a polis of ancient Greece, the parliamentary democracies of the eighteenth century or the cosmopolitan systems of governance that characterize late modernity – is the often partial accommodation of competing forms of moral, political and scientific authority. At least since the seventeenth century and the stuttering emergence of a modern scientific and technological enterprise, a central problematic for democratic politics has been the potential conflagration between scientific expertise, on the one hand, and popular representation on the other. In theory (and we must stress the word 'theory' here) democracies function not as a system of rule, or indeed government, but as a procedural solution to the problem of authority, with elaborately designed systems of delegation, accountability and oversight. Of course, in practice, democracy has remained anything but a static system of procedural administration. The transformations of democratic systems of government throughout the latter half of the twentieth century – that were precipitated by the failures in, and attempted recoveries and resuscitations of, these systems of bureaucratic resolution – have set the stage for a more reflexive pattern of relations between science and the projects of political ordering.

This volume is situated in the context of fundamental challenges to the accommodations that have, to date, constituted the kinds of institutional fixes deployed to proscribe and balance competing forms of authority and expertise. In a recent review Jasanoff suggests that though 'scientific ways of knowing have given rise to a politics of demonstration that modern nation states found supremely useful … the optimistic alliance between science, technology and democracy proved short-lived' (Jasanoff 2012, 2). The comforting assurance that the development of scientific understanding would result in socially progressive discoveries, and that in times of crisis bold assertions of scientific process and methodological rigour would carry the day, have seemingly been supplanted by a more questioning and ambivalent outlook. Technoscientific controversies – around objects such as chemical fertilizers, nuclear power plants, thalidomide and the production of chlorofluorocarbons resulting in holes in the ozone layer – revealed the downsides of science and innovation. At the same time concerns about the politicization of science in debates over anthropogenic climate change, data security, global pandemics and HIV research – together with iconic cases of scientific fraud (both alleged and actual) – have challenged the presumed neutrality of expert judgement. As Bijker *et al.* (2009, 1) have argued, we are living in profoundly paradoxical times where 'the cases in which scientific advice is asked most urgently are those in which the authority of science is questioned most thoroughly'.

A characteristic response to these shifting relations between science and democratic politics, and the emergence of a more reflexive attitude towards notions of scientific and technological progress, is the now commonplace argument that science needs to be made publically accountable in order that its accomplishments will be viewed as more socially acceptable. While economic competitiveness, environmental sustainability and advances in medical intervention are all said to depend on scientific and technological development, the future, as Marilyn Strathern (2005, 465) has eloquently argued, is increasingly 'forecast as fragile'. When it comes to matters of science, technology and the environment, it is increasingly apparent that it is no longer possible to operate in closed or secluded settings where public interest or social utility can be simply presumed. At the same time, contemporary systems of science policy-making and environmental regulation are no longer guaranteed by formal procedures of political representation. Over the past four decades or so, science *and* democracy have been increasingly opened up to diverse forms of public engagement and participation and wider civic scrutiny. Indeed, some have begun to suggest that the peculiar cultural and political sensitivities that characterize a distinctly late modern attitude towards science, technology and the environment represent an 'age of participation' (Einsiedel and Kamara 2006; Gottweis 2008; Blowers and Sundqvist 2010; Delgado *et al.* 2011), epitomized by the redistribution of expertise and attempts to incorporate a range of alternative actors and knowledges into processes of techno-political decision-making (Leach *et al.* 2005; Hagendijk and Irwin 2006; Irwin 2006; Callon *et al.* 2009).

From a position of relative obscurity, public relations with and understandings of science – accompanied by institutionalized practices designed to understand, move and govern publics – have become a characteristic feature of modern statecraft. So much so that it is now difficult to find processes of decision-making on science and environment-related issues that are not accompanied by forms of public participation and engagement. Take, for example, the case of national energy systems. In contrast to the relative degrees of public support, or at least lack of overt societal resistance, that accompanied the post-World War II development of modern energy systems in Western democracies and the initial establishment of nuclear power generation (Welsh and Wynne 2013), today the 'political situation' (Barry 2012) surrounding energy and low carbon transitions animates a multiplicity of public engagements with science, policy-making and innovation. These range from institutional moves to both characterize public opinion and motivate public behaviour change through deliberative consultation processes, opinion polls, e-democracy, citizen science, open innovation and the co-design of energy technologies, through to explicitly 'citizen-led' engagements in the form of protests, activism, alternative social movements, community energy initiatives and other forms of distributed innovation. The list goes on. What is striking about this particular example is the sheer *diversity* of contemporary forms of participation engaged by publics and civil society organizations in domains that have traditionally been the preserve of science, industry and the state. In these conditions it is often argued that science and democracy need to be rendered *more* democratic, *more* politically accountable, *more* publicly transparent and *more* socially responsive in order that they might be more successfully scientific and democratic.

In the context of this proliferation of participatory politics it is striking that much recent commentary continues to distinguish 'science' from 'politics' as two separate realms of human endeavour and collective action. While science tends to be viewed in objective terms, democratic forms of public participation are presented as necessary in limiting the transgressive potential of technological innovation and as providing an accountability mechanism that functions as a counterweight to systems of technical expertise.[1] However, viewed from a historical vantage point, this separation between science, politics and democracy appears to be a myopic image at best. In his seminal study of the interweaving histories of liberal democracy and experimental science, Ezrahi (1990) argues that the mutual entanglement of science and contemporary forms of social and political order have made it possible to resolve the competing demands of modern governance. Ezrahi identifies the ways in which an emergent 'attestive' visual culture gradually came to replace the 'celebratory' visual aesthetics of monarchical power. Rooted in the practices of public witnessing, a mode of scientific truth telling that emerged with the modern nation state, this attestive culture is premised on a notion of the popular equality of perception – a nascent political philosophy that would become a rhetorical hallmark of distinctly liberal modes of democratic participation. In the centuries that ensued, science came to adopt a central and privileged position in the establishment of democratic modes of social and political order, as a driver of social progress and as a source of political legitimacy. Working in the same historical terrain as Ezrahi, Shapin and Schaffer (1985, 332) conclude their groundbreaking study of the interweaving of political thought and emergence of experimental modes of scientific research with the striking claim that 'solutions to the problem of knowledge are solutions to the problem of social order', while the 'history of science occupies the same terrain as the history of politics'. For Shapin and Schaffer the development of new conventions for authorizing scientific knowledge were tied to the emergence of a new kind of political space characterized by the performance and public witness of experimental knowledge.

As has been well documented elsewhere (Latour 1987; Jasanoff 2004), Ezrahi's and Shapin and Schaffer's insights regarding the mutually reinforcing constitution of scientific and political authority provide the basis for co-productionist and pragmatist accounts of the relationship between science and democracy. Drawing on this body of work, in this volume we view the development of new modes of public engagement with science and the environment as deeply imbricated in the idiosyncratic, and at times uncomfortable but always mutually reinforcing, relationship between science and social order. It is the phenomenon of 'participation' writ large that forms the focus of this volume. However, rather than see 'participation', 'engagement' or indeed 'democracy' as outside or external to science or politics, with Shapin and Schaffer and Ezrahi we view these developments as an index of the contemporary social and cultural conventions that shape relationships between science and democracy, and the means through which societies legitimate claims to both political and epistemic authority. At the same time we argue that attempts to deal with the aforementioned ruptures, tensions and disconnects between science and democracy have been productive in bringing forward whole areas of (social) scientific scholarship, professional communities, expert bodies, industries, (non-) governmental organizations and social movements involved in constituting diverse forms of 'public participation' (see Chilvers 2008; Laurent 2011).

Developing the insights of co-productionist readings of the interweaving of science and the political, in this volume our goal is to both reflect on and move beyond the imperatives and dominant imaginaries of participation evident in much existing scholarship and practice in this mushrooming area of interest and activity. Simply put, contemporary analyses of public participation have been dominated by work focused on the development and extension of participatory methods and their evaluation and critique. This methodologically focused work has, in turn, adopted pre-given (often highly specific) normative models of participation that assume a correspondence theory of an external 'public' existing in a natural state waiting to be discovered and mobilized by participatory techniques and procedures. As we will outline below, this 'residual realist' perspective is shared by normative accounts of the value of public participation in technoscientific decision-making and more critical assessments of public participation practice.

Our ambition in this volume, then, is to rethink the taken-for-granted methodological and theoretical realism of this work, and the notion that public participation practices might be adjudicated with reference to pre-given and external democratic norms. The chapters that comprise this volume develop a picture of public participation as emergent and in the making. Instead of seeing publics as external to processes of public mobilization, the chapters take forward analyses of the ways in which the publics of science and democracy are actively brought into being through matters of concern and the various instruments, tools and forms of mediation deployed to know and move them. Similarly, rather than insist on a dualism between science and politics, in which democracy is figured as a pre-given and relatively timeless norm, we are interested in exploring the practical enactment of contemporary democratic politics and the creation of participatory spatialities. Questions and judgements of democratic legitimacy, in this sense, emerge within the practices and projects of democratization and participation. This focuses our attention on the shifting means through which societies arrive at collective determinations about science and democracy, arbitrate between competing forms of expertise and knowledge, and legitimate particular forms of political performance.

In other words, the volume aims to develop, articulate and explore a *co-productionist* approach to public participation and democratic engagement with science, technology and the environment. A tradition of co-productionist analyses at the interface between science and technology studies (STS), political theory, geography and anthropology has provided abundant evidence of the intimate interweaving and mutual constitution of science, technology and the projects of social and political ordering (Jasanoff 2004). Recent work in this tradition has extended co-productionist analyses in a number of productive directions – including at the interface between economic sociology (Callon *et al.* 2007), feminist science studies (Franklin 2013), citizenship studies (Benjamin 2009), post-colonial history and cultural theory (Hecht 2011; Amir 2012). Co-productionist understandings of science also constitute the conceptual hinterland of many participatory practices and projects designed to render science policy and environmental decision-making more open, transparent and accountable (Stirling 2008). In light of the extraordinary innovation across this field of inquiry, it is striking therefore to note that the pragmatics of public participation with science and the environment have received relatively little concerted analyses in ways that deploy the tools of situated interpretive and co-productionist analyses. The goal of this volume is therefore to deploy a *co-productionist idiom* (Jasanoff 2004) to open up a new terrain of theory, empirical inquiry and practice that rethinks participation (and 'the public') as relational and emergent. This is not only an interpretive-analytic exercise of deconstructing the *making* of public participatory practices and their relations with 'science'. It is also a constructively critical project where the authors take seriously the implications and possibilities of a co-productionist approach for *remaking* participatory practices, institutions and constitutions.

In the following sections of this introductory chapter we set developments in public participation against the background of changing relations between science, politics and society, and locate the dominant realist perspectives that characterise much of the existing scholarship and practice of participation. We then outline key features of the alternative co-productionist and relational conception of participation developed in this volume, before providing an overview of the chapters that follow.

Science and democracy in an age of participation

Three interlocking – and, at times, mutually reinforcing – axes are central to the changing relationships between science, democracy and public legitimation in which participation is continually reconfigured. The first concerns transformations in what might be thought of as the 'territory' or spaces of government. Barry (2001) argues that the spatiality of government has traditionally been 'conceived in terms of a relation between a national population and a national territory' (p2), underscored by concepts of society and economy as 'contained within the territorial boundaries of the nation-state' (p3; see also Elden 2013). However, as technological innovations – along with their promises of human progress, economic and social development and the perils associated with new novel risks and profound social, moral and ethical challenges – have become an increasing preoccupation of contemporary social and political debate, so too have the spaces of government expanded beyond the nation state. Barry (2001) continues by suggesting that in recent years government has come to operate 'not just in relation to spaces defined and demarcated by geographical or territorial boundaries but in relation to the zones formed around the circulation of technical practices and devices' (p3). Such transformations have been intensified by the twin dynamics of globalization and liberalization which – despite being premised on notions of openness, transparency, speed and global connectivity – have also posed considerable challenges for established modes of scientific and political representation. One of the consequences of these developments is that while traditional modes of democratic mobilization and representation face what many regard as a 'legitimacy crisis', we are also witness to the proliferation of new scales of government and participation, both *above* and *below* the nation state, that have challenged the presumed hegemonic unity of the nation as locus of political power and administrative coordination (Castells 1996).

The second axis concerns the changing relationship between scientific knowledge, technological artefacts and everyday subjectivities. In his analysis of the historic constitution of a modern conception of individual selfhood, Taylor (1989) argues that the advent of disenchanted understandings of the natural world paradoxically created the conditions for a model of selfhood defined by notions of individual preference and sentiment. He reasons that a 'modern feeling for nature which starts in the eighteenth century presupposes the triumph of the new identity of disengaged reason over the premodern one embedded in an ontic logos' (p301). For Taylor, a consequence of this development is that in modernity 'our own nature is no longer defined by a substantive rational ordering of purposes, but by our own inner impulses and our place in the interlocking whole' (p301). The paradox here is that the *scientization* of contemporary public and political discourse – and the extension of scientific expertise to all manner of social, moral and cultural problems – has also precipitated the individualization of contemporary modes of public reason and resistance to traditional modes of political hierarchy and social legitimation (Beck *et al.* 1994). The irony is that while technological innovation has become a source of social, ethical and moral dynamism – prompting new questions for collective deliberation – technoscience is increasingly geared towards the promises of individualization, personalization and genetic universalism, thereby challenging traditional notions of social solidarity, kinship and community (Pellizzoni and Ylönen 2012).

This apparent paradox underscores the third axis: the emergence of new modes of public reason and legitimation associated with what has been variously described as the 'new governance of science' and a 'deliberative turn' in contemporary technological politics (Hagendijk and Irwin 2006; Irwin 2006). A striking feature of the ways in which industrialized societies are coordinating the relationship between science and social organization is a bifurcation in the cultural and institutional conventions concerning the social meanings of technological change. In this context hegemonic notions of scientific authority – and particularly the notion that scientific expertise operates as an objective arbiter of moral meaning – sit uneasily alongside the notions of 'authenticity', 'transparency', 'trust' and 'responsibility' that underpin institutional commitments to public participation (Wynne 2006b). This disjuncture in the 'social life of science' (Mulkay 1979) is in part rooted in the risk-based politics that characterized the latter half of the twentieth century. During this period a political constitution tuned toward individual sentiments and the cultural aesthetics of modern 'life-politics' collided with (and to some degree disrupted) the high-end techno-political complex (Beck 1992). Underscoring the profound inadequacy of the institutionalized practices of risk assessment and public reassurance (Wynne 1988), the controversies that surrounded (for example) the commercialization of genetically modified crops, bovine spongiform encephalopathy (BSE, or mad cow disease), the siting of nuclear power facilities and disposal of radioactive waste, served to call into question what Jasanoff (2005) terms the 'intimate collaboration' between science and the state.

The principal effect of these controversies has been a set of schizophrenic institutional reconfigurations designed to both reaffirm the authoritative role of 'sound science' in 'evidence-based' decision-making and knowledge-based innovation whilst at the same time presenting scientific institutions as the transparent, accountable and authentic brokers of public deliberation (Brown and Michael 2002; Irwin 2006). This new mode of technological governance is most obvious in the conspicuous acts of public legitimation engaged around new and novel technologies where the use of deliberative forms of public discourse and consultation are coupled with programmes designed to improve the technical knowledge and dispositions of individual citizens (Irwin and Michael 2003; Kearnes and Wynne 2007). The paradox of this turn in contemporary science governance is that while institutions present themselves as more open and responsive to direct public participation, powerful and largely obscured underlying dynamics of closure – not least through driving forces of neo-liberalization, globalization and science-led progress – mean that participating publics have never been more removed from the centres of power and calculation (Stirling 2008; see also Wynne, Chapter 5 in this volume). At the same time, it is important to note that citizens and the general public are now, and perhaps more than ever, 'engaged' with science and technology. The proliferation of activist groupings, social media commentary and public discourse on a plethora of science-based issues is indicative of the multifaceted forms of public participation that characterize contemporary technological politics (Brown and Mikkelson 1990; Epstein 1996). Indeed, a common feature of recent scientific and environmental controversies is the almost anarchic emergence of participatory formations that are resistant to institutionally sanctioned processes of political closure and public consultation. These insurgent formations serve to call into question the overtly normative commitments to institutional accountability and transparency, while also enacting novel participatory spatialities that are situated across a multitude of socio-political sites (Barry 2001; Marres and Rogers 2008).

These dynamic relations between science and democracy have prompted the emergence of, and experimentation with, new forms of public participation and science–society interaction. The succession of alternative conceptual frameworks and methodologies for participatory technology assessment, science policy and the 'social shaping' of innovation – such as 'post-normal science' (Funtowicz and Ravetz 1993), 'mode II science' (Nowotny et al. 2001), citizen science (Irwin 1995), 'analytic-deliberative' risk assessment (Stern and Fineberg 1996), constructive technology assessment (Rip et al. 1995), 'upstream public engagement' (Wilsdon and Willis 2004), 'anticipatory governance' (Guston 2000) and 'responsible innovation' (Owen et al. 2012) – is an index of the vibrant spirit of innovation and institutional experimentation that have characterized attempts at 'opening up' the governance of science and technology to wider public values and internal reflection. Such moves have been matched only by the development of a vast array of techniques for coordinating the relations between science and publics – and particularly the development and deployment of deliberative and participative forms of decision-making and initiatives designed to create avenues for the active engagement of publics in science-related issues and innovation processes (Jasanoff 2005; Chilvers 2008). More broadly, notions of public participation and direct citizen consultation have also figured in the emergence of new models of public service delivery, environmental sustainability and policy development.

In this proliferation of techniques and practices designed to foster public engagement with and participation in 'science', we also see the marking out of a political and discursive space, defined by the construction of 'opinionated publics' – the enactment of publics who possess unique preferences and attitudes concerning developments in science and technology (Kearnes et al. 2014; Macnaghten et al. 2015).[2] Where citizens remain 'uninvited', the emergence of new participatory spaces often overflow into multivalent forms of activist, civic or citizen science and 'bottom-up' grassroots or distributed innovation (Epstein 1995; Wynne 2007b; Callon et al. 2009). In turn these developments have been accompanied by oftenfierce debate and critique (Irwin et al. 2013), centred on the epistemic problems of demarcation and 'extension' (Collins and Evans 2002; Wynne 2003) and political critiques of the underlying structures of popular participation and the evident depoliticization of contemporary democratic discourse (Mouffe 2000; Mahony et al. 2010; Swyngedouw 2010).

Residual realist perspectives of participation

In light of this vibrant and lively contestation of the purposes and objectives of contemporary techno-political decision-making, it is striking that many existing scholarly considerations of public participation are characterized by what we term 'realist' accounts of the nature of publics and the practices of public participation. This *residual realism* is evident in both normative accounts of the democratization of science and democracy, and critical analyses of the underpinning political commitments of contemporary modes of social legitimation. While often informed by post-positivist and social constructivist perspectives, in broad terms both areas of work share realist accounts of an objective public and a political epistemology that renders democracy as external to and independent of scientific and technological change. Before setting out the grounds for a more thoroughly co-productionist and performative account of the enactment of technological democracy, in this section we briefly chart key aspects of this residual realism evident in contemporary scholarship and practices of public participation.

Across a range of disciplinary perspectives the dominant discussion of public participation has tended to be couched in terms of methodological improvement and optimization through developing, experimenting with and evaluating the effectiveness of new participatory methods and designs. In pragmatic terms much of this work has concentrated on conceptual and methodological tools that enable and encourage public participation and deliberation with technoscientific and socio-political orders which, to date, have remained relatively secluded from broader scrutiny. Often rooted in normative theories of deliberative democracy and communicative rationality (Benhabib 1996; Habermas 1996; Dryzek 2000), such moves form part of an evaluative paradigm that seeks to adjudicate practices of public participation against theoretically defined and pre-given procedural standards, for example of inclusiveness, representation, competence, social learning and so on (Habermas 1984; Fiorino 1990; Renn *et al.* 1995; Bohman and Rehg 1997; Rowe and Frewer 2000; Fung and Wright 2003). Social scientific studies that have deployed public engagement methods have often 'imported in' these normative–deliberative models from political theory largely unproblematically, presenting public participation as a discursive and linguistic practice set in quasi-public spaces.[3] This has resulted in a methodological essentialism, where the normative ideals of deliberative participation are translated into procedural commitments, commonly through notions of openness, representation and the 'policy relevance' of participatory initiatives.

Partly in response to these drives towards methodological revisionism in participatory practice, a body of more critical work has highlighted the (seemingly paradoxical) potential of participation to exclude, disempower and oppress. Such critiques have emerged in disciplines such as development studies (Cooke and Kothari 2001), political science (Cruikshank 1999), planning (Tewdwr-Jones and Allmendinger 1998), geography (Swyngedouw 2005) and radical democratic theory (Mouffe 2005b). These studies often stress the persistence of uneven power relations and strategic behaviour throughout participatory processes that both undermine and are obscured by Habermasian-inspired consensual deliberative ideals (Pellizzoni 2001). Many scholars writing from this standpoint view the rise of public participation methods as an extension of modes of social control – particularly through the instantiation of relatively prescriptive subject positions evident in deliberative democratic practice – that maintain a distinction between, for example, 'publics' and 'experts' in ways that circumscribe and delimit 'public input' to participatory processes within technocratic discursive formations (Swyngedouw

2010; Thorpe and Gregory 2010; Jones *et al.* 2011). More broadly, this work has pointed to the often profoundly anti-democratic implications of public participation, where opportunities for substantive public contestation are evacuated by a consensual post-political populism (Rancière 2006).

While it would be unwise to talk of these expansive bodies of work in the same light, we argue that in these two diverse areas of commentary and scholarship it is possible to trace a predominant, though by no means exclusive, residual realist account of participation. In conceptual terms the debate between these models of participation hangs largely on the relationship between 'the democratic' and 'the political'. Though there has been substantial development in deliberative theory (see, for example, Dryzek 2000, 2010), in broad terms this body of work conceives of democracy as an outcome of processes of collective deliberation and decision-making. In contrast, analyses of participation inspired by cultural theories of hegemony (Laclau and Mouffe 2001; Mouffe 2005a) have powerfully highlighted that supposedly democratic politics – and particularly the formalized and consensual politics of deliberative practice – might be reconceived as the outcome of political struggle and power. Central to both models is a conception of difference and the asymmetric relations between contending political constituencies. The assumption is that contemporary knowledge-politics are propelled by the differential interests, understandings and perceptions possessed and mobilized by contending political subjects (for example, between citizens affected by an issue and technical experts). While both deliberative and agonistic models of participation deploy alternative theorizations of difference – either its resolution through deliberative reflection and communicative rationality (Habermas 1984) or the more critical contention concerning the productive role of social conflict and power relations in determining participatory outcomes – both positions share a commitment to pre-given normative models of democratic politics. 'The democratic' is, in both models, ready-made and external to the situated, material performances of democracy and participation (Marres 2012), and both accounts have relatively little to say 'about the existence and importance of materials and objects, which frequently come to animate public knowledge controversies' (Barry 2013, 8).

As a consequence, both of these accounts tend to assume a separation between the resolutions and accommodations arrived at by the contending parties of a political dispute and the technicalities of contemporary democratic participation. Matters of science, technology, expertise and the environment are only rendered political when they are brought into definitively political spatialities (either agonist or deliberative) by human agents. The 'remarkable technicality of politics' (Barry 2002, 269) and the performative, and always partial, enactment of 'atmospheres of democracy' similarly receives little attention in each of the accounts of participatory politics.

Despite their differences, then, we can identify in both affirmative and critical discourses a *residual realist* conception of participation. The key features of this dominant realist imaginary of participation and the public can be identified as entailing the following commitments:

1   *Publics as external to participation.* The dominant image in much existing scholarship and practice is of a public pre-exiting in a natural state external to the practices of participation. The methodological tenor of much of the existing literature on public participation attends to publics through a series of proxies – such as 'public opinion', 'public values' or as 'deliberating publics'. At the same time, much of this literature is also focused on questions of design and process – how avenues for participation might be designed to mobilize publics and represent externalized public 'truths' around preformed political and policy problems. This correspondence theory of representation broadly conforms to the view that 'governmental decisions should correspond to the pre-existing reality of either popular will or the objective public interest' (Brown 2009, 6). While cultural theories of hegemony focus on the disciplining of public sentiment by processes of public participation, and the often profoundly anti-democratic implications of deliberative ideals, both accounts maintain a conception of publics as external to processes of public participation.

2   *Publics as an aggregation of autonomous individuals.* The critical unit of representation is typically presented as individual human subjects, either through attempts to reach consensual agreement and aggregate individual preferences across wider populations *or* through emphasizing difference and divergence between autonomous individual subjects (Proctor 1998).

3   *Participatory democracy as pre-given and ready-made.* Participation is shaped by and evaluated against pre-existing models and norms (democratic, deliberative, discursive, communitarian, pluralist, agonistic and so on) that themselves are rarely reflected on. A dominant model and normative framework that has become established across many domains of participation and policy practice, including until recently its largely unquestioned import into STS scholarship, is the Aristotelian notion of participation as discursive and deliberative, situated in formally sanctioned public spaces (Marres and Lezaun 2011).

4   *Participation as technologized procedures.* Central to realist accounts of participation is the development, and standardization, of participatory techniques, instruments or tool-based procedures. Guides to 'good practice' in public engagement and participation suggest that such tools might be unproblematically deployed across different issues, contexts and cultures (such as the oft-cited examples of the consensus conference and citizen panel techniques).

5   *Participation as discrete and ephemeral events.* A key feature of many realist understandings of participatory democracy is an emphasis on one-off exercises coordinated at the service of particular moments of institutionally sanctioned decision- or knowledge-making. This is further reinforced in critical studies of participation, which often focus on front-stage spaces of participatory performances.

6   *Inclusion as a key quality of successful participation.* Dominant models of participation emphasize the inclusion of all relevant human actors in processes of public engagement and ensuring that participatory procedures are 'representative'. Here public participation practices are closely aligned with the development

and refinement of social science methods for producing representative samples of social wholes (Law and Urry 2004).
7   *A linear model of participation and engagement*. Much of the evaluation of public participation practices is based on linear 'cause-and-effect' understandings that processes of public engagement will (or at least *should*) directly precipitate actions and outcomes. This is evident in the ambition of 'impact on decision-making', which is often taken to be a criterion of good participation (e.g., Rowe and Frewer 2000) where knowledge closure and the generation of 'public truths' is expected to influence policy decisions and shape wider actions.
8   *Participation as separate from science and democracy*. The practical implication of the separation between the technical and the political, implicit in much participatory theory and practice, is that participation is viewed as discrete and external to science, technology and environment. In practice, participation is often conceived as a project to quantify and capture public understandings of new scientific and technological developments, or preferences concerning pre-framed policy problems, without reflecting more broadly on the social, moral and political commitments implicit in systems of technological innovation and innovations in participation itself. In this view, participation becomes an end in itself, and debates tend to centre on issues of extension and 'scaling up'.

In summarizing these implicit orientations in theory and practice, our argument is that realist conceptions of participation have become a dominant imaginary through which public participation and engagement is understood and conceived. While the proliferation of participatory initiatives says much about the idiosyncratic cultural politics of late-modern industrialized societies, this realist framework is itself tied up with a broadly affirmative and methodological paradigm. It is in this paradigm, and the circulation of methodologies, principles and evaluative frameworks, where much of the existing effort in the democratization of science and democracy has been channelled. Moves to institutionalize, technologize, and even professionalize and commercialize public engagement with science and the environment also serve to systemically reinforce these realist perspectives, and their underpinning of neo-liberalized science policy cultures (see Wynne, Chapter 5 in this volume). This realist understanding of participation also depends on, and is propelled by, critical interventions which are in turn translated into methodological and procedural commitments (Irwin *et al.* 2013).

The striking irony here is that while public participatory processes concerned with science, technology and the environment have drawn inspiration from the sociology of science and technology – and indeed are commonly situated at the interface between STS research and science policy processes (Webster 2007; Wynne 2007a) – the implications of co-productionist understandings of the social dynamics of science and invention have, for the most part, not been thoroughly worked through the practice of public participation (Macnaghten *et al.* 2015). While there is no doubt that work in both the affirmative and critical traditions has been both productive and useful, the approach developed in this volume represents a significant departure from – and, we would argue, an advance on – established theories and practices of public participation.

Participation as co-produced, relational and emergent

The chapters that follow develop and explore an alternative constructivist conception of participation as co-produced,[4] relational and emergent. Our approach is grounded in co-productionist accounts of participation that have begun to emerge in STS and cognate disciplines in recent years, pioneered by many of the authors in this book. Such work has begun to show that, far from being pre-given categories external to participatory practice, the subjects (publics and their concerns), objects (issues and material commitments) and formats (political ontologies and participatory procedures) that comprise the constituent elements of participation can more accurately be seen as both constructed through and emergent in the performance of carefully mediated, open-ended participatory experiments (Irwin 2006; Lezaun 2007; Lezaun and Soneryd 2007; Marres 2007; Chilvers 2008; Felt and Fochler 2010; Laurent 2011). This perspective helps to explain how participatory practices shape (and are shaped by) social and political orders by highlighting the ways in which they mediate the construction of subjects and public issues in the making of participatory processes. In other words, the realities of participation, the public and public knowledge-commitments do not pre-exist, but are instead the outcome of, collective participatory practices. In this view, rather than simply being composed of discrete formations of mini-publics, with linear relations with centres of power and calculation, multiple situated sites and forms of participation are continually being made, unmade and remade. So much for 'the participatory turn' then. Participation has always been, and is continually, turning!

Inspired by co-productionist theoretical resources that have been developed across contemporary STS research – in actor network theory (ANT) (Callon 1986; Latour 1987), in assemblage theory (Irwin and Michael 2003), in studies of the constitution of social and political orderings (Jasanoff 2004) and in processes of economic formation (Callon *et al.* 2007; MacKenzie *et al.* 2007; Pinch and Swedberg 2008; Muniesa 2014) – this volume begins from the proposition that public participation and engagement might be viewed as contingent and heterogeneous collectives of human and non-human actors, devices, settings, theories, social science methods, public participants, procedures and other artefacts. These diverse collectives can at times be established at particular sites to form localized and ephemeral time-spaces, while in other arrangements participatory formats become more durable, standardized and routinized. By invoking and developing the notion of co-production as an idiom for conceptualizing multiple modes of participation in questions concerned with science, technology and the environment, our intention is to highlight the active and contingent processes involved in the enactment or performance of participation. The objects (or issues), subjects (or publics/participants), instrumentalities (or methods and devices) and procedures (or political philosophies) of participation, in addition to the enrolment of actors, are not predetermined. Rather, in the mediation – or the assemblage – of participation, these diverse elements are held in tension. Furthermore, in our conception of the contingent constitution of participation, we do not presume that a central coordinating agency (the state, scientific institutions, civil society, etc.) is the locus of power and capability. Viewed in these terms, participation – and indeed the constitution of publics themselves – appears as an emergent and, at best, provisional accomplishment.

In developing a co-productionist account of participation, our approach takes seriously 'the proposition that the ways in which we know and represent the world (both nature and society) are inseparable from the ways in which we choose to live in it' (Jasanoff 2004, 2). In this sense we emphasize how socio-material collectives of participation always involve a mutual interweaving of social, normative, cognitive and material elements. Furthermore, the approach developed in this volume, and the chapters that comprise it, emphasize how emergent *participatory experiments and practices* are both shaped by – and, in turn, shape – technoscientific, political and social orders. This volume is therefore situated in an emerging area of work that extends co-productionist studies of the social life of science to the analyses of democratic engagement, processes of public engagement with 'science', and broader processes of public sense-making and legitimation. These studies have usefully highlighted the role of existing configurations of political economy, science policy cultures and sociotechnical imaginaries in shaping participatory practices (Irwin 2001; Jasanoff 2005; Goven 2006; Wynne 2006a) and have demonstrated how the co-productions of participatory practices shape meanings of science and science-related issues, forms of participatory democracy, public identities, and so on (Irwin and Michael 2003; Marres 2007; Felt and Fochler 2010).

A distinctive feature of many, but by no means all, existing co-productionist accounts of participation has often been a focus on situated studies of public engagement 'events' and what Michael (Chapter 4, this volume) terms the 'eventuation' of public participation (see also Michael and Rosengarten 2013). While the locus of empirical enquiry in STS has more often than not centred on the particularities of participatory processes, as we go on to argue (and lay out in more detail in Chapter 2), it is important for co-productionist studies of participation to develop an analytical gaze that extends beyond these discrete instances of participation – to utilize situated modes of analyses to uncover the overlapping sites and spatialities of participatory co-productions. We thus argue that it is important to extend social studies of participation in at least three ways. The first is centred on the portability, standardization and circulation of *technologies and expertise* of participation across time, place and culturalpolitical settings, and how powers of participatory prescription become lodged in expert communities, institutions and infrastructures. A second line of enquiry concerns the *spaces of controversy and negotiation* within which diverse participatory experiments intermingle – whether the space of negotiation is centred on the form of participation and democracy itself, or else around the definition of the public issues at stake or trajectories of socio-technical change. Finally, a richer co-productionist account of participatory democracy in the making needs to extend to the level of *political culture and constitutional relations* between citizens, science and the state, within which certain participatory collective practices and knowledge-ways become seen as authoritative and are endowed with legitimacy and meaning (Jasanoff 2005, 2011).

Extending out to understand these underexplored and interrelating spaces of co-productive coherence is an important and novel feature of many chapters that make up this volume. In further clarifying the alternative account of public participation offered by the authors in this collection, it is useful to identify key features of this co-productionist conception and approach that mirror the eight proposed for the realist perspective presented in the last section. In these terms a co-productionist, relational and emergent approach to participation emphasizes the following aspects:

1. *Publics as mediated and emergent.* Publics do not pre-exist in a natural state external to participatory practices. Nor are publics – or their proxies (public opinion, public values and public sentiment, for example) – simply mobilized methodologically. Rather, publics are actively mediated and brought into being by public issues, participatory devices and human interventions (including by publics themselves). This approach widens the definition of participation to encompass a conception of both the active configuration and mediation of 'participating publics' and the meanings of what counts as participation.
2. *Publics as collectives.* 'Individual' publics never act alone but always as part of socio-material collectives through which they know and accomplish things in the world. In developing situated accounts of participation in practice, the challenge is to account for the emergence of, and relations between, heterogeneous public collectives rather than simply persist with an image of publics as the aggregation of autonomous individuals.[5]
3. *Participation as collective experimental practices 'in the making'.* As with publics, the formats, philosophies, normativities and techniques of participation do not exist *a priori* but shape and are produced through the performance of participatory practices. This draws attention to their construction, negotiation and emergence. In this view, all forms of participation and public representation are contingent, exclusive, partially framed and subject to 'overflows'.
4. *Participatory collectives as co-produced, material and diverse.* The makeup of participatory collectives is shaped by and is constitutive of technoscientific, social and political orders. The interweaving of the social, normative, cognitive and material emphasizes the materiality of all forms of participation and the importance of material settings and objects/devices in participatory co-productions (cf. Marres and Lezaun 2011). This, in turn, opens up analyses to the sheer diversity and multiplicity (and in principle an infinite range) of definitions, meanings and configurations of public engagement with science, technology and the environment. This extends well beyond the discursive, deliberative, public and contentious to encompass material, embodied, visceral, private, everyday and mundane forms of public engagement, and more.

5 *Relational ecologies of participation.* In place of a focus on discrete participatory events, the chapters that make up this volume emphasize multiple, diverse, entangled and interrelating collectives of public involvement within particular political constitutions, systems or issue spaces. One implication of this approach is that the dynamics of collective participatory practices can never be fully understood in isolation, but should always be read in terms of their relational interdependencies and connectedness.

6 *Reflexivity and humility as key qualities of successful participation.* In place of the emphasis on methodological refinement that characterizes realist accounts of participation – the assumption that 'better methods' will produce 'better publics' and public knowledge – this volume suggests that framing effects, exclusions, contingencies and uncertainties are a characteristic feature of all participatory collectives. The challenge is therefore to render the specific closures and openings precipitated by participatory formats as matters of collective reflection and openness. This involves critical examination of preconditions that frame one's own participatory knowledge-commitments and those of others, as well as an anticipatory perspective.

7 *Participation as non-linear and multiply productive.* Participatory collectives form part of distributed agencies that make up contemporary knowledge societies, and relate to them in iterative and recursive ways. Our approach suggests that it is not possible to predetermine or prejudge the productions or qualities of participation – all forms of public engagement co-produce meanings, knowledges, forms of social organization and material effects. The multiple values of participation encompass these emergent qualities.

8 *Participation as constitutive of science and democracy.* Rather than being somehow separate, 'bolted on' or integrated into processes of scientific and technological development, participation and deliberation has always been part of how science operates and its public relations. At the same time, emergent participatory collectives have always been a central feature of democratic politics, which uphold, reproduce and challenge existing political formations.

Critical to this co-productionist approach are analyses of the ways in which public participation processes are intertwined with the transformations in contemporary democratic governance, new technologies and institutions of public representation, and new modes of contemporary public reasoning around scientific and technological developments. While these points offer a useful juxtaposition to realist conceptions of participation outlined in the previous section, the implications of understanding participation as both contingent and emergent will be more fully explored in the following chapters. Yet, as noted at the beginning of this introduction, our ambitions are not only interpretive-analytic in emphasis. In addition to embarking on a project that seeks to build theoretical, empirical and methodological advances in deconstructing the *making* of participation and participatory realities in contemporary democracies, this book takes seriously the need to also consider the implications of such analyses for reconstructing and *remaking* participation. In many respects this is a significant challenge and one that co-productionist accounts of participation have often sidestepped. We address such normative and instrumental questions head on toward the end of the book, rejecting a dismissive critique in favour of cultivating a critical standpoint that attempts to be constructive, additive and transformative in relation to already existing participatory developments and progress.

Remaking participation: an overview of the volume

In keeping with this overall ambition, the volume is structured into three main parts. Part I includes chapters that seek to set out what it means to *rethink public participation* and engagement with science from a co-productionist perspective. In Chapter 2 the editors lay out the foundations for such a perspective and the book as a whole through a theoretical review of the STS and critical social science literatures pertaining to participation and the co-production of science and social order. We first consider co-production at the level of situated participatory practices, identifying competing explanations of what brings participatory experiments into being and arguing that it is always to some extent an interplay between issues/objects, procedural formats and forms of human intervention. We then go on to set out other important spatialities of co-production and participatory coherence that extend beyond the immediate 'events' of participation – namely: spaces of standardization and prescription marked by technologies and expertise of participation; spaces of negotiation demarcated by the circulation of different meanings of the issue or political situation at stake within which diverse forms of engagement intermingle; and the level of political cultures and constitutions where particular participatory knowledge-ways become seen as credible and authoritative. It is suggested that interpreting the dynamics of participation in science and democracy depends on a thorough understanding of the interrelations within and between these topological spaces of participation.

In Chapter 3, Alan Irwin and Maja Horst develop a relational and emergent perspective on public engagement through the notion of (de)centredness. Set against earlier developments in the public understanding of science (PUS) and public engagement with science (PES) – which developed a view of participation 'centred' on formal institutions and policy-makers – Irwin and Horst articulate an alternative 'decentred' perspective of engagement as distributed, multiple, ambiguous and assemblage like. They develop this perspective through the case of a citizen summit on climate adaptation in the Danish town of Kalundborg, showing that this seemingly centralized process was simultaneously decentred – not only in its hybridity, fragmentation and overflows, but also its intermingling with other diverse forms of engagement with the climate adaptation issue by local and national groups (ranging from landowner associations through to national environmental organizations). Rather than being a source of dismissive critique of participatory practice and the archetypal 'policy-maker', Irwin and Horst conclude that a relational and decentred perspective on engagement can be productive in promoting continual improvement and greater attentiveness to the openings and closings of participation, rather than falling back on the simplicities of realist and fixed approaches to public engagement.

The focus on rethinking participation as co-produced and emergent radically alters taken-for-granted meanings of 'public engagement', as eloquently demonstrated by Mike Michael in Chapter 4. He contends that in place of the traditional focus of public engagement initiatives on technoscientific controversies in the public sphere, from a relational perspective even the most mundane, everyday and seemingly uncontroversial technologies might be the subject of 'public engagement with science'. Michael draws on the work of Alfred Whitehead and Isabelle Stengers to put forward a process philosophy approach that emphasizes the inherent relationality of all objects and provides a way of interrogating the (often forgotten) politics of emergence that attach to mundane technologies. This relationality extends to the argument that methodology is itself constitutive of its objects of study, thus highlighting how all elements of public engagement with science emerge in unforeseen ways. Michael outlines how a 'participatory politics of the mundane' calls for speculative approaches to engagement as a means of exploring the unfolding relations between publics and everyday technologies, like Velcro, rolling luggage and the band-aid plaster.

Moving out to more systemic and constitutional arrangements of politics and power, Brian Wynne builds on his past work to set out a thoroughly co-productionist and relational perspective of public engagements and relations with 'science' in Chapter 5. Despite popular experience to the contrary, he powerfully demonstrates how the 'empty signifier' of a singular public (in terms of 'the public interest', 'a public mandate' and so on) is continually reproduced as a necessary fiction of contemporary democracies – a ghost in the machine that is intensifying under the hermeneutically imposed meanings of science, the political economic driving forces of neo-liberalization, globalization and the requirements of incumbent powers to maintain credibility and control. Wynne argues that mainstream social science approaches to understanding science–public relations largely perpetuate this reductionist, singular imaginary of 'the public' while simultaneously falling into (and ignoring) these constitutional driving forces of science and democracy. Starting from a recognition of emergence and relationality as a fundamental human-ontological condition, Wynne explains key features of his own interpretive-relational approach to 'understanding publics and their sciences' developed with colleagues in the CSEC group at Lancaster University over the years. This more reflexive approach calls for social scientists to accept responsibility and be accountable for their own public representations and the ways in which they bring publics and public issues into being. This demands being reflexive about our own professional practices/normativities and about the wider collective-experimental dynamics where actors become aware of the assumptions of others and of the wider institutional-constitutional conditions that shape and are shaped by situations of public engagement.

In Part II, authors take forward more empirically oriented accounts of, and reflections on, *participation in the making*, drawing on and further elaborating the conceptual perspectives introduced in the first part of the book. In Chapter 6, Véra

Ehrenstein and Brice Laurent focus on state experiments in public participation, arguing that the making of participation is also a useful analytical entry point to explore the (re)making of different kinds of political organizations called states. Eschewing the tendency in mainstream participation literatures to focus on discrete engagement events that form in opposition to representative democracy and state institutions, the authors adopt an agnostic perspective to demonstrate how in cases of national importance participatory practices and governmental action are mutually at stake and co-produced. They lucidly illustrate this through a comparative analysis of two contrasting state experiments in participation, one a national public debate on nanotechnology conducted in France and the other a series of participatory initiatives in the Democratic Republic of the Congo to deal with the issue of national deforestation. Through this in-depth empirical analysis, Ehrenstein and Laurent conclude that while each national setting has distinct political constructs, in both cases the intense work invested in making participation happen at national levels contributed to the possibility of a 'state demonstration' that made the democratic quality of both French and Congolese government action visible and demonstrable to external audiences.

In Chapter 7, Linda Soneryd moves from this focus on the co-production of participation and the constitutional dynamics of national systems of governance to consider the spaces within which technologies of participation and forms of participatory expertise become (to some extent) standardized and circulate between cultures, times and places. Soneryd focuses on understanding the processes and work invested in making participatory designs portable, arguing that notions of the democratization of science and deliberative democracy need to be understood in relation to the transnational spread of ideas and standardized forms of governance. She develops her account through a combination of STS approaches on technologies of participation and new institutionalist approaches to analyse two cases where formalized designs for public deliberation have travelled – the first being a Swedish approach to stakeholder engagement in radioactive waste management that was translated to the Czech Republic, and the second being a visioning exercise that gathered citizens in different European countries to explore future developments in science and technology. Through both cases Soneryd describes how these technologies of participation are made to travel and the importance of 'organizational carriers' in these processes. While designs in both cases were to some extent 'de-scripted' and conducted in ways which were not standardized, they also adhered to strong scripts for how the events should be interpreted, ensuring the successful travel of the participatory formats and thus capacities to technologize issues of public concern.

In contrast to this focus on the dynamics of closure, in Chapter 8 Sarah Davies takes the notion of science communication to help 'open up' what she sees as taken-for-granted meanings, imaginations and formats of participation with science and technology. The chapter is a reflection on the ways in which different forms of science communication – such as science festivals, museums, broadcast media or grassroots activities such as hackerspaces and forms of public protest – might help to re-imagine participation. Davies suggests that it is useful to study such activities for two reasons. First, they foreground emotion and thus emphasize the material and affective dimensions of practices of public engagement with science. Second, they encourage us to reconsider the nature of scientific citizenship as extending way beyond discrete moments in deliberative-discursive spaces to include interdependencies between diverse material and embodied and affective forms of participation in wider systems.

In Chapter 9, Ulrike Felt takes time as a lens through which to understand participation in the making. Drawing on social theory and STS work on the sociology of time – that depicts time as situated and constructed but also as a universal and naturalizing force deeply entangled with questions of control and power – Felt argues that questions of time have been somewhat left out of debates about the democratization of science and technology. Drawing on empirical insights from her own extensive fieldwork to engage publics with emerging technosciences in Austria over the past decade, Felt explores the temporal patterns and choreographies by which time and participation are co-produced through the performance of public engagement with science. She demonstrates how engagements with science and technology are shaped and moulded by four forms of time, namely clock time, trajectorism, emplacement of time, as well as the multiplicities and inconsistencies of time. This perspective brings attention to the interrelations between specific situations of participation, the temporalities of emergent technoscience, and the rhythms of extant orders in relation to which they are co-produced. It alerts us to the need for a more care-oriented approach to participatory exercises that attend to the ways in which temporal choreographies frame distributed responsibilities in science and democracy.

Adopting a co-productionist stance on participation and democracy itself urges us to accept that even the most distant and removed theoretical and interpretive analyses of participation have a normative dimension in saying, in some way, how the world (including forms of participation within it) ought to be. The final part of the book thus moves on to more explicitly and deliberately consider what a co-productionist and relational perspective means for *remaking participation* in more practical terms.

Claire Waterton and Judith Tsouvalis take on this challenge in Chapter 10, in Part III, by exploring their own experience of carrying out a process of collective experimentation that sought to open up and reconfigure the framings and associations around the issue of eutrophication in a lake in the English Lake District. Responding to co-productionist critiques of the unacknowledged framing effects and power relations of many participatory projects, Waterton and Tsouvalis explain how a 'new collective' – called the Loweswater Care Project – was formed that sought to actively attend to the radical rationality and ongoing emergence of people (farmers, local residents, scientists and so on), things (such as blue-green algae in the lake) and heterogeneous associations involved in this particular participatory experiment. The experiment is an interesting and ongoing attempt to put the relational ideas of STS – most directly Bruno Latour's *Politics of Nature* and Karen

Barad's *Meeting the Universe Halfway* – into action to form new ways of questioning, forming knowledge and practising participation, which in turn brings forward a new ethical sensibility characterised by experimentation with intensities.

In Chapter 11, Cynthia Selin and Jathan Sadowski consider what it might mean to reconfigure participation in processes of technology assessment if obduracy were more fully taken into account. They argue that rather than a 'blank slate' approach evident in much future-oriented participatory practice, public engagement with emerging technologies should make explicit and material connections with the historic, social, material and economic durabilities and stabilities of socio-technical systems. Selin and Sadowski explore the possibilities for this through their own work to develop a place-based urban walking tour in which the constitution and interrogation of obduracies became a key focus. Through probing the relevance, role and risks of focusing on obduracy and diverse timescapes in participatory technology assessment, this chapter opens up questions about the politics of the future that are inscribed into public engagement processes and how to incorporate temporal considerations into their framing and performance.

Developing the 'technologies of participation' theme from earlier sections of the book, in Chapter 12 Jan-Peter Voß shifts our interest to a different kind of 'black boxing' of complex socio-material realities – relating to that of participation and democracy itself. This moves the interest from one of participation *in* technology assessment to one of the technology assessment *of* participation. Voß focuses on a particular democratic innovation – that of 'citizen panels' as a method of public participation – and provides a view of its ongoing transnational innovation journey. He argues that attempts to technicalize and expand this particular participatory technique have been countered by reflexive engagements – from critical academic discourse, direct protest actions and dedicated assessments – which work together as forms of 'informal technology assessment'. Voß goes on to describe a novel anticipatory assessment exercise on the future development of 'citizen panels', undertaken in Berlin in April 2014, which serves as one possibility for building reflexive engagement, anticipation and responsibility into the development of participatory practices and other areas of social innovation.

In the final chapter of the book the editors draw together perspectives from across the volume and move towards a set of concluding statements. After summing up key interpretive-analytical themes put forward by the different authors, the focus of the chapter then moves to outlining a more extensive set of programmatic statements about the implications for remaking participation. The co-productionist and relational ways of viewing the realities of participation set out in the book open up important paths for remaking participatory practice. Four are proposed as being particularly important – namely: the need to forge *reflexive participatory practices* that attend to their uncertainties, emergence, overflows and effects; the need to *ecologize participation* through attending to the interrelations and patterns of diversities of participatory collectives that are entangled in any given system or issue space; the need to catalyse processes and practices to bring about *responsible democratic innovations* through opening up spaces for reflexive engagement with and anticipatory governance of 'innovations in participation'; and, finally, an ongoing requirement for reimagining and *reconstituting participation* as constitutive of science and democracy, which brings forward new ways of seeing, being with, doing and caring for public participation with science and the environment in the twenty-first century.

Notes

1 Classic formulations of the relationship between science and democracy speak of an implied – and, at times, explicit – 'social contract for science' based on the promise that scientific research will produce novel technologies in return for continued public patronage and political autonomy (Guston 2000). While much recent commentary has focused on the shifting terms of this contract – with calls for public investment in socially, economically and environmentally 'relevant' research tied to new cultures of audit and accountability (Demeritt 2000; Strathern 2000; Gallopín et al. 2001; Slaughter and Rhoades 2005) – the scientific community has been relatively successful in maintaining a clear distinction between the often fraught political negotiations entailed in securing ongoing allocations of public funds for research (Hart 1998) and the socially protected role science plays in 'speaking truth to power' (Jasanoff 1990).
2 See Osborne and Rose (1999) and Law and Urry (2004) on the ways in which conceptions of public opinion are enacted by and are congruent with the development of sociological research technologies.
3 This appropriation of concepts derived from deliberative political theory persists in recent attempts to move the debate forward. For a critical review see Lövbrand et al. (2011).
4 The term 'co-production' has at least two meanings pertaining to our current discussion. First is the popular instrumental-organizational meaning of co-production as new forms of *interaction* between science, policy and society where different actors come together in discrete spaces of collaboration to 'make things together' – knowledge, innovations, decisions and so on. Forms of public participation have come to be seen as an (or in some cases the) exemplar of this meaning of 'doing co-production', which holds similar assumptions to the realist perspective of participation just outlined. The second meaning of co-production is the interpretive-philosophical one developed in STS and social theory that views science, politics and society as inherently intertwined and mutually constituted (Latour 1993; Jasanoff 2004). All science is social, the social is material, and everything (including participation and democracy) is co-produced. It is this *co-productionist* meaning of co-production that we adopt in this book, elaborating what it means to think about participation in co-productionist terms, starting in this section.
5 Here our approach is close to interpretive approaches to public sense-making, which have usefully highlighted the methodological individualism evident in much research concerned with 'public attitudes' to scientific and technological issues. Rather than persist with an image of the public as simply an aggregate of formally independent actors, this work has usefully explored interpretive readings of the intersubjective processes of collective sense-making (Proctor 1998).

References

Amir, S. 2012. *The Technological State in Indonesia: The Co-Constitution of High Technology and Authoritarian Politics*. London: Routledge.
Barry, A. 2001. *Political Machines: Governing a Technological Society*. London: The Athlone Press.
Barry, A. 2002. The anti-political economy, *Economy and Society* 31(2): 268–284.
Barry, A. 2012. Political situations: Knowledge controversies and transnational governance, *Critical Policy Studies* 6: 324–336.
Barry, A. 2013. *Material Politics: Disputes Along the Pipeline*. Oxford: Wiley-Blackwell.

Beck, U. 1992. *Risk Society: Towards a New Modernity*. London: SAGE.
Beck, U., Giddens, A. and Lash, S., eds. 1994. *Reflexive Modernisation: Politics, Tradition and the Aesthetic in Modern Social Order*. Cambridge: Polity Press.
Benhabib, S., ed. 1996. *Democracy and Difference: Contesting the Boundaries of the Political*. Princeton, NJ: Princeton University Press.
Benjamin, R. 2009. A lab of their own: Genomic sovereignty as postcolonial science policy, *Policy and Society* 28(4): 341–355.
Bijker, W. E., Bal, R. and Hendriks, R. 2009. *The Paradox of Scientific Authority: The Role of Scientific Advice in Democracies*. Cambridge, MA: MIT Press.
Blowers, A. and Sundqvist, G. 2010. Radioactive waste management–technocratic dominance in an age of participation, *Journal of Integrative Environmental Sciences* 7(3): 149–155.
Bohman, J. and Rehg, W., eds. 1997. *Deliberative Democracy: Essays on Reason and Politics*. Cambridge, MA: MIT Press.
Brown, M. 2009. *Science in Democracy: Expertise, Institutions, and Representation*. Cambridge, MA: MIT Press.
Brown, N. and Michael, M. 2002. From authority to authenticity: The changing governance of biotechnology, *Health, Risk and Society* 4(3): 259–272.
Brown, P. M. and Mikkelson, E. J. 1990. *No Safe Place*. Berkeley: University of California Press.
Callon, M. 1986. Some elements of a sociology of translation: Domestication of the scallops and the fishermen of St Brieuc Bay. In: J. Law (ed.) *Power, Action, Belief*. London: Routledge & Kegan Paul, pp196–223.
Callon, M., Lascoumes, P. and Barthe, Y. 2009. *Acting in an Uncertain World: An Essay on Technical Democracy*. Cambridge, MA: MIT Press.
Callon, M., Millo, Y. and Muniesa, F., eds. 2007. *Market Devices*. Oxford: Blackwell.
Castells, M. 1996. *The Rise of the Network Society: The Information Age: Economy, Society and Culture Vol. I*. Oxford: Blackwell.
Chilvers, J. 2008. Environmental risk, uncertainty, and participation: Mapping an emergent epistemic community, *Environment and Planning A* 40(12): 2990–3008.
Collins, H. and Evans, R. 2002. The third wave of science studies: Studies of expertise and experience, *Social Studies of Science* 32(2): 235–296.
Cooke, B. and Kothari, U., eds. 2001. *Participation: The New Tyranny?* London: Zed Books.
Cruikshank, B. 1999. *The Will to Empower: Democratic Citizens and Other Subjects*. Ithaca, NY: Cornell University Press.
Delgado, A., Kjolberg, K. L. and Wickson, F. 2011. Public engagement coming of age: From theory to practice in STS encounters with nanotechnology, *Public Understanding of Science* 20(6): 826–845.
Demeritt, D. 2000. The new social contract for science: Accountability, relevance, and value in US and UK science and research policy, *Antipode* 32(3): 308–329.
Dryzek, J. S. 2000. *Deliberative Democracy and Beyond: Liberals, Critics, Contestations*. Oxford: Oxford University Press.
Dryzek, J. S. 2010. *Foundations and Frontiers of Deliberative Governance*. Oxford: Oxford University Press.
Einsiedel, E. and Kamara, M. 2006. The coming of age of public participation. In: G. Gaskell and M. W. Bauer (eds) *Genomics and Society: Legal, Ethical and Social Dimensions*. London: Earthscan, pp95–112.
Elden, S. 2013. *The Birth of Territory*. Chicago, IL: University of Chicago Press.
Epstein, S. 1995. The construction of lay expertise: AIDS activism and the forging of credibility in the reform of clinical trials, *Science, Technology & Human Values* 20(4): 408–437.

Epstein, S. 1996. *Impure Science: AIDS, Activism and the Politics of Knowledge*. Berkeley: University of California Press.

Ezrahi, Y. 1990. *The Descent of Icarus: Science and the Transformation of Contemporary Democracy*. Cambridge, MA: Harvard University Press.

Felt, U. and Fochler, M. 2010. Machineries for making publics: Inscribing and de-scribing publics in public engagement, *Minerva* 48(3): 219–238.

Fiorino, D. J. 1990. Citizen participation and environmental risk: A survey of institutional mechanisms, *Science, Technology & Human Values* 15(2): 226–243.

Franklin, S. 2013. *Biological Relatives: IVF, Stem Cells and the Future of Kinship*. Durham, NC: Duke University Press.

Fung, A. and Wright, E. O. 2003. *Deepening Democracy: Institutional Innovations in Empowered Participatory Governance*. London: Verso.

Funtowicz, S. O. and Ravetz, J. 1993. Science for the post-normal age, *Futures* 25(7): 735–755.

Gallopín, G. C., Funtowicz, S., O'Connor, M. and Ravetz, J. 2001. Science for the twenty-first century: From social contract to the scientific core, *International Social Science Journal* 53(168): 219–229.

Gottweis, H. 2008. Participation and the new governance of life, *BioSocieties* 3: 265–286.

Goven, J. 2006. Processes of inclusion, cultures of calculation, structures of power: Scientific citizenship and the Royal Commission on genetic modification, *Science, Technology & Human Values* 31(5): 565–598.

Guston, D. 2000. *Between Politics and Science: Assuring the Integrity and Productivity of Research*. Cambridge: Cambridge University Press.

Habermas, J. 1984. *Theory of Communicative Action – Volume 1: Reason and the Rationalization of Society*. Boston, MA: Beacon Press.

Habermas, J.1996. *Between Facts and Norms: Contributions to a Discourse Theory of Law and Democracy*. Cambridge, MA: MIT Press.

Hagendijk, R. P. and Irwin, A. 2006. Public deliberation and governance: Engaging with science and technology in contemporary Europe, *Minerva* 44: 167–184.

Hart, D. M. 1998. *Forged Consensus: Science, Technology and Economic Policy in the United States, 1921–1953*. Princeton, NJ: Princeton University Press.

Hecht, G., ed. 2011. *Entangled Geographies: Empire and Technopolitics in the Global Cold War*. Cambridge, MA: MIT Press.

Irwin, A. 1995. *Citizen Science: A Study of People, Expertise and Sustainable Development*. London: Routledge.

Irwin, A. 2001. Constructing the scientific citizen: Science and democracy in the biosciences, *Public Understanding of Science* 10(1): 1–18.

Irwin, A. 2006. The politics of talk: Coming to terms with the 'new' scientific governance, *Social Studies of Science* 36(2): 299–320.

Irwin, A. and Michael, M. 2003. *Science, Social Theory and Public Knowledge*. Maidenhead: Open University Press.

Irwin, A., Jensen, T. E. and Jones, K. E. 2013. The good, the bad and the perfect: Criticizing engagement practice, *Social Studies of Science* 43(1): 118–125.

Jasanoff, S. 1990. *The Fifth Branch: Science Advisers as Policymakers*. Cambridge, MA: Harvard University Press.

Jasanoff, S., ed. 2004. *States of Knowledge: The Co-Production of Science and Social Order*. London: Routledge.

Jasanoff, S. 2005. *Designs on Nature: Science and Democracy in Europe and the United States*. Princeton, NJ: Princeton University Press.

Jasanoff, S., ed. 2011. *Reframing Rights: Bioconstitutionalism in the Genetic Age*. Cambridge, MA: MIT Press.

Jasanoff, S. 2012. *Science and Public Reason*. London: Routledge.

Jones, R., Pykett, J. and Whitehead, M. 2011. Governing temptation: Changing behaviour in an age of libertarian paternalism, *Progress in Human Geography* 35(4): 483–501.

Kearnes, M. and Wynne, B. 2007. On nanotechnology and ambivalence: The politics of enthusiasm, *Nanoethics* 1(2): 131–142.

Kearnes, M., Macnaghten, M. and Davies, S. R. 2014. Narrative, nanotechnology and the accomplishment of public responses, *Nanoethics* 8: 241–250.

Laclau, E. and Mouffe, C. 2001. *Hegemony and Socialist Strategy: Towards a Radical Democratic Politics*. London: Verso.

Latour, B. 1987. *Science in Action*. Cambridge, MA: Harvard University Press.

Latour, B. 1993. *We Have Never Been Modern*. Cambridge, MA: Harvard University Press.

Laurent, B. 2011. Technologies of democracy: Experiments and demonstrations, *Science and Engineering Ethics* 17(4): 649–666.

Law, J. and Urry, J. 2004. Enacting the social, *Economy and Society* 33(3): 390–410.

Leach, M., Scoones, I. and Wynne, B., eds. 2005. *Science and Citizens: Globalization and the Challenge of Engagement*. London: Zed Books.

Lezaun, J. 2007. A market of opinions: The political epistemology of focus groups, *The Sociological Review* 55(S2): 130–151.

Lezaun, J. and Soneryd, L. 2007. Consulting citizens: Technologies of elicitation and the mobility of publics, *Public Understanding of Science* 16(3): 279–297.

Lövbrand, E., Pielke, R. J. and Beck, S. 2011. A democracy paradox in studies of science and technology, *Science, Technology & Human Values* 36(4): 474–496.

MacKenzie, D., Muniesa, F. and Siu, L., eds. 2007. *Do Economists Make Markets? On the Peformativity of Economics*. Princeton, NJ: Princeton University Press.

Macnaghten, P., Davies, S. R. and Kearnes, M. 2015 (forthcoming). Understanding public responses to emerging technologies: A narrative approach, *Journal of Environmental Policy & Planning*, www.tandfonline.com/doi/full/10.1080/1523908X.2015.1053110 (accessed 24 September 2015).

Mahony, N., Newman, J. and Barnett, C., eds. 2010. *Rethinking the Public: Innovations in Research, Theory and Politics*. Bristol: Policy Press.

Marres, N. 2007. The issues deserve more credit: Pragmatist contributions to the study of public involvement in controversy, *Social Studies of Science* 37(5): 759–780.

Marres, N. 2012. *Material Participation: Technology, the Environment and Everyday Publics*. Basingstoke: Palgrave.

Marres, N. and Lezaun, J. 2011. Materials and devices of the public: An introduction, *Economy and Society* 40(4): 489–509.

Marres, N. and Rogers, R. 2008. Subsuming the ground: How local realities of the Fergana Valley, the Narmada Dams and the BTC pipeline are put to use on the Web, *Economy and Society* 37(2): 251–281.

Michael, M. and Rosengarten, M. 2013. *Innovation and Biomedicine: Ethics, Evidence and Expectation in HIV*. Basingstoke: Palgrave.

Mouffe, C. 2000. *The Democratic Paradox*. London: Verso.

Mouffe, C. 2005a. *On the Political*. London: Routledge.

Mouffe, C. 2005b. Some reflections on an agonistic approach to the public. In: B. Latour and P. Weibel (eds) *Making Things Public: Atmospheres of Democracy*. Cambridge, MA: MIT Press, pp804–807.

Mulkay, M. 1979. *Science and the Sociology of Knowledge*. Boston: George Allen & Unwin.
Muniesa, F. 2014. *The Provoked Economy: Economic Reality and the Performative Turn*. London: Routledge.
Nowotny, H., Scott, P. and Gibbons, M. 2001. *Re-Thinking Science: Knowledge and the Public in an Age of Uncertainty*. Cambridge: Polity Press.
Osborne, T. and Rose, N. 1999. Do the social sciences create phenomena? The example of public opinion research, *British Journal of Sociology* 50(3): 367–396.
Owen, R., Macnaghten, P. and Stilgoe, J. 2012. Responsible research and innovation: From science in society to science for society with society, *Science and Public Policy* 39(6): 751–760.
Pellizzoni, L. 2001. The myth of the best argument: Power, deliberation and reason, *The British Journal of Sociology* 52(1): 59–86.
Pellizzoni, L. and Ylönen, M., eds. 2012. *Neoliberalism and Technoscience: Critical Assessments*. Farnham, Surrey: Ashgate.
Pinch, T. and Swedberg, R., eds. 2008. *Living in a Material World: Economic Sociology Meets Science and Technology Studies*. Cambridge, MA: MIT Press.
Proctor, J. D. 1998. The meaning of global environmental change: Retheorising culture in human dimensions research, *Global Environmental Change* 8(3): 227–248.
Rancière, J. 2006. *Hatred of Democracy*. London: Verso.
Renn, O., Webler, T. and Wiedemann, P., eds. 1995. *Fairness and Competence in Citizen Participation: Evaluating Models for Environmental Discourse*. Dordrecht: Kluwer.
Rip, A., Misa, T. and Schot, J. 1995. *Managing Technology in Society: The Approach of Constructive Technology Assessment*. London: Thomson.
Rowe, G. and Frewer, L. 2000. Public participation methods: A framework for evaluation, *Science, Technology & Human Values* 25(1): 3–29.
Shapin, S. and Schaffer, S. 1985. *Leviathan and the Air-Pump: Hobbes, Boyle and the Experimental Life*. Princeton, NJ: Princeton University Press.
Slaughter, S. and Rhoades, G. 2005. From 'endless frontier' to 'basic science for use': Social contracts between science and society, *Science, Technology & Human Values* 30(4): 536–572.
Stern, P. and Fineberg, H., eds. 1996. *Understanding Risk: Informing Decisions in a Democratic Society*. Washington, DC: National Academy Press.
Stirling, A. 2008. 'Opening up' and 'closing down': Power, participation and pluralism in the social appraisal of technology, *Science, Technology & Human Values* 33(2): 262–294.
Strathern, M., ed. 2000. *Audit Cultures: Anthropological Studies in Accountability, Ethics and the Academy*. London: Routledge.
Strathern, M. 2005. Robust knowledge and fragile futures. In: A. Ong and S. Collier (eds) *Global Assemblages. Technology, Politics and Ethics as Anthropological Problems*. Oxford: Blackwell, pp464–481.
Swyngedouw, E. 2005. Governance innovation and the citizen: The Janus face of governance-beyond-the-state, *Urban Studies* 42(11): 1991–2006.
Swyngedouw, E. 2010. Apocalypse forever? Post-political populism and the spectre of climate change, *Theory, Culture & Society* 27(2–3): 213–232.
Taylor, C. 1989. *Sources of the Self: The Making of the Modern Identity*. Cambridge, MA: Harvard University Press.
Tewdwr-Jones, M. and Allmendinger, P. 1998. Deconstructing communicative rationality: A critique of Habermasian collaborative planning, *Environment and Planning A* 30(11): 1975–1989.

Thorpe, C. and Gregory, J. 2010. Producing the post-Fordist public: The political economy of public engagement with science, *Science as Culture* 19(3): 273–301.

Webster, A. 2007. Crossing boundaries: Social science in the policy room, *Science, Technology & Human Values* 32: 458–478.

Welsh, I. and Wynne, B. 2013. Science, scientism and imaginaries of publics in the UK: Passive objects, incipient threats, *Science as Culture* 22(4): 540–566.

Wilsdon, J. and Willis, R. 2004. *See-Through Science: Why Public Engagement Needs to Move Upstream.* London: Demos.

Wynne, B. 1988. Unruly technology: Practical rules, impractical discourses and public understanding, *Social Studies of Science* 18(1): 147–167.

Wynne, B. 2003. Seasick on the Third Wave? Subverting the hegemony of propositionalism. Response to Collins and Evans (2002). *Social Studies of Science* 33(3): 401–417.

Wynne, B. 2006a. Afterword. In: M. B. Kearnes, P. Macnaghten, and J. Wilsdon (eds) *Governing at the Nanoscale*. London: Demos, pp70–78.

Wynne, B. 2006b. Public engagement as a means of restoring public trust in science: Hitting the notes, but missing the music? *Community Genetics* 9(3): 211–220.

Wynne, B. 2007a. Dazzled by the mirage of influence? STS–SSK in multivalent registers of relevance, *Science, Technology & Human Values* 32: 491–503.

Wynne, B. 2007b. Public participation in science and technology: Performing and obscuring a political-conceptual category mistake, *East Asian Science, Technology and Society: An International Journal* 1(1): 99–110.

PART I
Rethinking participation

2

PARTICIPATION IN THE MAKING

Rethinking public engagement in co-productionist terms

Jason Chilvers and Matthew Kearnes

Introduction

The turn toward public participation in both the rhetoric and practice of contemporary science governance might be understood as one element of a broader shift in the relationship between science, the market and the practice of political power. Jasanoff (2011) situates this recent enthusiasm for public participation in a series of 'constitutional changes' in the relationship between science, society and political power. She argues that these transformations have entailed a renegotiation of 'the manner in which states and other authoritative institutions employ the power of expertise, and contests over those processes have become a fixture of modern democratic politics'. In this light 'public engagement is but the latest discursive rubric under which that contestation is played out' (p624). While predicated on the implied promise of science and innovation, practices of public participation are paradoxically designed to remake the conditions of scientific work in order to ensure the realization of this potential.

There is, of course, much to document here, and the thorough scholarly work that unpicks the attendant histories of these developments and prospects for recasting the relationship between science and society for progressive and democratic purposes has only recently commenced in earnest.[1] In this chapter our ambition is a more modest and conceptual one: to outline a theoretical vantage point for a performative and co-productionist account of public participation 'in the making'. To do this we draw on insights from a tradition of research that has insisted on the always-contingent and compositional nature of the social world. For example, studies informed by pragmatist social thought in the sociology of science have insisted on the '*institutive capacities* of scientific or technical knowledge and ... the types of realities that are brought into existence or modified throughout scientific ventures' (Muniesa 2014, 10, emphasis added) and, more broadly, that 'both the natural and life sciences, along with the social sciences, contribute toward enacting the realities that they describe' (Callon 2007, 315). Work in this tradition has vividly articulated the enactive quality of scientific knowledge making (Latour 1987), the construction and maintenance of financial systems (Callon 1998; MacKenzie *et al.* 2007) and the creation of global knowledge exchange networks (Edwards 2010). It is, however, striking to note that while writing inspired by science and technology studies (STS) has amplified the ways in which technoscientific artefacts – and their attendant social, political and economic orderings – are practical and contingent accomplishments sustained by an assemblage of translations between social and material worlds, this insight has not been systematically taken up in studies of public participation. A core goal of this chapter, and the volume as a whole, is to address this lacuna by developing avenues for symmetrical analyses of public participation that attend to the performative enactment of public collectivities around, against and in tension with scientific claims and technological developments.

Through reviewing and building on work in STS, political theory, geography and anthropology we set out the conceptual foundations for a more expansive co-productionist account of participation in the making, which to some extent mirrors enduring themes in STS scholarship on the mutual construction of technoscientific and social orders. The chapter is organized in three substantive sections. The first draws together and develops co-productionist understandings of participation in the making at the level of situated *participatory experiments and practices*. This forms a basis for understanding the emergence of all forms of participation and public representation as socio-material collectives that are both shaped by – and in turn shape – technoscientific, political and social orders. Not only does this open up the analytical horizon beyond formal event-based 'in vitro' studies of relatively controlled processes to also encompass diverse forms of public collectivities 'in the wild'. It also means that, rather than viewing publics as external social actors or democracy as defined by pre-given norms, publics and democracy are made (and indeed re-made) in the performative work entailed in constituting public collectives, whether brought into being by issues/objects, procedural formats and/or forms of human intervention.

The second main section moves beyond questions of the emergence of participatory collectives in particular situations or 'events' to consider the co-production of participation in wider spaces of coherence and prescription. Here our focus turns to the portability, standardization and circulation of *technologies and expertise* of participation across time, place and cultural-political settings, and how powers of participatory prescription become lodged in expert communities, institutions and infrastructures.

In the final main section we argue that understanding the dynamics of participation in science and democracy depends on considering the interrelations of participatory experiments and technologies of participation with respect to two further spaces of co-productive coherence: *spaces of negotiation and controversy* within which diverse participatory experiments intermingle; and *spaces of political culture and constitutional relations* within which certain participatory collective practices and knowledge-ways become seen as authoritative and are endowed with legitimacy and meaning. The co-productionist framework we outline suggests that interpretive analyses of the dynamics of participation in science and democracy should be extended to more fully understand relations within and between the topological spaces of participation explored in this chapter.

Participatory experiments and practices

A central theme in any move to re-conceptualize participation as co-produced, relational and emergent is the adoption of an experimental, empirically oriented and practice-based perspective on democratic engagement. As we have outlined in Chapter 1, this approach marks a departure from realist and positivist traditions in public attitude research – which depict public opinion as an aggregate of individual attitudes, preferences and desires (Proctor 1998) – and 'correspondence' theories of democracy that tend to represent 'the public' as largely autonomous, naturally occurring and external to technoscientific and democratic practice (Brown 2009). In contrast, a relational perspective – inspired by co-productionist work in STS, which has demonstrated the ways in which technological and scientific artefacts might be viewed as admixtures, composed of a hybrid assemblage of material, social and discursive relations (Latour 1993; Haraway 1997) – depicts all forms of participation as open-ended social experiments and collective practices (Irwin 2006). Rather than being somehow external or pre-given, practices of public participation actively produce publics, public issues, material commitments and forms of democratic engagement through the ways in which they are configured, mediated and performed.

However, we must be clear that this does not simply imply a form of *social* construction – that publics are simply constructed rhetorically, or inscribed discursively. Rather, participatory experiments and practices can be defined as heterogeneous socio-material collectives formed through the enrolment of material objects, devices, infrastructures, social practices, human subjects and theoretical and cognitive investments (Laurent 2011). This demarcates a specific participatory space and experimental setting that articulates a distinct 'atmosphere' of democracy (Latour and Weibel 2005), specific public identities, definitions of public issues, and other productive effects.[2] Public involvement in science and politics might therefore be defined as the enactment of socio-material collectives through which publics address collective public problems and in which the emergence of 'a public' (a democratic public, a deliberative public, an affected public and so on) is a contingent accomplishment and (at times) the focus of contestation.

In conceptualizing participation in co-productionist terms at the level of situated participatory experiments and practices, the approach outlined in this section sensitizes analyses to the following concerns:

- How do participatory experiments and practices come into being, what shapes them and how are they articulated?
- Who/what enrols heterogeneous actors into participatory collectives, mediates them and with what significant exclusions?
- What resistances and challenges do participatory experiments face?

- What are the productive dimensions and effects of collective participatory practices in terms of the ways in which they construct the issues or objects of participation, identities of the actors involved, material commitments and models of participation or democracy itself?

Co-producing participation: in vitro versus in vivo settings

Public experiments have always been a feature of science and democracy (Shapin and Schaffer 1985). Over time they have varyingly taken on epistemological, discursive and ontological forms, yet the notion of 'participatory experiments' forms a further category centred on participatory collectives and devices in and of themselves (Marres 2012). In practical terms, participatory experiments take on a variety of forms – from those that are *in vitro* (i.e., performed in closed, controlled, laboratory-like settings such as research projects developing new participatory techniques or applying established ones like citizen panels or focus groups) through to more spontaneous *in vivo* experiments conducted in the 'wild' where practices emerge more spontaneously around particular issues and concerns (Callon *et al.* 2009; also cf. Bogner 2012).

In-depth studies of specific collectives of public involvement that have tended to emphasize *in vitro* settings and 'invited' small group deliberative forms of engagement – or so-called 'mini-publics' (Goodin and Dryzek 2006) – have usefully demonstrated the ways in which participatory practices are both shaped by, and shape, technoscientific and social orders. An important theme in this work has been uncovering the ways in which participatory processes, and particularly the articulation and definition of issues and matters of collective concern, are framed by extant powers and orders, including incumbent scientific, political and economic institutions (Irwin 2001; Goven 2006; Stirling 2008). More recent work has emphasized the ways in which issues themselves are actively produced through the performance of participatory practices (Marres 2007). These situated studies have also demonstrated how the subjectivities and identities of 'the public' – for example, as innocent citizens, affected publics, interest group stakeholders and so on – are shaped by established democratic procedures and scientific-institutional assumptions (Irwin 2006; Wynne 2006; Braun and Schultz 2010) *and* the situated agency and active resistances of participating subjects themselves (Callon and Rabeharisoa 2004; Felt and Fochler 2010). What this means is a dual commitment that sees that the participatory normativities and political ontologies of participatory experiments are inscribed in the array of pre-exiting institutional arrangements for representing science and democracy (Brown 2009) *and* produced through the particular socio-material configurations, settings and arrangements that make up collective participatory practices (Braun and Whatmore 2010; Marres and Lezaun 2011).

These situated studies of participatory practice have helped to clarify the origins and locus of power in the formation of participatory experiments, who/what orchestrates participatory collectives and how, and the exclusions and effects that occur. Such studies have also brought forward critical and reflexive perspectives concerning the potential for participatory practices to reconstruct so-called 'technocratic' dynamics of closure and control that they seek to overcome (Rayner 2003; Chilvers 2008a; Stirling 2008). At the same time, the situated and symmetrical nature of these studies resists socially determinist interpretations of human and material forms of mediation, emphasizing instead the multiple and emergent productive effects and material commitments produced through the performance of participatory practices (Chilvers and Longhurst 2015). This breadth of discussion represents an avenue for remaking theories of participation from a relational and co-productionist standpoint in ways that enable richer analyses of participation *in vivo*, outside the constraints of formalized, mini-public and event-based participatory processes. It is important to stress here that we are not arguing that there is a simple opposition between formalized processes of public participation that now commonly form a key part securing bureaucratic legitimacy and collectively defined processes of participation 'in the wild'. All forms of participation – whether formal or informal – might be viewed as enacting forms of public collectives and their attendant political ontologies. However, as we will describe below, a key element in developing a more relational sensibility concerning public participation is a recognition of the multiple ways in which publics, public issues and political ontologies are enacted in both formal and informal settings.

Objects and issues

In existing studies of the enactment of publics there are at least three different ways of conceiving the dynamic process of performing participatory collectives, each of which foregrounds the relative role of: first, objects and issues; second, procedural formats; and, third, the human agency of mediators and participating subjects. To take the first of these themes, co-productionist accounts of technoscience have long emphasized the constitutive nature of problematization – and, indeed, the material, moral and discursive *objects* – around which heterogeneous actor-networks are enrolled in the production of scientific knowledge-commitments (Callon 1986; Latour 1987). In its desire to provide a symmetrical analysis of the social life of science and technology, research in STS has demonstrated the often highly partial ways in which public participation practices are framed by scientific and regulatory institutions and the often thinly veiled techno-managerial ambitions that underpin the state sponsorship of public engagement and science communication initiatives (Irwin and Wynne 1996). At the same time, this research has demonstrated the broader range of objects and issues that motivate public concern, and the inadequacies of formal engagement processes in generating more cosmopolitan understandings of the relationship between science, technology and society (Wynne 2007; Stirling 2011).

More recent work, inspired by STS and political theory, has taken forward and significantly deepened an *object and issue-centred* perspective on participation through bringing together a 'socio-ontological' approach (Marres 2007) with the work of earlier pragmatist political philosophers, most notably John Dewey

(1927), and materially sensitive accounts of democracy, citizenship and the public (Barry 2001; Bennett 2010; Braun and Whatmore 2010; Marres 2012). Such accounts foreground the ways in which issues or objects bring publics and forms of public involvement into being. As Marres (2007) states, following Dewey, this 'suggests that people's involvement in politics is mediated by problems that affect them' and focuses attention on the '"attachments" that people mobilise (and that mobilise people) in the performance of their concern with public affairs' (p759). The enactment of public participation is fundamentally entwined with the composition of publics and what Latour (2004) evocatively terms 'matters of concern'. For Latour, though 'scientific, technological, and industrial production has been an integral part of their definition from the beginning', matters of concern 'have no clear boundaries, no well-defined essences, no sharp separation between their own hard kernel and their environment' (p24). In areas of technological and environmental change as diverse as climate change, genetically modified (GM) technologies, food security, sustainable energy and public health, we see the rhizomatic articulation of matters of concern as a network that links material configurations with emergent public collectives. This enactment of public participation typically challenges existing institutional framings and capacities to deal with them. Emergent participatory practices not only overflow institutional issue definitions, but also established, formalized democratic procedures for representing publics – be they deliberative processes, opinion polls, elections and so on (Marres 2007).

This object-oriented perspective therefore provides an insight into the sheer multiplicity, variability and diversity of participatory practices. Participation is in this sense *continually emergent*. Viewed in these terms, participation might be understood as the always-partial process of defining objects of political concern – in which the objects of public participation, the constituency of affected 'publics' and what is legitimated as 'political' are themselves always a contingent outcome of the processes of participation. In addition, the pragmatist orientation of this work also helps to highlight the ways in which the objectives and purposes of public participation are always being transformed by processes of public participation itself. This includes participatory experiments 'in the wild' that are often ephemeral, fleeting or seemingly mundane. In this sense, the material entanglement of publics with objects and issues is therefore not bounded within or limited to formal spaces of public deliberation. This material entanglement is equally apparent in the more informal, private and mundane spaces of the home and the workplace, including, for example, public engagements in behaviour change initiatives, the domestication of smart energy technologies, social media commentary and consumption practices – potentially anything!

As Marres and Lezaun (2011) point out, this vision of participatory experimentation sits in stark contrast to traditions of social and political thought which, from Aristotle through to Habermas, have defined public engagement in politics largely in procedural, linguistic and discursive terms, where public action requires removing publics from everyday material attachments to allow 'proper' expression in public fora. Rather than being fixed in this way, an object-centred approach attempts the double move of recognizing the contingent construction of both the objects of public participation and the definition of what constitutes 'the public'. This intentionally experimental approach is located in pragmatist political thought, as first initiated by Dewey (1927) just under a century ago.

Procedural formats

While taking a similar view of participatory experiments and practices as socio-material collectives, a parallel area of contemporary scholarship foregrounds the role of *procedural formats and modes of organizing* in bringing instances of participation and 'participating publics' into being. This work has focused on the ways in which democratic concepts and norms (deliberative, liberal, agonistic and so on – see Durant 2010, 2011), participatory methods, and 'technologies of participation' play a crucial role as instruments and blueprints in shaping the configuration of participatory processes (Lezaun and Soneryd 2007; Laurent 2011). Most existing studies foregrounding this dimension have focused on *in vitro* experiments in public participation and have highlighted the ways in which participatory methods – such as citizen panels (Lezaun and Soneryd 2007), consensus conferences (Horst and Irwin 2010) and focus groups (Lezaun 2007), together with social science methods and research techniques (Osborne and Rose 1999; Law and Urry 2004; Ruppert et al. 2013) – are characterized by standardized mechanisms for enrolment, pre-given political epistemologies and modes of mediating collective participatory configurations.

In broad terms, this work highlights the traffic between social science method and normative political theory, and particularly the ways in which public participation practices function to enact 'ideal publics'. For example, Lezaun's (2007) enquiry into the political epistemology of focus groups sees the replication of this qualitative social science method as a kind of laboratory polity; an experimentally assembled and transient community. Importantly, Lezaun highlights the work invested in order to make this socio-material configuration work and the continual trials faced in order to produce tradable public opinions. This includes the disciplining of research subjects by managing their reflexivity and the materialization of a particular political philosophy (i.e., an isegoric situation based on equality of expression but where talk is always steered to the goals of the moderator), through the physical setting of the focus group (e.g., configuring a constrained space for discussion) and the arrangement of devices (such as digital recorders, used to assist the elicitation).

While such studies have yielded important insights into the configuration and formatting of forms of public participation, a key inspiration for the current volume is to develop a sensitivity for the contingencies of this performative work. While it is clear that public participation methods enact the kinds of publics envisaged in political theory (neoliberal, deliberative, dialogic, etc.), our interest is in the necessarily co-productive nature of this process. In part this is inspired by the observation that the importance of implicit democratic norms – and their associated technologies of participation – is not limited to *in vitro* experiments and formalized participatory processes. Rather, the movement, translation and standardization of techniques of public participation play a significant role in shaping a range of collective participatory practices, such as counter-experiments in resistance to formalized democratic formats (Laurent 2011) and forms of activism and public protest (Barry 2001). Indeed, in many cases it is precisely these normativities of participation themselves that become the object of both political controversy and experimentation (Chilvers 2008b; Chilvers and Burgess 2008). Through the enactive work of forming participatory collectives, these implicit normative commitments are often challenged, revised and creatively reproduced (Van Oudheusden and Laurent 2013).

The crucial point we want to emphasize here is that the enactment of participatory collectives entails the expression of distinct political ontologies – or following Latour and Weibel (2005), what we might we term 'atmospheres of democracy' – in the definitional work required in collectively articulating the nature of issues and their normative and material dimensions. Democratic norms are not simply pre-given or imported into the participatory process. Rather, democratic normativities – atmospheres of democracy – are made in practice. The formation of public collectives is also underpinned by a culturally and historically situated grammar of the common good (Boltanski and Thévenot 2006) that shapes the ways in which issues 'become issues', and the investment in particular kinds of procedural, bureaucratic and delegatory formats (Thévenot 1984).[3] The structures of democratic representation and deliberation (whether they are organized through formal processes of political delegation or more insurgent forms of public participation and protest) shape the development of and are produced through participatory experiments in the making.

For this reason, a relational approach to the making – and remaking – of public participation therefore serves to both problematize and extend the project of democratizing technoscience. While recent work in STS and associated interventions in policy and political practice have done much to render claims to scientific authority more publically and democratically accountable, much of this work has been underpinned by pre-given political normativities. Somewhat ironically, this has been a particular feature of what has been termed the 'participatory turn' in contemporary STS research and practice, where deliberative democratic and other models of participation, drawn from political science, have been appropriated largely unproblematically (de Vries 2007), with limited questioning about their conditions of validity and effects. A parallel body of work attempts to resolve the relationship between science and democracy through procedural innovation. Here we might think of Callon *et al.*'s (2009) 'hybrid fora', Latour's (2004) 'parliament of things' or Funtowicz and Ravetz's (1993) notion of 'extended peer communities' in post-normal science.

While constituting useful heuristics in their own terms, these procedural interventions have not successfully extended the symmetrical analyses of science 'in the making' to democratic practice and its objects and procedures. It is critical, then, that democratic, dialogic and procedural 'givens' – fora, parliaments and communities – are figured in broadly bureaucratic and procedural terms as a means of bringing order to a world made complex by the (de)construction of technoscience. As Marres (2007) observes, when Callon and Latour 'describe democratic processes in terms of "the composition of the common world", they commit themselves to a republican conception of democracy: they adopt a sociologised and ontologised notion of the common good' (p764). Rethinking participation as co-produced and 'in the making' thus stands, as we will outline below, to have far-reaching implications for understanding relations between science and society, and attempts to reconfigure or 'democratize' these relations.

Human agency of mediators and participating subjects

A final category of studies that view participation as experiments and practices in the making foreground, or at least place more emphasis on, the role of human agency and intentionality, particularly in the form of mediators, participating subjects or human imagination and affect. Much has been made of the important roles that facilitators or mediators can play in orchestrating, enrolling, guiding and publicizing the productions of participatory experiments (Elam *et al.* 2007; Lezaun and Soneryd 2007; Chilvers 2008b; Moore 2012). Indeed, recent work has demonstrated that mediators constitute a new category of expertise in late-modern knowledge societies (Osborne 2004; Meyer and Kearnes 2013). Osborne (2004) identifies the proliferation of mediation as characteristic of a new intellectual attitude related to the emergence of discourses of a knowledge society. Propositions for a new era of capital accumulation – under the sign of an emergent 'knowledge economy' – are typically accompanied by the paradoxical suggestion that the promises of technoscientific innovation will not be delivered by science, or scientists, alone. We might see this idiosyncratic political accommodation that demands both the liberalization of government regulation of science and the coordination of state investment in research as an effect of the 'neoliberalization' of contemporary research policy. In this context the mediator resembles an 'enabler', 'catalyst' and 'broker' of knowledge, always 'in the middle of things' (Osborne 2004), who functions to realize the potential of science (Meyer 2010). In the same way that it is common to suggest that science needs brokers in order to be truly innovative, it is also commonly accepted that democracy needs to be coaxed into becoming more democratic.

While these forms of mediation and facilitation are especially apparent in formalized and 'invited' *in vitro* participatory processes, they are also evident in more *in vivo*, citizen-led and organic forms of engagement (Chilvers 2010). Recent work on public participation processes has also emphasized the potential agency of participating subjects through their 'de-scription' of procedural models imposed by facilitators. Participants can often struggle with, shift or reject the script of participatory procedures (Felt and Fochler 2010; see also Akrich 1994), while formalized public participation processes at times become the focus of public controversy and protest. As Irwin and Michael (2003) capture with their notion of ethno-epistemic assemblages (EEAs), in more spontaneous, organic and citizen-led forms of collective public involvement, participants themselves typically self-organize and co-construct participatory procedures and the issue in question in a more rhizomic fashion – blurring the boundaries between mediator and participant.

Finally, in the same way that materialist accounts of the composition of science, technology and society have been critiqued for 'downplay[ing] the role that human consciousness plays, in exercising action, credibility, beliefs and ideas' (Chilvers and Evans 2009, 385; see also Jasanoff 2004), an object-oriented perspective on public participation can have the effect of underplaying the role of emotions, beliefs and affective dimensions in the co-production of collective participatory practices. To this end, we emphasize the small but growing number of studies highlighting the role of emotions, feelings, beliefs and imaginaries in making participation, and the importance of embodied, emotional, imaginative, sensory and affective elements in the emergence of participatory experiments (Harvey 2009; Felt 2015; Davies this volume).

Participatory co-productions

In conceptualizing participation as co-produced and relational – and by focusing on specific participatory experiments and practices – we have emphasized three related sets of theoretical positions that foreground the agency of, first, objects/ issues; second, procedural formats; and, third, human action in mediating emergent collectives of participation. The co-productionist conception of participation developed in this book emphasizes the *mutual constitution* of all of these elements that make up emergent collectives of participation, rather than ascribing analytical privilege to any one of them. The degree of emphasis on each of these relational elements – both in shaping and being produced through participatory experiments – will vary in practice and becomes in this view a profoundly contingent matter. Attending to the particular *modes of construction* evident in particular participatory practices is therefore a largely empirical question, rather than something that can be articulated *a priori*. Indeed, rather than seeing evaluative frameworks and normative commitments as being 'imported into' participation, the perspective we advocate here is sensitized to the performative enactment of quasi-democratic atmospheres and their reconfiguration in practice.

In this way, a co-productionist perspective on participation views participatory experiments as *multiply productive*. The productivity of participation goes way beyond discursive or linguistic outcomes to include material commitments (e.g., in bringing forward new technological commitments through 'grassroots' and distributed innovation in community energy projects, hackerspaces, design collectives, etc.), alternative visions and imaginaries (e.g., through forms of activism and artistic engagement), the potential for transformed social practices beyond the setting of specific participatory experiments (e.g., in relation to pro-environmental behaviour change initiatives), and so on (Chilvers and Longhurst 2012).

In sum, then, having outlined a co-productionist and relational understanding of participation in the making at the level of emergent socio-material collective practices, we can summarize the distinctive features of this perspective, and what we consider its advantages, compared to realist and correspondence theories of democracy that treat forms of participation as pre-given or natural categories that participatory practice is evaluated against. Three points are particularly pertinent. First, rather than presume that normativities of participation are projected *a priori* and materialized through established techniques (e.g., Crosby *et al.* 1986; Chambers 1994; Joss and Durant 1995), a co-productionist perspective sees these dimensions as being actively constituted through the ways in which participatory practices are constructed and performed. This alerts us to the careful work and struggle involved in achieving particular closures, representations and outcomes, and emphasizes the contingencies, power relations and politics of public engagement. Second, participation can no longer be considered a specific form of institutional or democratic practice, characterized by particular political commitments and organizational formats. Taking seriously the materiality and experimental normativities of collective participation opens the analytical horizon up to the multiplicity in public engagement practice – from the narrow meanings of public deliberation as a linguistic enterprise into mundane, domestic, private spaces and beyond. This is not a small change. It orients analysis to the potentials and possibilities of participation in terms of what it is, was, and could be (cf. Arendt 1958). It offers a symmetrical theory of participation that might be developed and applied across a range of diverse settings, and is open to emergent participatory formats and publics. Finally, understanding the construction and production of participatory experiments in this situated sense moves us away from an emphasis on evaluating participatory practices against pre-given normative principles that define the 'best' form of democratic engagement in advance (e.g., Webler 1995; Rowe and Frewer 2000) to an interest in analysing relationships between the way in which heterogeneous collectives of participation are configured and the political openings and closings that occur as a result (Barry 2001; Stirling 2008).

Technologies and expertise of participation

One of the key features of many existing constructivist studies of public participation is an over-reliance on situated case-based accounts of particular public engagement initiatives. This focus on the relatively ephemeral time-spaces of participatory initiatives, we argue, carries the risk of (re)producing popular meanings of participation as discrete one-off 'events'. An exclusive focus on the specific sites where participation is practised stands to mask and make invisible systemic power relations and co-productions occurring at sites beyond these 'front stage' performances and in other spaces and levels of social aggregation. In this section we therefore move on to outline a second important theme in considering public participation as co-produced and relational, that of technologies and expertise of participation, together with the ways in which public participation tools and meanings become standardized, codified and mobile across space and time through growth of expert communities, institutions and infrastructures.[4]

In moving to consider the *science and technologies of participation*, key questions brought forward under this theme include:

- How do participatory collectives become 'technicalized'? That is, how the traffic between localized experiential practices becomes more-or-less stabilized as technologies of participation.
- What does it mean to understand processes of participation and democratic engagement as socio-material technologies (rather than as simply the means by which publics engage in technoscience)?
- How do technologies of participation and engagement expertise move and travel from place to place and across different cultures of production and implementation?
- What is the nature and form of public participation expertise? How is it constructed, performed and demarcated? (Who counts as an expert of participation?)
- What effects do technologies and expertise of participation have, for example, on other democratic practices, technoscience and politics?

Technologies of public making

In beginning to explore such questions, social studies of participation in STS and other critical social science approaches have taken formalized techniques, tools and instruments for public involvement as their main entry point. For Rose (1999, 189) these 'technologies of community' that have been 'invented to make communities real' include things like public attitude surveys, market research, opinion polls, focus groups and citizens' juries. Lezaun and Soneryd (2007) prefer the term 'technologies of elicitation' to describe these instruments produced in transient, ephemeral and experimental time-spaces to generate public views on the issues in question and feed them into the policy process. Broadening the connotations of this approach further still to include participatory procedures alongside other forms of democratic practice – such as elections and parliaments – Laurent (2011) has introduced the term 'technologies of democracy', defined as: 'instruments based on material apparatus, social practices and expert knowledge that organise the participation of various publics in the definition and treatment of public problems' (p649).[5] We build on these formulations in this volume, developing the term 'technologies of participation' in ways that go beyond the focus on extractive methods of eliciting public views while remaining focused on diverse instruments that organize, shape and consolidate public engagement with science, technology and the environment.

Our goal here is to base a theory of participatory practice on an understanding of technology as the assemblage and standardization of skills and know-how in the form of artefacts that can be used by others and translated across a range of different contexts (Latour 1991). In this sense, technologies of participation can be conceptualized as coming into being where the formats, configurations and skills for enrolling heterogeneous collectives that make up specific participatory experiments become more or less stabilized, standardized and 'blackboxed' in the form of established designs that are able to travel from place to place. Techniques designed to enrol heterogeneous actors become 'stabilised in lasting networks only to the extent that the mechanisms of enrolment are materialised in various more or less persistent forms' (Rose and Miller 1992, 183–184; see also Latour 1987). For technologies of participation this can include participatory techniques, tools, instructions, guidelines, handbooks, facilitation training courses, professional accreditation systems and so on. Such inscriptions help to stabilize mobile and loosely affiliated networks around particular technologies of participation.

However, participatory practice is not simply a case of the manifestation of a set of codified designs or tools in material settings. Sociological studies focused on technologies of participation reveal the intense effort and material/cognitive investments required to materialize and reproduce particular participatory blueprints and collective arrangements across cultures and political settings. As Laurent (2011) has shown in his study of public involvement in deliberations over nanotechnology research and policy in France, technologies of participation – such as the Ile-de-France Citizen Conference process in 2006, organized by the market research company IFOP – require material and cognitive investment to stabilize the format of a 'citizen panel', produce particular publics, fix definitions of collective public issues, and stabilize a particular atmosphere of democracy. Even though IFOP was a market leader with a well-established procedure, making the citizen conference 'work' demanded careful selection of 'neutral' or innocent citizens with no prior interest in the issue, disciplining participants' discussion to fit with the constraints of the technical procedure and issue framing, and where resistances came forward from dissenting 'critical citizens', editing out their voices from the final report that represented public views. This replication of an instrument of participation is challenging, not always successful and subject to resistances. Success depends on the stabilization of the boundary between the issue and the (now decontextualized) technical procedure in order to reproduce the technology of participation (Laurent 2009), which in turn allows actors to make the procedure circulate and be reconstituted at other sites.

Standardization and circulation

As public participation formats have become increasingly mobile and standardized, a number of high-profile participatory techniques have become internationally renowned. A well-known example is participatory rural appraisal (PRA) – which while first developed at the Institute of Development Studies in Sussex, UK, now circulates in transnational networks and participatory development settings across the 'global south' (Chambers 1994; Singh 2001). The diffusion of the consensus conference format is a similar case: from its humble beginnings at the Danish Board of Technology, to its global circulation and application in science policy processes across most Western democracies (Joss and

Durant 1995; Seifert 2006). This move from localized practices to technologies of participation has become a key dynamic in the public participation field. The evident mobility of such technologies is an index of the growth of a globally connected public engagement industry involved in the transnational circulation of expertise, people, skills and techniques in a marketplace of methods. This process of professionalization and standardization is evident across a range of global settings: in science, technological and environmental policy development in the UK and Europe (Chilvers 2008b; Saretzki 2008), in deliberative public planning and infrastructure provision processes in Australasia and the US (Hendriks and Carson 2008; Lee 2015), in participatory development in the 'global south' (Kothari 2005), and the formation of international networks and organizations such as the International Association for Public Participation (IAP2).

The standardization and professionalization of public participation expertise has also been characterized by the establishment of trans-local expert networks. These 'epistemic communities' cohere around distinct normativities and epistemologies of participation, or break down into more specific 'communities of practice' around particular participatory instruments and formats (Chilvers 2008b). Such 'communities' also help to stabilize particular normative principles and to standardize participatory tools and practices, measures, validity criteria and so on. Ranging from local to transnational in extent, participatory expert communities span sites residing in universities, research institutes, think tanks, consultancies, companies, professional associations, government departments, civil society organizations and other social groups (Chilvers 2013). Human and material intermediaries that hold expert networks together can play an important role in the circulation of particular technologies of participation (Latour 2005b) and therefore the 'social life' (Law and Ruppert 2013) or 'innovation journey' (Polley *et al.* 1999; Voß, Chapter 12 this volume) of particular technologies of participation.

'Participation experts' – or so-called facilitators and moderators – also offer specific forms of mediation. They intervene in developing, organizing and assessing public involvement processes and linking them with various forms of action (Elam *et al.* 2007; Chilvers 2008b). As such, participation experts – and indeed the professionalization of an engagement 'industry' – play a critical role in making technologies of participation work in particular settings. Herein lies an inherent tension in the facilitator's role, between the necessity of intervention to make participation work and the demand for neutrality and independence in order to achieve credibility and legitimacy (Lezaun 2007; Chilvers 2008b; Moore 2012). The ability for facilitators to disappear from writings on democratic theory and from the products of participatory processes depends on a chameleon-like quality to simultaneously intervene in and maintain distance from issues and participants, and in publicizing public representations. This, coupled with the often experiential and embodied character of participation expertise, is one way in which participation experts endow themselves and their knowledge claims with credibility as they move between different settings, times and domains of implementation (Chilvers 2010).

The transnational qualities and circulation of technologies of participation and their connections and stabilizations across diverse sites, as explained here, can be understood in terms of what Barry (2006) calls 'technological zones'. For Barry the notion of 'technological zones' operates as a figure for the 'trans-territorial' production of science, expertise and material artefacts. He argues that a 'technological zone can be understood, in broad terms, as a space within which differences between technical practices, procedures or forms have been reduced, or common standards have been established' (p239). Viewing public participation in these terms allows us to see the ways in which new realities of participation and democracy become established in trans-local, and sometimes transnational, spaces. This can help to explain, for example, why institutionally sanctioned public engagement processes focused on emerging technologies – such as biotechnology, nanotechnology and geoengineering – across Europe and North America over the past decade, have often (re)produced relatively narrow socio-material collective configurations (small group deliberations) comprising supposedly representative 'innocent citizens' across diverse cultural-political settings (Irwin 2006; Lezaun and Soneryd 2007; Wickson *et al.* 2010). While this is partly due to the nature of the objects of engagement, and their relative novelty and limited existing forms of public expression and mobilization, this dynamic can also be explained in terms of a zone of participatory standardization, which involves the translation and reproduction of evaluation metrics, auditing techniques and learning infrastructures oriented towards achieving participatory 'best practice'.[6] Techniques of enrolment and normative prescriptions also become lodged in institutions, professional participation organizations, and science and policy institutions.[7] In this way, innovation in technologies of public dialogue can become narrowed down to particular deliberative-dialogic normativities and formats of participation to the exclusion of alternatives (Chilvers 2010, 2012).

Opening up technologies of participation

Studying the science and technologies of participation in this way is therefore crucial in revealing wider, more systemic, trajectories of democratic innovation and how participatory technologies and expertise can become 'locked in' (Arthur 1989) to particular formats and normativities of democratic engagement. These dynamics cannot be fully understood or mapped through an exclusive focus on situated participatory events at particular sites. For this reason, STS-informed approaches to the study of the technologies of participation have revealed the powers that get lodged and invested in infrastructures, institutions and spaces of participatory prescription. They offer possibilities for looking inside the 'black box' of technologized forms of participation, making visible these 'imputable mobiles' together with their attendant political and material commitments (Latour 1987). Such studies therefore offer the prospect for opening public participation up to wider scrutiny and reflexive questioning, focused on the purposes, accountability and transparency of particular forms and practices of participation (as we will move on to consider in the final chapter in this volume). Finally, a further value of a technologies of participation approach is in the ways it reveals the foreclosure of alternatives in the design and execution of participatory practices. This provides a vantage point for developing alternative participatory normativities and formats as well as understanding the exclusions, invisibilities, uncertainties and contingencies of knowledge-commitments inherent within specific collective participatory practices.

While socially determinist perspectives emphasize the mediating and controlling capacities of instruments and devices of participation as 'technologies of government' offering the possibility of governing 'at a distance' (Rose and Miller 1992; Rose 1999), our notion of technologies of participation emphasizes the intense effort entailed in making collective participatory practices work, their fragility and the fact that they can and do often fail. There will always be resistances. These may be dissenting participants within authorized processes of public participation, the organization of 'counter-experiments' in public participation or insurgent disruptions and interventions.[8] It is for this reason we emphasize that any participatory practice can become codified and reproduced in different places, and this includes the collective participatory practices of activism, protest, behaviour change, citizen science, grassroots innovation and so on (Chilvers and Longhurst 2015). Transnational social movements and international activist networks (Routledge et al. 2007) and the global spread of citizen-led engagements associated with the transition town movement that originated in the United Kingdom (Smith 2011) are characteristic cases of this traffic in seemingly more organic or less formalized participatory processes. All participation is technical and material, just as the social is inherent to all technologies (Barry 2001). What this does is break down the closed notion of participatory expertise evident in many of the above examples, viewing it in more distributed terms and raising questions about the status and character of 'lay' or 'vernacular' forms of mediation and participation expertise.

There are limits to the notion of technologies of participation, however. Notwithstanding the sensitivities just mentioned, it carries the risk of reifying the particular technologies of participation at the expense of other possibilities and imaginations of participatory politics. It is therefore essential that studies of technologized forms of participation also keep in view relations with wider diversities of participatory experiments and practices. In addition, while a focus on technologies of participation provides important advances, it also has its limits in understanding the co-production of technoscience and society. The significant advance is in how it transforms our understandings, from the common notion that participation works in opposition to the innovation system, or operates as a separate procedural add-on that potentially impacts upon technoscience, to an understanding that sees participation and innovation as inherently intertwined and co-evolving. However, the technologies of participation perspective has a tendency to foreground the role of social science methods in enacting quasi-democratic mini-publics rather than a thoroughly co-productionist account which would explore scientific and social scientific modes of constructing and legitimating epistemic and political authority in public (Callon et al. 2009; Jasanoff 2012). This potential lacuna demands a broader – even systemic or ecological – perspective that brings into view the relational interdependencies between multiple collectives that make up the governance of science and democracy in wider spaces of negotiation and political culture. It is this requirement to which we now turn.

Political spaces and constitutions

In this section we move on from co-productionist accounts located in situated collective participatory practices and technologized zones of participation, toward an understanding of the relational entanglement of both within the wider political and cultural worlds of science and democracy.[9] In doing this we consider the relational roles of participatory collectives in more extensive and complex spaces of socio-material coherence, whether that be in terms of spaces of *controversy* or *political culture*. In terms of the former, our interest is in how autonomous collectives of participation are relationally connected with each other (and with other technoscientific collectives) in making up *spaces of negotiation* around the framing and resolution of controversies: whether that is over issue definitions, trajectories of socio-technical system change, or contestations over models of democracy itself. In terms of political culture we go on to consider the *spaces of collective public reason*, and constitutional relations between citizens, science and the state, in which certain participatory collective practices and knowledge-ways become seen as authoritative and are endowed with legitimacy and meaning.

When it comes to issues such as anthropogenic climate change, food security, chemical pollution, and the disposal of nuclear waste, participatory experiments operate simultaneously at a range of scales, in conjunction with the projection of each issue as a site of citizenry scale. The enactment of participatory subjectivities is therefore encoded with the logics of geographical locality while also being plugged into circuits of international civil society and globalist citizenship. For this reason questions of scale and spatiality – particularly the mobility of participatory technologies between scales and the broader composition of participatory assemblages, lodged at multiple scales simultaneously – are critical for a co-productionist account of public participation.

As we have argued earlier, while co-productionist analyses of contemporary technoscience have developed accounts of processes of 'collective experimentation' that weave associations between science and society (Latour 1998), these accounts often leave procedural and democratic categories 'as given'. This can have the effect of upholding the imaginary of a single collective or fora that bounds experimental practice or a notion of an expanding collective or fora that encompasses all relevant actors implicated in the experiment – a 'parliament of collectives', to paraphrase Latour (2004). In much of the literature on public participation, including public engagements with science and the environment, this is conceived as a problem of 'scaling up' the kinds of mini-publics formed in deliberative settings.

What we see here is a procedural solution to questions of scale and extension, where modes of public participation are scaled up to function in sites of global governance (Worthington *et al.* 2013). If, as we have argued earlier, the procedural-democratic formats and configurations that make up 'collective experiments' are themselves experiments in the making (rather than all-encompassing sites of collective coherence), the problem of scale and scaling up might therefore be recast in terms of the composition of multiple, diverse and interrelated socio-material collectives and participatory experiments. Adopting a relational understanding of public participation as the experimental enactment of socio-material collectives therefore entails attending to how we might 'make an *assembly* out of all the various *assemblages* in which we are already enmeshed' (Latour 2005a, 27). What such an assembly is and how we can conceive of a 'collective of collectives' are therefore critical questions in developing a foundation for a more systemic reading of public participation practice; one that moves from an imaginary of participation as single one-off 'events' or experiments that form a linear relationship between knowledge and power, to one that sees multiple participatory collectives relationally connected as part of diverse distributed agencies that make up wider systems and constitutions of science and democracy (cf. Hagendijk 2004; Felt and Wynne 2007; Parkinson and Mansbridge 2012; Macnaghten and Chilvers 2014; Stilgoe et al. 2014).

Key questions guiding enquiry at these levels of co-productive coherence include:

- How do multiple and diverse collectives of participation interrelate and relationally connect with each other (and with other technoscientific collectives)?
- In what ways do such 'ecologies of participation' relate to and form part of wider spaces of coherence – for example, controversies, issue spaces, systems or constitutions?
- What are the conditions of political culture and power by which certain collectives of participation and 'participatory knowledge-ways' become authoritative and credible, at the expense of others?

Assemblies of participatory collectives

As we have described earlier, Latour's proposition of an assemblage of assemblages invites a procedural solution for the question of whether it is possible to conceive a collective of collectivities. Brown (2009) has taken up this challenge, locating such an assembly within an 'ecology of institutions', an internally differentiated system of practices and institutions for representing nature and society. Following Iris Marion Young's (2000) call for 'pluralising the sites and modes of representation', Brown goes on to argue: 'The degree to which citizens enjoy democratic representation, therefore, should be judged with respect to the ecology of institutions to which they have access, rather than with regard to any single institution' (p237). This is a thoughtful response to the perplexing question of how to proceed once democracy, in addition to science, is viewed as relational and constructed.

An alternative way of addressing Latour's question is to take a more analytical and *empirical* stance, which would view a collective of collectives as a topological space formed through the enactment of multiple, autonomous, diverse, interrelating, participatory (and other) socio-material collectives. Such a perspective has a long tradition in STS, particularly in the area of controversy studies, and has been developed in more recent 'issue-oriented' work on the formation of participatory collectives (Marres 2012). As we described above, this issue-oriented perspective draws Dewey's pragmatist political philosophy into conversation with socio-ontological work in STS. The space of coherence in this perspective is the 'issue space' made up of multiple socio-material collectives (i.e., partly exclusive, partial and divergent associations that are entangled in an issue). As Marres (2007) states, 'articulating a public affair renders explicit … the mutual exclusivities between associations that different constituencies bring to a controversy, and which are caught up in the matter at stake, and de-publicising articulations can render such exclusivities obscure' (p773). This perspective lays a foundation for understanding coherence in the co-production of science and participation at the level of spaces of negotiation or controversy, made up of multiple, jostling and interrelating socio-material collectivities. This goes beyond Brown's institutionally centred approach, as it allows for and acknowledges diversities in issue articulation *and* democratic-procedural formats that overflow established institutional jurisdictions and capacities. Of course, whether institutions, or other actors, have the will and capacity to listen and respond relationally to such overflows is another matter – one that we turn to in the concluding chapter of this volume.

While we are sympathetic to this approach, a relational and emergent perspective regarding participatory practices suggests an even more open reading of the possible 'spaces of negotiation' (Murdoch 1998) in which multiple participatory collectives jostle, intermingle and to some extent cohere. Critical here are the complexities and interrelations between 'issue spaces' in wider 'political situations' (Barry 2012). By emphasizing the complex entanglement of discrete and autonomous collectives of participation within multiple issue spaces, such a perspective moves beyond the limits of controversy studies and a focus on set-piece analyses of specific objects of participation. What we have in mind are, for example, the ways in which a local transition town group, involved in developing grassroots innovations in community energy, might also be entangled in controversies over the siting of wind energy infrastructure, resisting fracking technologies, and opening up alternative pathways for tackling climate change. The multiply productive dimensions of participatory experiments therefore form part of what we might think of as swarming 'collective vitalities' (Bennett 2010) that constitute spaces of negotiation around, for example, trajectories of techo-scientific change (Stirling 2008) or participatory-democratic innovation (Smith 2009). These can constitute alternative spaces of coherence within which participatory collectives intermingle. The material, cognitive and normative associations that are drawn into the definition and contestation of issues extend well beyond the formal boundary of the substantive issue itself.

An account of the collective vitalities that circulate in spaces of negotiation therefore offers a vantage point for conceptualizing the ways in which democracy itself becomes an object of contestation in processes of public participation, where questions centre on the trajectories of democratic and political change more generally (Chilvers and Burgess 2008). At the same time, this sense of the plurality of participatory practices helps to recast dominant analyses of political withdrawal and public alienation. In recent years it has become commonplace to speak of a 'legitimacy crisis' affecting many Western democracies, characterized by popular dissatisfaction with and withdrawal from political participation and representation. Indeed, the notion that publics are increasingly retreating from formal political processes, which has become something of an unquestioned 'social fact', provides a key incentive for institutional experimentation in more direct forms of citizen participation. Democracies need to be renewed, the argument goes, by making them more democratic. However, as Wynne (Chapter 5, this volume) evocatively demonstrates, public inaction, withdrawal and indeed silence might be reconceived as forms of participation (see Eliasoph 1998) in which delegated forms of representation have become a site of contestation. By attending to the multiple spatialities of participation, a relational and co-productionist account might therefore offer a means for rewriting dominant narratives of democratic stagnation and popular withdrawal.

Systems of participation

The notion of socio-technical systems has become an important strand of work in STS and innovation studies, including the co-evolutionary perspective that understands system change in terms of multilevel interactions between socio-technical niches, regimes and the landscape (Geels 2002). Such frameworks have until recently been relatively quiet on the democratic implications of taking a systems approach to the coordination of technological change and the place of participation and engagement within them (although for contrast see Hendriks 2008; Chilvers and Longhurst 2012; Smith 2012).[10] Yet it is the area of political and democratic theory where the notion of systems of participation – or 'deliberative systems' – is arguably being taken most seriously (Parkinson and Mansbridge 2012). This marks a move by some deliberative theorists to recognize the limits of understanding deliberative democracy in terms of the 'mini-publics' model of a formal, structured, small group deliberative process, aligned with the imaginary of participation as one-off invited events, that predominated earlier deliberative democratic ideals (Habermas 1984; Goodin and Dryzek 2006). In recognizing the simplicity and limits of this model, deliberative democrats have first moved to consider 'macro-scale' deliberation in the wider public sphere (Hendriks 2006), and more recently developed a notion of 'deliberative systems'. Mansbridge *et al.* (2012, pp1–2, authors' emphasis) define this more systematic understanding of deliberative practice in the following terms:

> To understand the larger goal of deliberation, we suggest that it is necessary to go beyond the study of individual institutions and processes to examine their interaction in the system as a whole. We recognise that most democracies are complex entities in which a wide variety of institutions, associations, and

sites of contestation accomplish political work – including informal networks, the media, organised advocacy groups, schools, foundations, private and non-profit institutions, legislatures, executive agencies, and the courts. We thus advocate what may be called a *systemic approach to deliberative democracy*.

This is a useful intervention that opens deliberative democratic theory and practice up to diverse forms of engagement, understanding how multiple spaces are interconnected in making up a 'deliberative system'. The approach assesses the notion of 'deliberative quality' at the level of the system rather than individual processes; one deliberative process may have suboptimal performance but make a positive contribution to system performance (and *vice versa*).

The concept of deliberative systems constitutes an important development in deliberative democratic theory and practice that, as Miller *et al.* (2013) argue, have tended to concentrate on the development of 'innovative, deliberative models of public engagement in scientific and technological decision-making' and 'focus on individual technologies rather than complex technological and social systems and usually involve relatively small-scale public engagement, rather than seeking to understand and enhance the capacity for deliberative systems to enable society-wide conversations' (p146). However, current formulations of deliberative systems lack the co-productionist, relational and emergent perspective of participation outlined in this chapter. While emphasizing system complexities and patternings of 'deliberative ecologies' (Mansbridge *et al.* 2012), the approach remains faithful to a residual realist understanding of participation (as outlined in Chapter 1) by predefining normative procedural principles for arbitrating the effectiveness of deliberative systems rather than attending to their construction through the performance of deliberative practice. In addition, while opening up the meaning and purposes of deliberation, this approach remains avowedly 'talk-based', limiting participation to discursive and linguistic forms that underplay the multiplicity and diversities of participation we have emphasized under the above socio-material conception put forward in this chapter. The emphasis of deliberative systems thinking is on decision-making and legitimacy, which includes distributed societal decisions, rather than expanding the definition of participation to the diverse productions and distributed agencies of participatory collectives. Finally, deliberation within 'the system' is seen as somehow separate from or integrated into science and democracy, rather than being held in tension with them in processes of mutual co-constitution.

Ecologies of participation

In many respects the relational understanding of participation developed in this chapter sits uneasily with the implied holism of emerging systemic approaches. For example, Mansbridge *et al.* (2012) define a system as 'a set of distinguishable, differentiated, but to some degree interdependent parts, often with distributed functions and a division of labour, connected in such a way as to form a complex whole' (p4). In contrast, Marres (2012) warns that it is a mistake to try and build a holistic multilevel theory of participation. Rather than generating a conceptual map that seeks to clarify the complex interplay of overlapping aggregations and problematizations once and for all, the particular relations between specific engagement procedures and broader currents in political and policy practice for Marres remains an empirical and pragmatic question. The analytical task is both diagnostic and co-productionist – to map the relations between and within modes of public participation while documenting the ways in which democratic and political ontologies are themselves co-produced through processes of participation.

We are sympathetic to this argument. In keeping with the relational framework we have outlined in this chapter, however, we would go on to further emphasize the possible formation, imagination and interconnections of topological spaces of participation at different levels of complexity and extent in the form of socio-material collectives of participatory practice, technologized spaces of participatory prescription and spaces of negotiation. Viewed in this way, rather than constituting a 'complex whole' – an aggregate of smaller units of participatory practice – systems of participation might be better understood as socio-material assemblages in which the composition of collective elements is both contingent and being worked out in tension with normative concerns about how overlapping forms of participation might be assembled 'better'. For us, widening analyses of participation and emphasizing the dynamics and interdependencies of multiple, interacting socio-material collectives evokes a metaphor of 'ecologies of participation' (Chilvers 2010, 2012). An ecological conception of participation suggests that is not possible to properly understand any one collective of participation without understanding its relational interdependence with other collective participatory practices, technologies of participation, spaces of negotiation and the cultural-political settings in which they become established. Viewing participation as a web of connections highlights the interrelations between all these spaces of participation. It also emphasizes the multiplicity, diversities and variabilities of collectives of participation that make up spaces of negotiation, as well as their relevance/irrelevance or visibility/invisibility in relation to incumbent institutions – or any other situated perspective for that matter, including from the perspective of publics themselves (cf. Michael 2009).

Political cultures

In order to make sense of the dynamics of this ecological and topological reading of participation and its relations with arrangements for governing science and technology, a more explicit theorization of political culture is necessary. We use the terminology of 'political culture' advisedly. There is, of course, the tendency in accounts of 'political cultures' toward a prosaic reading of the coincidence between cultural formation and territorial boundaries – that different political systems might be distinguished by the different cultural expectations they impose. Here we use the language of culture much more specifically, to explore the importance of historical-cultural antecedences in understanding contemporary ecologies of participation and the political, cultural and institutional settings within which certain forms of participation cohere and are granted a kind of authority.

Work in STS and political science has highlighted how the meanings, performance and legitimacy of participatory practices vary cross-culturally and in different national contexts (Hagendijk and Irwin 2006; Dryzek and Tucker 2008; Horst and Irwin 2010). Jasanoff's (2005) comparative study of the political reception of biotechnology in the United States and Europe is particularly important in laying the foundations for understanding how practices and technologies of participation interact with collective forms of public reason in modern nation states. It shows how political culture (i.e., the systematic and routinized ways in which a political community validates knowledge and makes binding collective choices) plays out in technological governance and affects the production of public knowledge. Jasanoff introduces the term 'civic epistemology' to explain the collective and historically and culturally grounded 'tacit knowledge-ways through which [members of a given society] assess the rationality and robustness of claims that seek to order their lives' (p255). These knowledge-ways vary across political cultures. Through the biotechnology case, Jasanoff contrasts civic epistemologies between the *contentious*, pluralist and interest-based approach in the United States, the British *communitarian* service-based style grounded in embodied expertise, and the more *consensus seeking* culture of public knowledge making in Germany which is more corporatist and institution-based. Demonstrations or knowledge claims that do not meet these styles in each national setting may be dismissed as irrelevant. In emphasizing how publics assess claims by and on behalf of science, it follows from this that the notion of civic epistemology applies just as much to claims emanating from collective participatory practices, forms of social science and *participatory knowledge-ways*. In light of our discussion above, these entrenched cultural expectations, which are grounded in public life and discourse, can help to explain how some collective participatory practices become established and others struggle to achieve relevance or become endangered in particular settings.

Jasanoff's work on civic epistemologies centres on comparative analyses of nation systems for science policy articulation and decision-making,[11] as the scale at which institutionally authorized knowledge-ways achieve a measure of epistemological and practical coherence. The framework put forward in this chapter builds on this approach by exploring how such epistemic and socio-material coherences can become established at a range of spaces and scales. The analysis of this process of spatial coherence remains for us a profoundly empirical question. The work entailed in taking a relational perspective on participation is therefore to document the specific sites and institutional configurations in which participatory practices cohere and are rendered authoritative. In developing a pragmatist understanding of participation, we draw here on Barry's (2012) resistance to the claim that there exists an *a priori* domain of social life termed 'the political'. Drawing on the sociologist W. I. Thomas, Barry argues for a 'situational analysis' of the ways in which situations and issues are defined as political by the actors engaged in their articulation. Building on this concept, our argument is that participation is *both* a product of engrained patterns of institutional behaviour and institutionally sanctioned epistemologies – shaped by social, material and political commitments – and the pragmatic practices entailed in defining issues, objects and publics as sites for participatory intervention. If we were to venture a general rule – admittedly a risky business at best – we would suggest that spaces of participatory coherence are assembled (and are *being* assembled) at multiple interrelating scales simultaneously. We therefore posit that interpreting the dynamics of participation in science and democracy depends on more fully understanding the interrelations within and between these topological spaces of participation.

It follows from this that the ecologies of participation that become established in particular social and political situations will be shaped by, and will shape, the political cultures in which they are situated. For example, understanding institutional configurations in the UK can help to explain the mushrooming of particular institutionally mediated forms of public engagement with science over the past decade and a half in a country that has historically lacked experience in broad public involvement. In response to the overt breakdown of public trust in science advice over the BSE crisis and other controversies around emerging technologies such as GM crops through the 1990s (Jasanoff 1998; Hinchliffe 2001), the attempts by the British state to rebuild expert credibility included new institutional forms that were more pluralist in composition and open to public inputs. This could have taken different paths but in practice took a particular course emphasizing 'invited' small group deliberative events that sought out representative groups of 'innocent citizens' as opposed to diverse forms of interested, innovative or activist publics (Jasanoff 2005; Irwin 2006; Wynne 2007; Chilvers 2010).

These developments reflect a historically situated commitment to demonstration through empirical science, evidenced by the emphasis on established technologies of participation grounded in social science methods, and a political culture that emphasizes embodied and service-based expertise, a cultural setting conducive to established social scientists and other participation experts taking up these service roles and claiming to 'speak for the people' at this juncture in British techno-politics (see Chilvers 2008b). While these developments can be seen to have partially enriched ecologies of participation in certain science-related issues, they have in other ways diminished and limited possible participatory pathways, linked to an enduring imaginary of the public as a problem or threat (Welsh and Wynne 2013; Wynne, Chapter 5 this volume).

Constituting public participation

These insights are relevant to any space or setting where particular participatory knowledge-ways and credibility criteria have become stabilized, which includes nation states but can also equally apply to other spaces in which political cultures may cohere – including organizational, disciplinary and institutional arrangements, or the specific settings of participatory practices and technological zones discussed in the first two main sections of this chapter. In addition, rather than seeing ecologies of participation simply as contemporary formations, however, such understanding needs to be underlain by a deeper and more thoroughgoing understanding of their cultural and historical antecedents. Jasanoff (2011) has outlined the conceptual foundations for such thinking in terms of 'constitutional moments' which are 'brief periods in which, through the unending contestation over democracy, basic rules of political practice are rewritten, whether explicitly or implicitly, thus fundamentally altering the relations between citizens, [science] and the state' (p623). Jasanoff situates contemporary US interest in public engagement in the context of two historical cycles of constitutional developments in governing science and technology. The first (circa 1940 to 1980) saw an expansion in the scope of the public sphere, state action, as well as public participation being more pluralist in nature with liberalized rules of access. The second period (circa 1980 to the present) saw the institutionalization of some participatory practices but others contracted, thus reducing the possibilities for public input. This is explained by the relative withdrawal of the state, the increasing commercialization of science and technology transfer, and the privatization and professionalization of value debates through this period.

Not only does this analysis highlight the need for new public engagement practices to take account of these historical contours of the relationship between science, technology and political ordering. It is also significant in emphasizing the ways in which relations between citizens, experts and the state in particular democratic settings are held together in constitutional configurations – made up of established institutions, regulations, laws, political economic arrangements, socio-material infrastructures, policy cultures, participatory mechanisms and so on – which are transformed over time. Ecologies of participation are thus co-produced in relation with and as part of these constitutional configurations that are durable and exhibit stabilities but are also precipitated by moments and cycles of transformational change.

In this section we have added spaces of negotiation, political culture and constitutional formations to our understanding of participation as co-produced, relational and emergent. Socio-material collectives of participation can link up to form trans-local technologies of participation or diverse ecologies that make up political situations, both of which intersect with the political cultures of nation states and systems of governing science, technology and the environment. This perspective is useful in moving from relational notions of participatory collectivities as somehow separate entities to seeing them as interrelating with other socio-material collectives and as an inherent part of the constitutional relations that make up science and democracy in modern nation states.

Conclusions

As Ezrahi (2012, 1) puts it, a democratic society's

> precarious existence depends upon mutually reinforcing democratic ideas, political culture, political imaginaries, institutions, and practices ... A

democracy, like any other political regime, must be imagined and performed by multiple agencies in order to exist. Like a symphony, democracy has to be performed reasonably well in order to be realised as a political world.

As part of this symphony, collective participatory practices and spaces of participatory negotiation and prescription continually emerge as momentary ruptures and openings unsettle constitutional relations between citizens, science and the state. What appear on face value to be novel and emergent participatory experiments are thus part of the cyclical and continual readjustments in the democratic order of things. The so-called 'participatory turn' becomes more accurately seen as yet another reconfiguration and remaking of the elements and relations that make up science and democracy, out of the many others that have occurred through history. Participation is always turning, always 'in the making'.

Bringing co-productionist analyses into closer engagement with political and democratic theory takes us beyond situated studies of discrete participatory experiments, but also allows us to gain greater resolution and a more granular understanding of the multiple interrelating elements that constitute technical democracy. In attempting to set out the conceptual foundations for understanding participation in co-productionist terms, this chapter has opened up the analytical-interpretive horizons for social studies of participation. While co-productionist analyses of situated participatory experiments and practices will continue to be important, we have argued that a relational understanding of the dynamics of participation in technological democracies also depends on taking forward new situated studies into the spaces of standardization that form around technologies and expertise of participation; interrelating ecologies of participation that make up spaces of negotiation (whether conceived in terms of issue spaces, systems and so on); and, in turn, how these spaces of participation perform in relation to wider political cultures and constitutional relations between citizens, science and the state. These new interpretive research paths are explored in the following chapters, which also go on to consider how rethinking the reality of participation as co-produced and emergent in these ways in turn opens up possibilities for its remaking.

Notes

1 Aside from normatively oriented literatures, arguing *for* public participation in science, and accounts rooted in public engagement practice, there are few synthetic accounts that document the conditions of contemporary scientific knowledge-making. Notable exceptions here include Balough (1991); Epstein (1996); Barry (2001); Irwin and Michael (2003); Brown (2009); and Jasanoff (2012).
2 In this sense we take an enduring theme in STS scholarship concerning the emergence and stabilization of new objects and phenomena through scientific experiments and demonstrations (Collins 1975; Latour and Woolgar 1979; Latour 1993), shifting the focus from scientific artefacts to the production of representations, procedures and knowledges of/about democracy and the public.

3 Our use of the notion of 'normativities' also draws on Latour's (2009) work in the ethnography of the law. In a recent review of this work McGee (2014) argues that in Latour's legal ethnographies the dichotomy between 'facts' and 'norms' (or between 'facts' and 'values') is reconstructed as a starting point for empirically sensitized enquiry. He argues that for Latour it is 'no longer possible to invoke a common backdrop or shared normative horizon, an always already present lifeworld or interpretive community, a social context, or a general system of rights and duties, but normative forms proliferate in their absence' (pxvii). Our interest here is in the analogous process in which normative forms and claims proliferate through participatory practices.
4 A focus on how particular situated participatory arrangements and products can become more or less stabilized and move around the world evokes another important strand of STS co-productionist scholarship on the intelligibility, portability and circulation of the products of technoscience across time, place and different cultural-political settings (Latour 1987; Jasanoff 2004). Instead of studying natural and physical scientific practices in the making *per se*, under this theme in the current volume we turn our attention to social scientific and participatory ways of knowing, representing and mobilizing publics and societies.
5 See Callon *et al.* (2009) for a discussion of 'technical democracy'.
6 STS as a discipline is not outside of these co-productions but rather inherently implicated in them alongside other distributed agencies. For example, this discussion is not unrelated to the observation, made in the previous section, that until recently most STS scholarship on public engagement has tended to ship in models of participation from deliberative democratic theory that align with a small group, event-based imaginary of participation.
7 For example, such processes in the standardization of participatory practice have been linked with the development of government offices of science and technology and participatory technology assessment, such as the Danish Board of Technology in Denmark, the Rathenau Institute in the Netherlands and TA-Swiss in Switzerland. In the UK this has been evident in the establishment of the government-funded Sciencewise Expert Resource Centre for public dialogue on science and technology where a particular model of dialogue became established through institutional principles, guidelines, evaluation systems, resource structures and participatory expert networks.
8 See Joly and Kaufmann (2008); Laurent (2011); and Meyer (2013) for discussion of recent non-governmental organization (NGO) interventions and protests that have surrounded public engagement processes focused on nanotechnology and synthetic biology in France.
9 This is reflective of the move in co-productionist analyses from a focus on the production and circulation of technoscientific artefacts, practices and institutions to make more explicit connections with macro spaces of political power and culture (e.g., Shapin and Schaffer 1985; Ezrahi 1990; Jasanoff 1990). Here we focus more explicitly on how the production and circulation of diverse participatory practices forms part of the picture.
10 More systemic perspectives of participation have also been implied in recent frameworks to understanding the reflexive governance of science and technology, many of which have also included principles framed around notions of the common good (Stilgoe *et al.* 2013; Macnaghten and Chilvers 2014).
11 See also Jasanoff and Kim (2009, 2013) for a discussion of *national* socio-technical imaginaries.

References

Akrich, M. 1994. The de-scription of technical objects. In: W. E. Bijker and J. Law (eds) *Shaping Technology/Building Society: Studies in Sociotechnical Change*. Cambridge, MA: MIT Press, pp205–224.

Arendt, H. 1958. *The Human Condition*. Chicago, IL: University of Chicago Press.
Arthur, W. B. 1989. Competing technologies, increasing returns and lock–in by historical events, *Economic Journal* 99(394): 116–131.
Balough, B. 1991. *Chain Reaction – Expert Debate and Public Participation in American Commercial Nuclear Power, 1945–1975*. Cambridge: Cambridge University Press.
Barry, A. 2001. *Political Machines: Governing a Technological Society*. London: Athlone Press.
Barry, A. 2006. Technological zones, *European Journal of Social Theory* 9(2): 239–253.
Barry, A. 2012. Political situations: Knowledge controversies in transnational governance, *Critical Policy Studies* 6(3): 324–336.
Bennett, J. 2010. *Vibrant Matter: A Political Ecology of Things*. Durham, NC: Duke University Press.
Boltanski, L. and Thévenot, L. 2006. *On Justification*. Princeton, NJ: Princeton University Press.
Bogner, A. 2012. The paradox of participation experiments, *Science Technology & Human Values* 37(5): 506–527.
Braun, B. and Whatmore, S., eds. 2010. *Political Matter: Technoscience, Democracy and Public Life*. Minneapolis: University of Minnesota Press.
Braun, K. and Schultz, S. 2010. '… a certain amount of engineering involved': Constructing the public in participatory governance arrangements, *Public Understanding of Science* 19(4): 403–419.
Brown, M. 2009. *Science in Democracy: Expertise, Institutions and Representation*. Cambridge, MA: MIT Press.
Callon, M. 1986. Some elements of a sociology of translation: Domestication of the scallops and the fishermen of St Brieuc Bay. In: J. Law (ed.) *Power, Action, Belief*. London: Routledge & Kegan Paul, pp196–223.
Callon, M., ed. 1998. *The Laws of the Markets*. Oxford: Blackwell.
Callon, M. 2007. What does it mean to say that economics is performative? In: D. MacKenzie, F. Muniesa and L. Siu (eds) *Do Economists Make Markets? On the Peformativity of Economics*. Princeton, NJ: Princeton University Press, pp311–357.
Callon, M. and Rabeharisoa, V. 2004. Gino's lesson on humanity: Genetics, mutual entanglements and the sociologist's role, *Economy and Society* 33(1): 1–27.
Callon, M., Lascoumes, P. and Barthe, Y. 2009. *Acting in an Uncertain World: An Essay on Technical Democracy*. Cambridge, MA: MIT Press.
Chambers, R. 1994. Participatory Rural Appraisal (PRA): Analysis of experience, *World Development* 22(9): 1253–1263.
Chilvers, J. 2008a. Deliberating competence: Theoretical and practitioner perspectives on effective participatory appraisal practice, *Science, Technology & Human Values* 33(2): 155–185.
Chilvers, J. 2008b. Environmental risk, uncertainty and participation: Mapping an emergent epistemic community, *Environment and Planning A* 40(12): 2990–3008.
Chilvers, J. 2010. *Sustainable Participation? Mapping out and Reflecting on the Field of Public Dialogue on Science and Technology*. Harwell: Sciencewise Expert Resource Centre.
Chilvers, J. 2012. *Expertise, Technologies and Ecologies of Participation*. 3S Working Paper 2012–17. Norwich: Science, Society and Sustainability Research Group, University of East Anglia.
Chilvers, J. 2013. Reflexive engagement? Actors, learning and reflexivity in public dialogue on science and technology, *Science Communication* 35(2): 283–310.
Chilvers, J. and Burgess, J. 2008. Power relations: The politics of risk and procedure in nuclear waste governance, *Environment and Planning A* 40(8): 1881–1900.

Chilvers, J. and Evans, J. 2009. Understanding networks at the science–policy interface, *Geoforum* 40(3): 355–362.

Chilvers, J. and Longhurst, N. 2012. *Participation, Politics and Actor Dynamics in Low Carbon Energy Transitions*. Norwich: Science, Society and Sustainability Research Group, University of East Anglia.

Chilvers, J. and Longhurst, N. 2015 (forthcoming). Participation in transition(s): Reconceiving public engagements in energy transitions as co-produced, emergent and diverse, *Journal of Environmental Policy and Planning*.

Collins, H. 1975. The seven sexes: A study in the sociology of a phenomenon, or the replication of experiments in physics, *Sociology* 9: 205–224.

Crosby, N., Kelly, J. M. and Schaefer, P. 1986. Citizens panels: A new approach to citizen participation, *Public Administration Review* 46: 170–178.

de Vries, G. 2007. What is political in sub-politics? How Aristotle might help STS, *Social Studies of Science* 37(5): 781–809.

Dewey, J. 1927. *The Public and its Problems*. Athens, OH: Shallow Press.

Dryzek, J. S. and Tucker, A. 2008. Deliberative innovation to different effect: Consensus conferences in Denmark, France and the United States, *Public Administration Review* 68(5): 864–876.

Durant, D. 2010. Public participation in the making of science policy, *Perspectives on Science* 18(2): 180–225.

Durant, D. 2011. Models of democracy in social studies of science, *Social Studies of Science* 41(5): 691–714.

Edwards, P. N. 2010. *A Vast Machine: Computer Models, Climate Data and the Politics of Global Warming*. Cambridge, MA: MIT Press.

Elam, M., Reynolds, L., Soneryd, L., Sundqvist, G. and Szerszynski, B. 2007. *Mediators of Issues and Mediators of Process: A Theoretical Framework*. ARGONA FP6 Project Report. Göteborg: Göteborg University.

Eliasoph, N. 1998. *Avoiding Politics: How Americans Produce Apathy in Everyday Life*. Cambridge: Cambridge University Press.

Epstein, S. 1996. *Impure Science: AIDS, Activism and the Politics of Knowledge*. Berkeley: University of California Press.

Ezrahi, Y. 1990. *The Descent of Icarus: Science and the Transformation of Contemporary Democracy*. Cambridge, MA: Harvard University Press.

Ezrahi, Y. 2012. *Imagined Democracies: Necessary Political Fictions*. Cambridge: Cambridge University Press.

Felt, U. 2015. Keeping technologies out: Sociotechnical imaginaries and the formation of Austria's technopolitical identity. In: S. Jasanoff and K. Sang-Hyung (eds) *Dreamscapes of Modernity: Sociotechnical Imaginaries and the Fabrication of Power*. Chicago, IL: Chicago University Press.

Felt, U. and Fochler, M. 2010. Machineries for making publics: Inscribing and de-scribing publics in public engagement, *Minerva* 48(3): 219–238.

Felt, U. and Wynne, B. 2007. *Science and Governance: Taking European Knowledge Society Seriously*. Report of the Expert Group on Science and Governance to the Science, Economy and Society Directorate, Directorate-General for Research. Brussels: European Commission.

Funtowicz, S. O. and Ravetz, J. 1993. Science for the post-normal age, *Futures* 25(7): 735–755.

Geels, F. W. 2002. Technological transitions as evolutionary reconfiguration processes: A multi-level perspective and a case-study, *Research Policy* 31(8): 1257–1274.

Goodin, R. E. and Dryzek, J. S. 2006. Deliberative impacts: The macro-political uptake of mini-publics, *Politics & Society* 34(2): 219–244.

Goven, J. 2006. Processes of inclusion, cultures of calculation, structures of power: Scientific citizenship and the Royal Commission on genetic modification, *Science, Technology & Human Values* 31(5): 565–598.

Habermas, J. 1984. *Theory of Communicative Action – Volume 1: Reason and the Rationalization of Society*. Boston: Beacon Press.

Hagendijk, R. P. 2004. The public understanding of science and public participation in regulated worlds, *Minerva* 42(1): 41–59.

Hagendijk, R. P. and Irwin, A. 2006. Public deliberation and governance: Engaging with science and technology in contemporary Europe, *Minerva* 44: 167–184.

Haraway, D. J. 1997. *Modest Witness@Second_Millennium.FemaleMan©_Meets_OncoMouse™: Feminism and Technoscience*. London: Routledge.

Harvey, M. 2009. Drama, talk and emotion: Omitted aspects of public participation, *Science, Technology & Human Values* 34(2): 139–161.

Hendriks, C. M. 2006. Integrated deliberation: Reconciling civil society's dual role in deliberative democracy, *Political Studies* 54(3): 486–508.

Hendriks, C. M. 2008. On inclusion and network governance: The democratic disconnect of Dutch energy transitions, *Public Administration* 86: 1009–1031.

Hendriks, C. M. and Carson, L. 2008. Can the market help the forum? Negotiating the commercialisation of deliberative democracy, *Policy Sciences* 41(4): 293–313.

Hinchliffe, S. 2001. Indeterminacy in-decisions – science, policy and politics in the BSE (Bovine Spongiform Encephalopathy) crisis, *Transactions of the Institute of British Geographers* 26(2): 182–204.

Horst, M. and Irwin, A. 2010. Nations at ease with radical knowledge: On consensus, consensusing and false consensusness, *Social Studies of Science* 40(1): 105–126.

Irwin, A. 2001. Constructing the scientific citizen: Science and democracy in the biosciences, *Public Understanding of Science* 10(1): 1–18.

Irwin, A. 2006. The politics of talk: Coming to terms with the 'new' scientific governance, *Social Studies of Science* 36(2): 299–320.

Irwin, A. and Michael, M. 2003. *Science, Social Theory and Public Knowledge*. Maidenhead: Open University Press.

Irwin, A. and Wynne, B. 1996. *Misunderstanding Science? The Public Reconstruction of Science and Technology*. Cambridge and New York: Cambridge University Press.

Jasanoff, S. 1990. *The Fifth Branch: Science Advisors as Policy Makers*. Cambridge, MA: Harvard University Press.

Jasanoff, S. 1998. The mad cow crisis: Health and the public good, *Public Understanding of Science* 7(4): 354–356.

Jasanoff, S. 2004. The idiom of co-production. In: S. Jasanoff (ed.) *States of Knowledge: The Co-Production of Science and Social Order*. London: Routledge, pp1–12.

Jasanoff, S. 2005. *Designs on Nature: Science and Democracy in Europe and the United States*. Princeton, NJ: Princeton University Press.

Jasanoff, S. 2011. Constitutional moments in governing science and technology, *Science and Engineering Ethics* 17: 621–638.

Jasanoff, S. 2012. *Science and Public Reason*. London: Routledge.

Jasanoff, S. and Kim, S.-H. 2009. Containing the atom: Sociotechnical imaginaries and nuclear power in the United States and South Korea, *Minerva* 47: 119–146.

Jasanoff, S. and Kim, S.-H. 2013. Sociotechnical imaginaries and national energy policies, *Science as Culture* 22(2): 189–196.

Joly, P. B. and Kaufmann, A. 2008. Lost in translation? The need for upstream engagement with nanotechnology on trial, *Science as Culture* 17(3): 225–247.
Joss, S. and Durant, J., eds. 1995. *Public Participation in Science: The Role of Consensus Conferences in Europe.* London: The Science Museum.
Kothari, U. 2005. Authority and expertise: The professionalisation of international development and the ordering of dissent, *Antipode* 37(3): 425–446.
Latour, B. 1987. *Science in Action.* Cambridge, MA: Harvard University Press.
Latour, B. 1991. Technology is society made durable. In: *Sociology of Monsters: Essays on Power, Technology and Domination.* London: Routledge, pp103–161.
Latour, B. 1993. *We Have Never Been Modern.* Harlow: Pearson Education.
Latour, B. 1998. To modernize or to ecologize? That's the question. In: N. Castree and B. Braun (eds) *Remaking Reality: Nature at the Millennium.* London: Routledge, pp221–242.
Latour, B. 2004. *Politics of Nature: How to Bring the Sciences into Democracy.* Cambridge, MA: Harvard University Press.
Latour, B. 2005a. From realpolitik to dingpolitik or how to make things public. In: B. Latour and P. Weibel (eds) *Making Things Public: Atmospheres of Democracy.* Cambridge, MA: MIT Press, pp4–31.
Latour, B. 2005b. *Reassembling the Social: An Introduction to Actor-Network-Theory.* Oxford: Clarendon.
Latour, B. 2009. *The Making of Law: An Ethnography of the Conseil d'Etat.* Cambridge: Polity Press.
Latour, B. and Weibel, P. 2005. *Making Things Public: Atmospheres of Democracy.* Cambridge, MA: MIT Press.
Latour, B. and Woolgar, S. 1979. *Laboratory Life: The Social Construction of Scientific Facts.* Beverly Hills, CA: SAGE.
Laurent, B. 2009. *Replicating Participatory Devices: The Consensus Conference Confronts Nanotechnology.* CSI Working Paper No. 18, www.csi.ensmp.fr/working-papers/WP/WP_CSI_018.pdf (accessed 28 August 2015).
Laurent, B. 2011. Technologies of democracy: Experiments and demonstrations, *Science and Engineering Ethics* 17(4): 649–666.
Law, J. and Ruppert, E., eds. 2013. *The Device: The Social Life of Methods.* Special issue of *The Journal of Cultural Economy* 6(3).
Law, J. and Urry, J. 2004. Enacting the social, *Economy and Society* 33(3): 390–410.
Lee, C. W. 2015. *Do-It-Yourself Democracy: The Rise of the Public Engagement Industry.* New York: Oxford University Press.
Lezaun, J. 2007. A market of opinions: The political epistemology of focus groups, *The Sociological Review* 55(S2): 130–151.
Lezaun, J. and Soneryd, L. 2007. Consulting citizens: Technologies of elicitation and the mobility of publics, *Public Understanding of Science* 16(3): 279–297.
MacKenzie, D., Muniesa, F. and Siu, L., eds. 2007. *Do Economists Make Markets? On the Peformativity of Economics.* Princeton, NJ: Princeton University Press.
Macnaghten, P. and Chilvers, J. 2014. The future of science governance: Publics, policies, practices, *Environment & Planning C* 32(3): 530–548.
Mansbridge, J., Bohman, J., Chambers, S., Christiano, T., Fung, A., Parkinson, J., Thompson, D. F. and Warren, M. E. 2012. A systemic approach to deliberative democracy. In: J. Parkinson and J. Mansbridge (eds) *Deliberative Systems: Deliberative Democracy at the Large Scale.* Cambridge: Cambridge University Press, pp1–26.
Marres, N. 2007. The issues deserve more credit: Pragmatist contributions to the study of public involvement in controversy, *Social Studies of Science* 37(5): 759–780.

Marres, N. 2012. *Material Participation: Technology, the Environment and Everyday Publics.* Basingstoke: Palgrave Macmillan.
Marres, N. and Lezaun, J. 2011. Materials and devices of the public: An introduction, *Economy and Society* 40(4): 489–509.
McGee, K. 2014. *Bruno Latour: The Normativity of Networks.* London: Routledge.
Meyer, M. 2010. The rise of the knowledge broker, *Science Communication* 32(1): 118–127.
Meyer, M. 2013. Debating synthetic biology: A necessity or a masquerade? www.csi.mines-paristech.fr/blog/en/?p=36 (accessed 16 December 2014).
Meyer, M. and Kearnes, M. 2013. Intermediaries between science, policy and the market, *Science and Public Policy* 40: 423–429.
Michael, M. 2009. Publics performing publics: Of PiGs, PiPs and politics, *Public Understanding of Science* 18: 617–631.
Miller, C., Iles, A. and Jones, C. 2013. The social dimensions of energy transitions, *Science as Culture* 22(2): 135–148.
Moore, A. 2012. Following from the front: Theorizing deliberative facilitation, *Critical Policy Studies* 6(2): 146–162.
Muniesa, F. 2014. *The Provoked Economy: Economic Reality and the Performative Turn.* London: Routledge.
Murdoch, J. 1998. The spaces of actor-network theory, *Geoforum* 29: 357–374.
Osborne, T. 2004. On mediators: Intellectuals and the ideas trade in the knowledge society, *Economy and Society* 33(4): 430–447.
Osborne, T. and Rose, N. 1999. Do the social sciences create phenomena? The example of public opinion research, *British Journal of Sociology* 50(3): 367–396.
Parkinson, J. and Mansbridge, J., eds. 2012. *Deliberative Systems: Deliberative Democracy at the Large Scale.* Cambridge: Cambridge University Press.
Polley, D. E., Garud, R. and Venkataraman, S. 1999. *The Innovation Journey.* Oxford: Oxford University Press.
Proctor, J. D. 1998. The meaning of global environmental change: Retheorising culture in human dimensions research, *Global Environmental Change* 8(3): 227–248.
Rayner, S. 2003. Democracy in the age of assessment: Reflections on the roles of expertise and democracy in public-sector decision making, *Science and Public Policy* 30(3): 163–170.
Rose, N. 1999. *Powers of Freedom: Reframing Political Thought.* Cambridge: Cambridge University Press.
Rose, N. and Miller, P. 1992. Political power beyond the state: Problematics of government, *British Journal of Sociology* 43(2): 173–205.
Routledge, P., Cumbers, A. and Nativel, C. 2007. Grassrooting network imaginaries: Relationality, power and mutual solidarity in global justice networks, *Environment & Planning A* 39(11): 2575–2592.
Rowe, G. and Frewer, L. 2000. Public participation methods: A framework for evaluation, *Science, Technology & Human Values* 25(1): 3–29.
Ruppert, E., Law, J. and Savage, M. 2013. Reassembling social science methods: The challenge of digital devices, *Theory, Culture & Society* 30(4): 22–46.
Saretzki, T. 2008. Policy-analyse, demokratie und deliberation: Theorieentwicklung und forschungsperspektiven der 'policy sciences of democracy'. In: F. Janning and K. Toens (eds) *Die Zukunft der Policy-Forschung: Theorien, Methoden, Anwendungen.* Wiesbaden: Verlag für Sozialwissenschaften, pp23–54.
Seifert, F. 2006. Local steps in an international career: A Danish-style consensus conference in Austria, *Public Understanding of Science* 15(1): 73–88.
Shapin, S. and Schaffer, S. 1985. *Leviathan and the Air-Pump: Hobbes, Boyle and the Experimental Life.* Princeton, NJ: Princeton University Press.

Singh, K. 2001. Handing over the stick: The global spread of participatory approaches to development. In: M. Edwards and J. Gaventa (eds) *Global Citizen Action*. Boulder, CO: Lynne Rienner, pp175–187.

Smith, A. 2011. The Transition Town network: A review of current evolutions and renaissance, *Social Movement Studies* 10(1): 99–105.

Smith, A. 2012. Civil society in sustainable energy transitions. In: G. Verbong and D. Loorbach (eds) *Governing the Energy Transition: Reality, Illusion or Necessity?* London: Routledge, pp180–202.

Smith, G. 2009. *Democratic Innovations: Designing Institutions for Citizen Participation*. Cambridge: Cambridge University Press.

Stilgoe, J., Lock, S. J. and Wilsdon, J. 2014. Why should we promote public engagement with science? *Public Understanding of Science* 23(1): 4–15.

Stilgoe, J., Owen, R. and Macnaghten, P. 2013. Developing a framework for responsible innovation, *Research Policy* 42(9): 1568–1580.

Stirling, A. 2008. 'Opening up' and 'closing down': Power, participation and pluralism in the social appraisal of technology, *Science, Technology & Human Values* 33(2): 262–294.

Stirling, A. 2011. Pluralising progress: From integrative transitions to transformative diversity, *Environmental Innovation and Societal Transitions* 1(1): 82–88.

Thévenot, L. 1984. Rules and implements: Investment in forms, *Social Science Information* 23(1): 1–45.

Van Oudheusden, M. and Laurent, B. 2013. Shifting and deepening engagements: Experimental normativity in public participation in science and technology, *Science, Technology & Innovation Studies* 9(1): 3–22.

Webler, T. 1995. Right discourse in citizen participation: An evaluative yardstick. In: O. Renn, T. Webler and P. Wiedemann (eds) *Fairness and Competence in Citizen Participation: Evaluating Models for Environmental Discourse*. Dordrecht: Kluwer, pp35–86.

Welsh, I. and Wynne, B. 2013. Science, scientism and imaginaries of publics in the UK: Passive objects, incipient threats, *Science as Culture* 22(4): 540–566.

Wickson, F., Delgado, A. and Kjølberg, K. 2010. Who or what is 'the public'? *Nature Nanotechnology* 5: 757–758.

Worthington, R., Rask, M. and Minna, L., eds. 2013. *Citizen Participation in Global Environmental Governance*. London: Routledge.

Wynne, B. 2006. Public engagement as a means of restoring public trust in science – Hitting the notes, but missing the music? *Community Genetics* 9(3): 211–220.

Wynne, B. 2007. Dazzled by the mirage of influence? STS–SSK in multivalent registers of relevance, *Science, Technology & Human Values* 32: 491–503.

Young, I. M. 2000. *Inclusion and Democracy*. New York: Oxford University Press.

3
ENGAGING IN A DECENTRED WORLD

Overflows, ambiguities and the governance of climate change

Alan Irwin and Maja Horst

> Shifting such path dependent patterns is not something that can be easily achieved. So while governments can make major differences almost certainly much of what has to happen is more localized, decentralised, happening at the margins.
>
> (John Urry: http://politicsofclimatechange.wordpress.com/priorities-for-low-carbon-transition/john-urry)

The global politics of climate change nicely embodies one of the central governance challenges of the twenty-first century. As John Urry has argued, addressing climate change means acknowledging the significant role of traditional democratic institutions (especially government and parliament) but also of embedded social practices and 'path dependencies' – not least reliance on cheap oil and the associated phenomenon of 'high carbon mobile lives'. As Urry expresses this:

> These practices are not 'mere choices' in the sense that they can be picked up and put down like individual items in a supermarket. They are embedded and ingrained, core elements of modern life around which affects, routines and futures are intricately woven.
>
> (Urry 2011, 88)

In such a situation, governance extends substantially beyond government as deeper-rooted modes of living, and what we will describe here as 'decentred' activities are opened up to discussion, appraisal and challenge.

Experience of the December 2009 United Nations Climate Change Conference, held in Copenhagen, vividly captures this dual character of contemporary governance. On the one hand, the Danish state had high ambitions for the event and national governments performed a central role in establishing the conference, framing discussions and reaching tangible agreements (or not). On the other, the limited success of the 15th session of the Conference of the Parties (COP 15) can only be understood in the context of a plurality of localizing and globalizing forces – including the impacts (and causes) of economic recession, scientific debate over key factors and consequences, ingrained patterns of production and consumption, the role of industrial organizations, lobby groups and non-governmental organizations (NGOs), and the emergence of new social movements and alliances. As the many demonstrators at the Copenhagen conference (themselves acting as an emergent and heterogeneous assemblage) served to underline, it is very possible to challenge the democratic accountability of such a gathering. However, one is also obliged to ask more profound questions about the nature of democratic governance in such a setting. What is even meant by 'democratic accountability' in this context: accountability by and to whom? As Julia Black has expressed the general issue with reference to questions of regulation:

> Complexity, fragmentation of knowledge and of the exercise of power and control, autonomy, interactions and interdependencies, and the collapse of the public/private distinction are the central elements of the composite 'decentred understanding' of regulation. Together they suggest a diagnosis of regulatory failure which is based on the dynamics, complexity and diversity of economic and social life, and in the inherent ungovernability of social actors, systems and networks.
>
> (Black 2002, 6)

In this chapter, we will address issues of 'decentredness' with particular regard to discussions over the purposes and impact of what has become known as 'public engagement with science' (PES). The very notion of 'engagement' seems to imply a traditional social and political structure with discrete and relatively fixed actors who can 'engage' with one another in a specific encounter and for a particular period of time, and then resume their separate, business-as-usual existences: government, the public, industry, scientific advisors. What happens to our understanding of 'engagement' once one begins to view social and political action in terms of fluidity, hybridity, fragmentation and (in Beck's terminology: Beck 1992) sub-politics – including a challenge to the very idea that 'policy' is formed in national government departments which can then direct (or set a framework for) subsequent events? *What* is then being engaged with and *by whom*?

Rather than simply taking a broad perspective on these issues, we will be especially concerned with the possibilities and challenges facing what will be presented as the general figure of the 'policy-maker'. If the issues constantly slip (as Beck would put it) through our conceptual fingers, then how can government officials or politicians 'get to grips' with them in terms of making meaningful political and policy interventions? If the most interesting action is happening 'at the margins', what does this suggest for the institutional 'centres'? We employ the 'policy-maker' as an ideal type with the specific purpose of identifying a series of policy-related dilemmas and challenges. Drawing on one particular example of the local discussion of the impact of climate change, we will then consider some of the implications of a decentred world.

Putting engagement at the centre

Once it appeared relatively simple for our 'policy-maker'. When it came to matters of science and technology policy, the dominant assumption was (and in many ways still is) that science objectively interpreted the world and that policy-makers should apply this knowledge in the best interests of the wider public. Meanwhile, decision-making took place within established institutions (typically, of government) and within the operational parameters and representative structures of the state. Sometimes, of course, the public could not recognize what was in its best interest: lack of scientific understanding, emotion and irrationality came into play. In such situations, the best course of action was either to ignore the public altogether or (as in the much-discussed deficit model: Wynne 1992; Bodmer 2010) to 'educate' them into a state of understanding – and therefore agreement. On that basis, the field of practice that became known as the 'public understanding of science' (or PUS) developed a vigorous emphasis on science communication and on persuading sceptical publics of the cultural, economic and practical benefits of scientific knowledge (Bucchi 2008). The important point here is that the 'deficit' model is not solely a matter of the status of scientific knowledge but also of the state of societal institutions – and, more broadly, of socio-technical relations (both present and future).

Although controversy and disagreement around science and technology go back substantially further, it seems that the 1990s offered a particular turning point in these discussions (House of Lords 2000; Irwin 2006). Certainly, one can trace from that period a growing sense of challenge to the 'deficit' assumption of an uninformed and ignorant public, a challenge partially driven by a series of controversial cases (BSE, civil nuclear energy, genetically modified foods, stem cell research: see, for example, Jasanoff 2005; Hagendijk and Irwin 2006; Horlick-Jones et al. 2007). A number of official documents and reports from that time – especially in the UK but also at a much broader European level – began to acknowledge that scientific and technological change could raise legitimate trans-scientific issues and that increased societal resistance to innovation could be anticipated if the public was not somehow 'engaged with' (for one landmark treatment of these issues see the 2000 Phillips Report). More broadly, engagement came to be seen as a means of winning back public trust in science and in the robustness of institutional (and epistemological) structures (see, for example, European Commission 2002). Public engagement represented a new addition to the toolkit of science policy mechanisms: engaging with the public might help to create a more constructive social debate than was possible through traditional techniques of information and education. The (admittedly partial and inconsistent) call now was for the policy-maker to be more transparent and more flexible, to acknowledge uncertainty and to consult with – and perhaps even learn from – citizen views and assessments (Irwin 1995; Felt and Fochler 2008).

From the 1990s also – and driven by a critique of the conventional treatment of 'public understanding of science' – qualitatively oriented social scientists began to complicate things (see Irwin and Wynne 1996). Rather than adopting the worldview of scientific institutions and enquiring about the nature and extent of public ignorance around science, a more 'symmetrical' approach was called for. The significance of social and cultural context was stressed. Societal encounters with science were presented as heavily mediated by the circumstances in which they took place – such that 'science' might not even be the central issue at all. In a number of studies, it was demonstrated that the PUS approach to dealing with the wider publics as if they were passive and ignorant tended to miss the sociological point and in practice often turned out to be politically provocative (Wynne 1995). At the same time, social scientists made the important argument that 'the public' as a single entity did not exist. Instead, policy-makers and others must acknowledge the presence of a plurality of publics (Leach *et al.* 2005; Kerr *et al.* 2007).

Put very simply, one can identify a convergence of these social scientific and institutional responses around a call for democratic engagement, consultation, openness and dialogue. PUS began with a commitment to 'top-down' communication but was increasingly replaced with a PES (public engagement with science) perspective which prescribed engagement as an important tool for the institutional policy-maker. Meanwhile, social scientists working especially in the science and technology studies tradition have often demonstrated both a scholarly orientation to contextualizing socio-scientific encounters but also a well-argued commitment to the 'democratization' of scientific decision-making (e.g., Jasanoff 2003; Felt and Wynne 2007).

However, for scientific institutions – such as government departments, industrial bodies and learned societies – the partial and still contested move from 'deficit to democracy' continues to raise many challenges. Certainly, one prevalent response to public engagement activities has been to view them in a rather critical light: as falling short of some democratic ideal, as failing to address fundamental issues of economic power or else as restricted in structure, intention and content (Wynne 2006; Callon *et al.* 2009; Horst 2010; Irwin *et al.* 2013). A common form of this has been to challenge the output from such activities – especially as this has taken shape in governmental response. There is also the abiding issue of *who* gets to engage in engagement activities – and whether participation can, or should, be representative of the wider population. Moreover, there are crucial questions of the *framing* of the issues for debate – are policy-makers implicitly (or possibly deliberately) missing the point of citizen concerns in their enthusiasm to construct these issues as largely 'technical' in character (Irwin 2006; Wynne 2006)? Such criticisms often focus on the marginality, fragility and conflictuality of engagement initiatives – and, consequently, their relative lack of transformative power. No matter how brave or worthy the intentions, what real hope is there for such small-scale, occasional and partial efforts?

Looked at from the perspective of our ideal–typical policy-maker, we can see how all this might offer some hope but also some decidedly gloomy news. On the one hand, criticism of top-down and one-way communication has been widely expressed over the last two decades (the very term 'deficit model' builds upon a criticism of institutional policy and practice). On the other, attempts at public engagement readily lead to accusations of superficiality, inconsequentiality and legitimation: effectively, an accusation of 'deficit by other means'. It seems that 'public engagement' offers a promise that often leads to disappointment – or perhaps that policy-makers have been presented with a gift which they might not entirely wish to accept but cannot easily give back. However, and as we will discuss in the next section, it may be that the complications for policy-makers and their institutions are only beginning.

Decentring science and democracy

> [W]hen uncertainties about possible states of the world and the constitution of the collective are dominant, the procedures of delegative democracy are shown to be unable to take the measure of the overflows created by science and technology. Other procedures of consultation and mobilization must be devised; other modes of decision-making must be invented.
>
> (Callon et al. 2009, 225)

In their account of *Acting in an Uncertain World*, Callon et al. (2009) present science and technology as overflowing the boundaries of existing frameworks of knowledge and governance. In line with Beck's risk society (Beck 1992), controversial areas of science and technology serve to contest technical-social barriers, and the traditional distinctions between 'experts' and 'lay', and 'science' and 'policy', disintegrate. In a double challenge to our knowledge of the world and to the composition of the collective, new alliances and assemblages emerge which question both existing assumptions about scientific advice and established forms of delegative democracy. Meanwhile, the (still generalized) policy-maker finds herself under distinct pressure: 'Decision makers think that the parameters of the questions to be dealt with have been suitably and properly defined, from both a technical and a political point of view, and now overflows identified by the actors demonstrate the opposite' (Callon et al. 2009, 29). Instead, Callon and colleagues see particular possibilities in what they term 'hybrid forums': new forms of direct engagement in which a broader array of political and technical possibilities can find expression and new actors emerge.

As Callon et al. present these, hybrid forums offer

> an enrichment of delegative democracy, and not a threat to it ... They replace a conception of the public space made up of detached, transparent actors lacking existential substance with a 'cluttered' public space in which individual wills are worked out and nourished by attachments that concerned groups have negotiated and discussed at length and in breadth.
>
> (Callon et al. 2009, 262)

On that basis, Callon *et al.* reflect on over 30 years of 'experimentation under real-life conditions' in this area, including consensus conferences, citizens' panels and juries, and collective organization around HIV/AIDS. Regarding the consensus conference format, for example, Callon and colleagues note that this represents a 'meaningful start' in recognizing the role of laypeople in scientific matters and in strengthening delegative democracy by rendering visible a broader array of public voices. However, 'the exercise does not allow a real exploration and formation of new identities, or the composition of the collective that could result from this' (Callon *et al.* 2009, 176). In that sense 'the group of ordinary citizens confines itself to a traditional vision of the collective and the general will' (p173). One reason for these limitations is, of course, the constraints imposed by the very structured form of consensus conferences, including the fact that the citizens are intended to focus on the common good and represent the 'general' public rather than specific concerned groups (Horst 2008). At least in Denmark, the participatory consensus conferences pioneered by the Danish Board of Technology build on a cultural tradition which aims at strengthening the shared identity of the community, rather than challenging it (Horst and Irwin 2010; Horst 2012).

The concept of ethno-epistemic assemblage (EEA) provides a further means of exploring how science–society relations can, in specific contexts, recombine and redefine previously fixed categories, especially that between 'expert' and 'lay' actors. The EEA concept offers a way of 'investigating how the blurring of science and society … might entail rather surprising new resources and methods' (Irwin and Michael 2003, 113). Put more concretely, and broadly in congruence with Callon *et al.*'s account, the term draws attention to the emergence of new hybrid entities and new 'admixtures of science and society' (Irwin and Michael 2003, 113). The EEA concept is intended above all as a heuristic device: '[I]t is a tool with which to explore how such heterogeneous groupings might be characterised. Ethno-epistemic assemblages are meant to aid us in examining how such "odd" mixtures come together, cohere and "work" as … unitary or singular actors' (Irwin and Michael 2003, 113). These 'odd mixtures' do not necessarily consist solely of human actors. Instead, 'non-humans' can play a significant part in the assemblage. In the case of climate change debate, the relevant EEA potentially encompasses complex mathematical models, patterns of biodiversity and agriculture, different visions of the future and of social organization, and threats to human and animal habitats. Rather than closing down or reinforcing the conventional boundaries between the human and the non-human, the social and the scientific, the lay and the expert, and the governmental and the public, the EEA concept attempts to be open to the transgression of boundaries and the flexible construction of actants. In that sense, EEAs 'are a means of *expanding the range* of entities, actors, processes and relations that get blurred and mixed up' (Irwin and Michael 2003, 114). This 'blurring and mixing' lies at the core of what was termed above the 'decentred' approach to scientific governance.

The particular implication of the EEA concept – and also of Callon *et al.*'s hybrid forum – in the current context is that 'engagement' cannot be seen as embodying a two-partner, science–society interaction, nor can the definition of the issue simply be assumed from the start. Instead, a diverse range of groups and organizations claim to speak both for 'science' and 'society', and the constitution of these groups will change over time. This in itself represents a profound challenge to the governance of science and technology. As we described earlier, critical approaches to engagement typically emphasize the marginality and fragility of one-off exercises. The assemblage perspective, with its emphasis on 'blurring and mixing', raises more basic questions: what is it that is being governed, how is it being governed, who is being governed in this way and why? Even the very concept of 'government' becomes open to question as we encounter the search for sustainable alliances that combine individuals and organizations traditionally defined as being both inside and outside the state (see also Dean 1999).

At this point, it is useful to bring into the discussion Marres' (2005, 2007) treatment of public involvement in controversy. One of the basic assumptions within the conventional model of public engagement is that there is indeed a 'public' (or 'publics') with which to engage. However, Marres – drawing on Walter Lippmann, in particular – argues that publics only emerge in response to specific issues. As she quotes Lippmann:

> The work of the world goes on continually without conscious direction from public opinion. At certain junctures, problems arise. It is only with the crisis of some of these problems that public opinion is concerned. And its object in dealing with a crisis is to allay that crisis.
> (Lippmann 2002 (1927) quoted in Marres 2005)

From this perspective, publics become engaged in controversial areas not because of some general enthusiasm for 'democracy for its own sake' but rather because of the particular content of an issue. Seen in this way, the policy-maker cannot simply tap into 'the public' at convenient points in the decision-making process since it is only the issues themselves which bring out 'the passions of the public' (Marres 2005). Equally, 'public engagement' in the full sense of that term becomes a matter for exceptional 'opening up' rather than for the more 'routine' operation of the bureaucracy. The 'centre' cannot connect with the public when the former so chooses because the public does not exist as a category other than in very atypical situations.

Perhaps unsurprisingly, Callon *et al.* are critical of Lippmann's approach. However, the debate over the role to be accorded to 'the public' is an important one in the current context. Put crudely, should we view heterogeneous assemblages as 'overflowing' the state so that it 'constantly has to be reinvented, shifted and taken in charge by a multitude of different, fragmented publics' (Callon *et al.* 2009, 241) or else consider the public as a more shadowy body, energetic when its 'passions' are stirred but otherwise largely passive? One possibility here is that while activists and committed social scientists tend to the 'overflow' perspective, our policy-maker may be at ease with a more 'ghostly' view of the public. If that hypothesis is correct, then it is not hard to see why activities such as consensus conferences tend still to operate within a relatively restricted perspective on the 'collective will' (to use Callon *et al.*'s term).

So far we have dealt in general terms with both the movement from PUS to PES and that from a traditional perspective on science and politics to one which offers at least the potential for transformation, emergence and change. However, we have also noted in passing some of the challenges this might represent for policy-makers who, having received what was earlier described as the 'gift' of public engagement, now discover that – far from providing an elegant elaboration and enhancement of current processes – this may have cuckoo-like (or ghost-stirring) consequences. Rather than helping with the daily task of policy-making, this 'gift' seems at least potentially to mess things up.

At this point, we move to a very particular case in a very particular context. In the following section, we will consider one specific engagement initiative within Denmark – a country with a strong international reputation for innovativeness in public engagement. The case concerns a citizen summit, a procedure with what Callon *et al.* term a 'family resemblance' to consensus conferences (hardly surprisingly when the same body – the Danish Board of Technology – has to a very large degree been parent to both). How does the previous 'decentred' perspective help us to interpret such a specific case and, rather crucially, does our discussion suggest any new possibilities in terms of tackling the entrenched challenges of climate change?

A citizen summit on climate change

The Danish town of Kalundborg is located on the western coast of Zealand. It is economically dependent on a few major industries, several of which have closed in recent years. The municipality is largely rural and possesses 160km of coastline, mainly very low lying. In 2006 the area experienced extensive flooding. A substantial proportion of the houses, summer houses, industrial premises and arable areas are at risk of being even more severely flooded in the future. Legally in Denmark, it is the individual landowner's responsibility to take action to prevent flooding, but the municipality has acted to develop a climate adaptation strategy in which the issue of flooding is prominent. Drawing upon a longstanding Danish tradition of consensus-building (Horst 2008; Horst and Irwin 2010) and especially through its participation in a European Union-funded collaborative project (baltCICA: www.baltcica.org/about.html), the municipality has worked to involve local stakeholders in the development of this adaptation strategy.

In March 2011, a citizen summit was organized by the Danish Board of Technology. According to the mayor of Kalundborg, the intention was to inform citizens about the situation and what climate adaptation might mean, but also to gauge citizens' opinions on the local climate adaptation plan since such a plan might contain controversial decisions – not least about abandoning certain parts of the Kalundborg area rather than seeking to protect them from the impacts of climate change (Lyngse 2011a). The citizen summit is a participatory technology developed by the Board of Technology which aims to include a large group of citizens in dialogue around an issue in order to create direct input to political decision-making. It is a one-day event which includes information-giving, dialogue and debate among citizens and subsequent voting.

The citizen summit in Kalundborg was part of a larger process organized under the auspices of the baltCICA project by the Danish Board of Technology in collaboration with the municipality. In autumn 2010 a workshop with local stakeholders developed four scenarios for a selected area around the Reersø peninsula. This area has a high recreational value and would be dramatically altered by rising sea levels. The four scenarios suggested different adaptation strategies: sea dikes, large dikes on land, gradual phasing out of buildings in vulnerable areas, and active conversion of vulnerable areas to open and recreational space. These scenarios were then employed in the background material presented to the participants before the summit.

The participants in the citizen summit were chosen on the basis of an invitation to 7,000 randomly selected people from the locality. Subsequently, 500 of those who responded positively to the invitation were selected in such a way as to reflect the demographic diversity of the municipality (Teknologirådet 2011a). In practice, this procedure led to a composition where just over half of the participants came from the town of Kalundborg and the other half were equally spread across the entire municipality. More than 80 per cent were over 45 years old, 17 per cent had a summer house in the area, and 80 per cent had previously experienced problems in relation to flooding.

As preparation for participants, a background paper explained how climate change might affect specific areas of the municipality (Kalundborg Municipality 2011). In this paper, the mayor of Kalundborg expressed the objective of the summit: 'To hear your opinions on local problems which might develop as part of climate changes. The results will be used in the subsequent municipality work on planning climate adaptation' (Kalundborg Municipality 2011, 1). One of the specific issues presented in the material was the distribution of responsibility and cost between the municipality, the citizens and local industry and whether 'the municipality should be authorized to make decisions for the common good' (Kalundborg Municipality 2011, 21).

The citizen summit itself was held (as often seems to be the case) in a large sports hall. At the summit, six to eight citizens were placed at each table with a local politician, civil servant or a university student as chair. The role of the chair was to mediate without engaging in the substance of the discussions. The summit was organized in different rounds in which citizens were given different tasks. Each round included debate and discussion which served as input for a general vote on three to five questions, with immediate projection of the results onto a big screen.

As an example of preferences regarding adaptation in the rural area, the citizens were asked to vote for one of the four scenarios presented. In total 42 per cent were in favour of a gradual phasing out of buildings in vulnerable areas, whereas 36 per cent were in favour of dikes either on land or sea (Teknologirådet 2011b). A large majority voted in favour of giving the municipality an active part in adaptation. The general conclusion from the voting was that most people were in favour of letting nature take its course in rural areas and concentrating resources on protecting the town of Kalundborg (Lyngse 2011b). The councillor in charge of infrastructure and environment was later quoted as being pleased with this outcome. He had feared that the citizens might vote for the large dikes, which in his eyes were unrealistic. At the same time, he was aware that half the participants came from the town itself, and therefore that a different composition of participants might have produced a different outcome: 'If you ask the people at Reersø and the summer cottage owners, then they probably would like some form of protection' (Lyngse 2011a).

In general, the local politicians stressed that they would take the results of the summit into account in the future development of the climate adaptation strategy: 'It is also a lot easier for us politicians to implement things if you have a sense that there is support for it' (Lyngse 2011a). The same councillor emphasized that it would especially help the municipality to make unpopular decisions:

> You cannot satisfy everybody ... we have to ask about what we, in the interest of the whole, can do as a society. What is easiest to do without? Some things we will have to leave to nature. I think that it is easier if we have a sense of what the citizens want.
>
> (Lyngse 2011a)

Bringing the (local) centre back into decentredness

So far, this chapter has moved from a broad exploration of decentredness to a very particular case of local action, from fluidity (social, cognitive, political) and multiplicity to one specific participatory assemblage. However, use of the term 'assemblage' in this context allows us immediately to make the point that the citizen summit, despite its more 'established' and 'centred' character, was also the product of hybridity, fragmentation and overflows. Nothing about this case was simply given. On the contrary, institutional, cognitive and political work was required to give it shape and substance. The legal issues could have been left with the landowners. The local mayor could have decided that there were no votes to be gained from these difficult and potentially divisive issues. The Danish Board of Technology could have chosen more high-profile ways of applying its creative energies. The very task of bringing together such a large range of questions within one specific context could have been viewed by the organizers as too open, too heterogeneous and too slippery to be even considered worthwhile.

In making this point that even 'centred' activities can (certainly in this case) possess the characteristics of an assemblage, we must note too that this 'engagement assemblage' did not sit alone but existed alongside (and was in many ways dependent upon) other networks and assemblages. Indeed, Denmark is a land of associations. One can argue that the strong Danish sense of national identity builds precisely upon a whole web of formal and informal organizations which allow issues to be considered and potential conflicts to be dealt with immediately and directly (Horst and Irwin 2010). In this case, especially important are the landowner associations (*grundejerforeninger*) which typically have special responsibility for roads, drainage and infrastructure. Also relevant (among many others) are the various farming associations, bird-watching and fishing organizations, local associations of national environmental organizations, political groupings, shopowner and industrial organizations, and religious bodies.

At least in this Danish setting (and here there may be important local and national differences), the 'centre' does not reside in fortress-like isolation but is to a large degree sustained by a diversity of other alliances and assemblages. And the municipality of Kalundborg is not a 'natural' unit but has been constituted through many decades of political reform and local reorganization. In making these points, we are also suggesting that the 'centre' might look rather different in different settings. The bicycle-riding official in a small Danish town – whose children attend local schools, who shops in the same supermarkets as everyone else and who is a member of many of the same associations – may in that sense be very different to her equivalent in Paris, Brasilia or Washington, DC. While it might be helpful at some level to generalize 'the policy-maker' – and not least in order to identify common dilemmas – we should also be extremely careful not to iron out these important differences in political culture and in the constitution of the policy-making process.

At this point also, we should note that ideas of (de)centredness depend, of course, on a spatial logic or metaphor. As has been noted elsewhere (e.g., Crang and Thrift 2000), care needs to be taken when adopting such a framing concept – and all such approaches carry unhelpful baggage. In previous work, we have referred to 'space' as denoting, first, the creation of new meeting points between institutions and wider stakeholders; second, the social relations occupying these points; and, third, the ideas and values, particularly with regard to knowledge, shaping the policy context (Jones and Irwin 2010). In line with other chapters in this volume, however, we take inspiration from STS explorations of the contextuality and situatedness of scientific knowledge (e.g., Latour and Woolgar 1979; Lynch 1993). Just as STS scholars have mapped the situated understandings, epistemological interstices and boundary work of science, notions of decentredness at least have the potential to open up the institutional embedding, meeting points and organizational practices of scientific governance.

In what follows, we deliberately operate in a spirit of modest pragmatism as we assess some of the policy and political implications of the Kalundborg case. Our version of modest pragmatism, and, indeed, our ideal type of the policy-maker, draw heavily on Charles Lindblom's concept of 'disjointed incrementalism' (Lindblom 1968). For Lindblom, policy-making is best seen as a process of 'muddling through', with radical change only taking place occasionally and not necessarily for the best.

We can start this discussion by acknowledging that critical and decentred perspectives do indeed provoke important questions about the purposes and limitations of 'engagement'. Notably, there is a strong tendency for (in this case, regional) policy-makers to set administrative boundaries around issues both in terms of their definition and the conditions for debate – as if the fate of Kalundborg could be separated from other regions and other climate-related challenges, as if such a large and complex set of discussions could be properly conducted in a highly restricted period of time, and as if the relationship between local administrative authorities and delegative democracy could simply be taken for granted. However, while all this can be presented as a problem, there are also good grounds for suggesting that it is only by delimiting (and in that sense 'centring' within the Kalundborg region) a problem that it becomes possible to 'get to grips' with it in a meaningful way. In that sense, 'centredness' is a way of turning slippery problems into tangible debates and action, even if such 'centredness' also risks distorting the very questions it sets out to tackle. Looked at from a 'disjointed incrementalist' perspective, where else can one start if not in the here and now, no matter how imperfect the 'here and now' might be?

Very importantly also, and as Urry and others have recognized, the existence of decentredness does not imply that national or local government can make no difference. In this case, the capacity of local government to mobilize resources in an organized and consistent fashion should not be under-emphasized. At the same time, however, our brief discussion of the Kalundborg case reminds us that engagement assemblages do not sit alone but must be sustained by other, perhaps pre-existing, assemblages and associations. At a very practical level, the process of mapping, tuning in to, nurturing and (as necessary) challenging existing collectives is likely to be crucial to the success of any engagement initiative.

Returning to Marres's discussion of the Lippmann–Dewey debate, it must be acknowledged that in this case both the issues and the public were in a fundamental sense 'manufactured' rather than emerging spontaneously or as a direct consequence of some 'crisis' (although flooding is certainly a matter of local concern). However, in governance terms, there seems little alternative to this if there is to be any kind of 'constructive technology assessment' around issues of climate change (and here, of course, we reveal our own normative preference for active policy debate and intervention). From a policy perspective, there is an institutional responsibility to anticipate as well as react – which suggests that, in Marres's terms, there will inevitably be a degree of artificiality around the definition of who should count as the 'concerned public'. Rather than operating with the notion that there is only one 'authentic' public, however, the more fundamental point may be to recognize the contingency and constructedness of all groups claiming to speak for 'the public'.

Going further, this case certainly raises important issues of representation and representativeness. The 500 citizens stand for 'the publics' in a manner which is very clearly open to challenge. The sample could easily be biased towards those with personal experiences of flooding, strong views about the underlying issues or simply the time and inclination to take part. These 'biases' might have influenced parts of the discussion – for instance, in deciding whether summer cottages or else the town of Kalundborg should be saved. But the 'impurity' – or one might say the 'ambiguity' – of engagement can also provoke debate and create energy. Indeed, far from representing something to be 'filtered out' from the engagement assemblage, ambiguity and impurity can help to make the issue into something with which people can engage and with which they wish to engage. As one satisfied participant put it: 'I received the invitation and I thought that, if you get the opportunity, it is a citizenly duty to participate. It is also an exciting issue and something that you have to do something about' (Teknologirådet 2011b).

Presented in this way (and here we offer a point also made by Callon *et al.* 2009, among others), public engagement – despite its acknowledged limitations – can be a way of invigorating and enhancing democracy. It brings people into one 'social and intellectual space' and allows a focus for discussion concerning a case that can otherwise seem very remote. From this perspective too, democracy is something to *perform* rather than *make perfect*. Moreover, a sense of imperfection and ambiguity can provide a powerful stimulus to engagement. The analytical and political challenge is then less a matter of how to remove doubt, decentredness and uncertainty from any engagement exercise, but instead one of deciding how much centredness, how many limitations and what degree of 'top-downedness' is both possible and optimal (and from whose viewpoint?).

Conclusion

> The notion of ambiguity must not be confused with that of absurdity. To declare that existence is absurd is to deny that it can ever be given meaning; to say that it is ambiguous is to assert that its meaning is never fixed, that it must be constantly won.
>
> (de Beauvoir 1948/1976, 129)

How then should we view the relationship between social scientific reflections on 'decentredness' and this very specific case? Taking some inspiration from Simone de Beauvoir's discussion of the ambiguity of meaning, we suggest not one but four responses.

One response is in terms of the language of *critical disappointment*. Certainly, it is not difficult to contrast the abundance of potentially overflowing democratic possibilities with the more mundane realities of this particular exercise and accordingly find the latter wanting. Analytical scepticism is an essential element within social scientific enquiry. And external criticism is important if policy and practice in this area are to improve. From our perspective, however, a sense of critical disappointment may be necessary (and even unavoidable) faced with the realities of individual engagement exercises – but is not in itself sufficient.

A second response has been argued in this chapter: a view of the governance process as an *assemblage* and not simply the 'attenuated other' (i.e., a one-dimensional or rigidly stereotypical category). Such an approach draws particular attention to the significance of new governance forms but also to the organizational work involved in creating 'centres' in the midst of such slippery and heterogeneous concerns. The assemblage perspective raises questions concerning the tolerance of ambiguity within governance systems. It may be that the Kalundborg case suggests a greater willingness to risk-taking and openness (and, hence, ambiguity) than can be found in other more 'institutionalized' approaches to environmental governance. Hypothetically, this acceptance of a more open and ambiguous approach may be associated with wider political culture – with a relatively high level of trust in political institutions within Denmark encouraging greater risk-taking (Horst and Irwin 2010).

A third response is based on a *dialectical viewpoint*. From this perspective, 'centredness' and 'decentredness' are both essential dimensions of the same relationship – both are therefore necessary since neither can exist without the other. This response suggests that the key question is not whether a particular engagement initiative is either 'centred' or 'decentred' but rather of how these elements interact and shape one another. From this perspective also, it is probably better to think of (de)centredness less as a fixed state and more as a process of becoming, as '(de)centring' rather than '(de)centredness'. Equally, even decentred movements must have a centre if they are to act in the world: how else could the protesters at COP 15 have converged? And, to make the opposite point, the divided response from senior political leaders at the Copenhagen conference sharply indicates the decentredness within what might otherwise resemble the 'global centre'.

A fourth and (for now) final response is more *reflexive* in character. It invites discussion of the relationship between contemporary social science and institutional action. There is much that could be said here. In this chapter, we have tried to be modestly but also constructively pragmatic concerning the relationship between especially STS scholarship and the trials and tribulations (and no doubt joys) of the – admittedly overgeneralized – 'policy-maker'. One brief proposal for now is that the kinds of scholarship presented here have much to offer in terms of democratic engagement with science and technology – not as a quick fix but as a critical companion, a source of occasional comfort, a challenge and a provocation. As one illustration of what we have in mind, social scientists have a potentially important role to play in encouraging experimentation in this area but also in identifying the unavoidable (and, we would say, necessary) ambiguities of engagement and governance practice. The point, as we have emphasized above, is not to view ambiguity as a weakness but as a stimulus and an opportunity for institutional reflection and development. This also calls for a social science which is open about its own uncertainties and limitations – and not least in terms of the need to develop a richer vocabulary for the analysis of the relationship between scientific governance and public engagement.

At this late point, we must return to our ideal–typical policy-maker. Her journey commenced with the greater certainties of the deficit model before being troubled, criticized and possibly inspired by talk of democracy and engagement. However, and as our metaphor of the unwanted gift suggests, discussions of decentredness, fluidity and multiplicity raise even greater uncertainties and can be read as implying that the traditional sense of policy- and decision-making no longer functions, since (as the poet Yeats has put it) the centre cannot hold.

That, however, is not this chapter's conclusion. It is true that we point to potentially greater multiplicities and ambiguities around the practices of scientific and environmental governance. One can also read this account as suggesting a number of policy dilemmas: not least, concerning the challenges of organizing for open engagement, inviting in the publics without ignoring the 'uninvited' (Wynne 2007), and framing the key issues without 'crowding out' alternative frameworks and scenarios. It seems inevitably true that public engagement promises more than it can ever deliver and in that sense is doomed to failure and disappointment (see also Horst 2010). Nevertheless, we have suggested that this imperfection and limitation can be a stimulus and provocation and not simply a matter for regret. Of course, this does not in any way remove the institutional responsibility to seek improvement in this area (not least since current initiatives are so very obviously short of perfection). However, our particular emphasis is on public engagement as a process rather than a point of arrival and closure. Attention to the dynamics of centring and decentring might be one crucial way in which our policy-maker can maintain this sense of process and improving rather than retreating into the greater simplicities and beguiling certainties of technocratic, fixed and formulaic approaches to engagement.

References

Beck, U. 1992. *Risk Society: Towards a New Modernity*. London: SAGE.
Black, J. 2002. *Critical Reflections on Regulation*. CARR Discussion Paper DP 4. London: Centre for Analysis of Risk and Regulation, London School of Economics.
Bodmer, W. 2010. Public understanding of science: The BA, the Royal Society and COPUS, *Notes and Records of the Royal Society* 64(suppl.1).
Bucchi, M. 2008. Of deficits, deviations and dialogues: Theories of public communication of science. In: M. Bucchi and B. Trench (eds) *Handbook of Public Communication of Science and Technology*. London and New York: Routledge, pp57–76.
Callon, M., Lascoumes, P. and Barthe, Y. 2009. *Acting in an Uncertain World: An Essay on Technical Democracy*. Cambridge, MA: MIT Press.
Crang, M. and Thrift, N., eds. 2000. *Thinking Space*. London: Routledge.
de Beauvoir, S. 1948/1976. *The Ethics of Ambiguity*. New York: Citadel.
Dean, M. 1999. *Governmentality: Power and Rule in Modern Society*. London: SAGE.
European Commission 2002. *Science and Society Action Plan*, http://ec.europa.eu/research/science-society/pdf/ss_ap_en.pdf (accessed 11 September 2015).
Felt, U. and Fochler, M. 2008. The bottom-up meanings of the concept of public participation in science and technology, *Science and Public Policy* 35(7): 489–499.
Felt, U., rapporteur, and Wynne, B., chairman. 2007. *Taking European Knowledge Society Seriously*. Report of the Expert Group on Science and Governance to the Science, Economy and Society Directorate, Directorate-General for Research, European Commission, http://ec.europa.eu/research/science-society/document_library/pdf_06/european-knowledge-society_en.pdf (accessed 11 September 2015).
Hagendijk, R. and Irwin, A. 2006. Public deliberation and governance: Engaging with science and technology in contemporary Europe, *Minerva* 44(2): 167–184.
Horlick-Jones, T., Walls, J., Rowe, G., Pidgeon, N., Poortinga, W., Murdock, G. and O'Riordan, T. 2007. *The GM Debate: Risk, Politics and Public Engagement*. Abingdon, UK: Routledge.

Horst, M. 2008. In search of dialogue: Staging science communication in consensus conferences. In: D. Cheng, M. Claessens, T. Gascoigne, J. Metcalfe, B. Schiele and S. Shunke (eds) *Communicating Science in Social Context: New Models, New Practices*. Dordrecht: Springer, pp259–274.

Horst, M. 2010. Collective closure? Public debate as the solution to controversies about science and technology, *Acta Sociologica* 53(3): 195–211.

Horst, M. 2012. Deliberation, dialogue or dissemination: Changing objectives in the communication of science and technology in Denmark. In: B. Schiele, M. Claessens and S. Shunke (eds) *Science Communication in the World: Practices, Theories and Trends*. Dordrecht: Springer, pp95–108.

Horst, M. and Irwin, A. 2010. Nations at ease with radical knowledge: On consensus, consensusing and false consensusness, *Social Studies of Science* 40(1): 105–126.

House of Lords. 2000. *Science and Society*. London: The Stationery Office.

Irwin, A. 1995. *Citizen Science: A Study of People, Expertise and Sustainable Development*. London: Routledge.

Irwin, A. 2006. The politics of talk: Coming to terms with the 'new' scientific governance, *Social Studies of Science* 36(2): 299–320.

Irwin, A. and Michael, M. 2003. *Science, Social Theory and Public Knowledge*. Maidenhead and Philadelphia: Open University Press.

Irwin, A. and Wynne, B., eds. 1996. *Misunderstanding Science?* Cambridge: Cambridge University Press.

Irwin, A., Jensen, T. E. and Jones, K. E. 2013. The good, the bad and the perfect: Criticizing engagement practice, *Social Studies of Science* 43(1): 119–136.

Jasanoff, S. 2003. Technologies of humility: Citizen participation in governing science, *Minerva* 41(3): 223–244.

Jasanoff, S. 2005. *Designs on Nature: Science and Democracy in Europe and the United States*. Princeton, NJ: Princeton University Press.

Jones, K. E. and Irwin, A. 2010. Creating space for engagement? Lay membership in contemporary risk governance. In: B. M. Hutter (ed.) *Anticipating Risks and Organising Risk Regulation in the 21st Century*. Cambridge: Cambridge University Press, pp185–207.

Kalundborg Municipality 2011. *Borgertopmøde om klimatilpasninger*, www.kalundborg.dk/Borger/Veje-_teknik_og_milj%C3%B8/Klima/Klimatilpasning/Borgertopm%C3%B8de.aspx (accessed 11 September 2015).

Kerr, A., Cunningham-Burley, S. and Tutton, S. 2007. Shifting subject positions: Experts and lay people in public dialogue, *Social Studies of Science* 37: 385–411.

Latour, B. and Woolgar, S. 1979. *Laboratory Life: The Social Construction of Scientific Facts*. London: SAGE.

Leach, M., Scoones, I. and Wynne, B., eds. 2005. *Science and Citizens: Globalization and the Challenge of Engagement*. London and New York: Zed Books.

Lindblom, C. E. 1968. *The Policy-Making Process*. New Jersey: Prentice Hall.

Lynch, M. 1993. *Scientific Practice and Ordinary Action: Ethnomethodology and Social Studies of Science*. New York: Cambridge University Press.

Lyngse, L. 2011a. Skal sommerhusene eller industrien drukne?, *Teknologidebat* 2, www.teknologidebat.dk/issue/article/skal-sommerhusene-eller-industrien-drukne (accessed 11 September 2015).

Lyngse, L. 2011b. I fremtiden må vi sejle til Føtex, *Teknologidebat* 2.

Marres, N. 2005. Issues spark a public into being: A key but often forgotten point of the Lippmann–Dewey debate. In: B. Latour and P. Weibel (eds) *Making Things Public*. Cambridge, MA: MIT Press.

Marres, N. 2007. The issues deserve more credit: Pragmatist contributions to the study of public involvement in controversy, *Social Studies of Science* 37: 759–780.
Phillips Report 2000. *The BSE Inquiry: Vol 1*. London: The Stationery Office.
Teknologirådet 2011a. Citizen summit in Kalundborg, www.tekno.dk/article/kalundborg (accessed 11 September 2015).
Teknologirådet 2011b. Deltagernes oplevelse af borgertopmødet, www.teknologidebat.dk/issue/article/deltagernes-oplevelse-af-borgertopm%C3%B8det (accessed 11 September 2015).
Urry, J. 2011. *Climate Change and Society*. Cambridge and Malden, MA: Polity Press.
Wynne, B. 1992. Public understanding of science research: New horizons or hall of mirrors?, *Public Understanding of Science* 1: 37–43.
Wynne, B. 1995. Public understanding of science. In: S. Jasanoff, G. E. Markle, J. C. Petersen and T. Pinch (eds) *Handbook of Science and Technology Studies*. Thousand Oaks, London and New Delhi: SAGE, pp361–388.
Wynne, B. 2006. Public engagement as a means of restoring trust: Hitting the notes but missing the music?, *Community Genetics*, 9: 211–220.
Wynne, B. 2007. Public participation in science and technology: Performing and obscuring a political-conceptual category mistake, *East Asian Science, Technology and Society: An International Journal* 1(1): 1–13.

4

ENGAGING THE MUNDANE

Complexity and speculation in everyday technoscience

Mike Michael

Introduction

The 'participatory turn' in the public understanding of science and the emergence of a field of 'public engagement with science' has been hotly debated over the last decade and a half or so. On the one hand, there is a view that this is an important development in enabling the voice of the public to enter into the process of scientific decision-making; on the other, that the various mechanisms of participation oversimplify the character of the public and diminish the purview of democratic politics. Shared across both these broad positions is the idea that participation is necessary in order to engage with the social, political, economic and environmental impacts of scientific and technological innovation. However, the emphasis on 'innovation' – complex, though this is – and the potential controversiality of such innovation is what is at stake in this chapter.

At its base, the questions that animate the present argument are: What does it mean to 'do engagement' about that which is not (or is no longer) marked by innovation and controversy? How can mundane technological artefacts and common-sensical knowledges be subject to participatory debate? What does this imply for our conceptions of politics, of science and technology, and of publics? What models of social science and society do we need in order to pursue these questions? What techniques are available to us that allow us to pursue these questions empirically?

These questions are, in part, sparked off by the idea that any object or event is comprised of many heterogeneous entities. Thus any object or event is connected to, and is a combination of, a huge array of elements. This is a version of the relationality of the object or event that finds an echo in actor–network theory, in the process philosophy of Whitehead, and even in Lefebvre's Marxian sociology of everyday life. If we accept this idea of relationality, then from even the most mundane technology, we can trace that technology's emergence (and mundanity) out of a range of entities and in this process pose a host of queries related to those constitutive entities. The mundane thus becomes a route for interrogating that which comprises it, including those elements that can be attached to the exposition of 'controversiality'. That is to say, the mundane can serve as a way of accessing a lost, or forgotten, politics. The other concern at the heart of this chapter derives from the argument that methodology is itself constitutive of its objects of study: method is part of what 'makes' the event of engagement or participation, and the ways in which the various participants emerge within such an event. The upshot of this is that there is a need to think of the engagement or participatory event in terms of an 'intervention', one through which the participants, the issues at stake and even the researchers themselves can emerge in unforeseen ways. As such, the point of an engagement event is not 'simply' to record or represent the process of debate and the development of new positions (sometimes leading to a consensus) around a given issue, but to contribute to – provoke even – a process of emergence that might redefine and re-demarcate what those positions might be, and what shape that issue might take. In the context of an engagement with the mundane, this also means moving beyond the 'interrogative' and the 'controversial' toward the speculative.

In what follows, I begin with a short outline of the 'participatory turn' in science–society relations, before addressing some of the limitations enacted through the focus on overt 'innovation' and 'controversy'. I then deploy a process philosophical analytic to argue that even the most mundane artefacts and events can constitute a fruitful line of enquiry into the dynamics of science–society relations. This point can be situated in relation to the sociology of everyday life, and to environmental sociology (as applied to mundane practices). After a brief explication of the constitutive role of methodology, I go on to illustrate a 'participatory politics of the mundane' first by looking at such everyday artefacts as Velcro and rolling luggage and, second, by drawing on the speculative design tradition to sketch out what a public engagement with the mundane – illustrated with the everyday technology, the band-aid – might look like.

A participatory turn

As is well known, early studies of science–society relations, particularly those that attempted to quantify the public's 'understanding' of science, have been characterized in terms of the deficit model. Rather than scientific knowledge, it is the public's deficiency in scientific literacy that was being measured. This deficiency was seen to be at the root of the public's apparent lack of support for science (e.g., Wynne 1991, 1992; also Gregory and Miller 1998). Against this, arising around the late 1980s, there was the critical or ethnographic perspective which drew on such qualitative techniques as semi-structured interviews, focus groups and participant observation. The aim here was to address the public's relevant knowledges and skills that were seen to mediate and reflect local cultural conditions and identities. It was also noted that this knowledge was routinely marginalized by scientific institutions, whose 'institutional body language' of certainty and self-confidence tended to privilege accredited scientific knowledge. Given that the application of such expert knowledge was, on occasion, problematic, critical PUS was also interested in how relations of trust between experts and lay publics were affected. Indeed, this turned on its head the earlier view that scientific knowledge led to support of science: the public's lack of trust in science (in part because of scientific institutional body language) led to its scepticism about scientific knowledge (Wynne 1996; Irwin and Michael 2003).

In the intervening years, this field has developed in a number of ways. The divide between science and society, lay and expert, that was once presumed has been redrawn at empirical, conceptual and political levels. Empirically, it has been shown that the divide is much more complex than first thought – breached by such figures as 'lay experts', for instance (Arksey 1998; Epstein 2000). Conceptually, it has been suggested that there is a need to theorize the complex interweaving of expert and lay constituencies as, for instance, assemblages (Irwin and Michael 2003) or hybrid forums (Callon *et al.* 2001). Politically, it has been argued that mechanisms need to be set in place that enable the voice of the public to enter into the scientific decision-making process. This is partly because, on the one hand, such a participatory initiative would enhance the social robustness of scientific decision-making and, on the other, would serve in the continued enfranchisement of the public – that is, in the 'democratization of democracy' (e.g., Wynne 2005; Hagendijk and Irwin 2006).

Under the general rubric of 'public engagement with science and technology' (PEST), numerous techniques have been developed (e.g., citizens' juries, consensus conferences, deliberative polling, card-based group discussion). Often, a key objective is to assess the appropriate parameters and efficacy of these methods for the 'doing' of participation (e.g., Chilvers 2008; Irwin *et al.* 2013). These various methods have also been subjected to a panoply of criticisms, ranging from a concern that they do not actually feed into any scientific decision-making process, through a worry that they detract from other, more radical, forms of participatory politics, and on to disquiet about the sorts of citizen that are being enacted and promoted (e.g., Michael 2012a). Further, PEST practitioners are interested in tracing the complex ways in which the views of lay people and experts might change in the process of (some sort of) a dialogical engagement around a specific scientific and technology controversy. To be sure, such controversy might only be a potential one, especially where the science or technology is at a relatively embryonic stage. Nevertheless, another key aim of the engagement process is to address a substantive issue, grounded in this or that scientific or technological innovation, which is likely to be worrisome in some degree. For instance, an engagement event might focus on a specific topic – nanotechnology, genetically modified (GM) crops, stem cell research, synthetic biology, climate change, fracking, geoengineering – that can be linked to contemporary or future risks to the environment, or to human health, or to particular communities, identities or folk knowledges.

Here, the important point is that these PEST methods – what have been called 'formalised mechanisms of voicing' (Michael and Brown 2005) or 'technologies of elicitation' (Lezaun and Soneryd 2007) – topicalize 'controversy'. That is to say, technical processes of designing, organizing, recruiting, implementing and analysing participation, deliberation and engagement become enmeshed with the processes of identifying, specifying and focalizing controversial topics. This comes into relief when we step outside of this nexus of technique and topic, and propose, for instance, a consensus conference on such mundane 'topics' as Velcro, or rolling luggage, or the paper clip. Within the framework of PEST, this is likely to be seen as a trivial waste of resources or some sort of cynical exercise in market research.

Yet, in the sociology or everyday life, and in recent environmental sociology, the mundane has been an object of critical analysis. Classically, Lefebvre (1947) writes that even a woman buying a bag of sugar can be subject to analysis that discloses not only such things as her class background or the 'state of the markets' but, indeed, 'the sum total of capitalist society, the nation and its history' (p57). For Ben Highmore (2002), the everyday is a rich source of critical insight not least when it can be de-familiarized via a 'plurality of approaches, a range of attentions that place [the everyday] within a framework of critical interdisciplinarity' (p143). Elizabeth Shove (2003) has shown how embedded and embodied everyday practices of cleanliness and comfort (and their embroilment with often 'invisible' technologies such as water, lighting and heating systems) can be critically regarded as contributing to all manner of environmental problems.

However, note that these approaches treat the everyday as a topic of *critical* analysis. The social scientist stands outside of these mundane events, with a particular analytic armoury in place – whether that be Marxian, or critical theoretical, or environmentalist. As such, the mundane can end up serving as an illustration of a particular critical version of the world, say broadly Marxian or environmentalist. By comparison, Gay Hawkins (2006; see also Gabrys et al. 2013) has, in relation to plastic, shown that a more nuanced analysis is possible in which plastic, as both product and waste, can lead into a nexus of complex issues which troubles any singular analytic frame, not least because empirically these issues also encompass the pleasures and possibilities of plastic. Put another way, the mundane 'actual entity' that is being analysed is comprised of a hugely heterogeneous and complex array of mutually affecting entities and relations, including those contributed by the researcher herself. This necessitates both a more circumspect and a more 'creative' treatment: more circumspect because it is important not to attach this mundane heterogeneity, complexity and emergence too enthusiastically to this or that theory; and more creative because we will need to address the 'processuality' of everyday events and entities in ways which access this 'openness'.

Mundane events and entities

The foregoing mention of 'processuality' and 'openness' indicates that this chapter takes inspiration from those perspectives that emphasize a version of the world that is relational, heterogeneous and in a process of becoming. Core authors here are Whitehead, Deleuze and Stengers. This processual sensibility yields two implications that are especially pertinent to the present discussion.

The first is that any event or entity emerges through the combination of myriad other events or entities (or what Whitehead might call the concrescence of prehensions in an actual occasion or an actual entity). We see this echoed in Donna Haraway's (1994) work. Her notion of a 'black hole' can be extended beyond her figuration of the cyborg to address the mundane as a complex heterogeneity in which 'the mythic, textual, political, organic and economic dimensions implode. That is, they collapse into each other in a knot of extraordinary density that constitutes the objects themselves' (p63). Heterogeneity also marks actor–network theory's figure of the 'actant', which in its complex relationality is itself comprised of a network (e.g., Latour 2005). Indeed, for several commentators, this is a great weakness of actor–network theory: How can one analyse an actant when its extensiveness is potentially infinite? How does the analyst know what to include and what to ignore (e.g., Button 1993)? For Callon and Law (1995), the resolution was to acknowledge that such choices came down to a matter of 'taste'.

This last issue seems only to make sense where there is a methodological commitment to an empiricism in which the event or entity that is being empirically accessed is 'stable', and where the process of empirical inquiry does not affect that event or entity. Law's (2004) later writings have been central in articulating the performativity of method – how empirical engagement is partially constitutive of its objects of study. Accordingly, method involves 'the crafting and enacting of boundaries between presence, manifest absence and Otherness' (p85) in which 'data' come to be excluded, not least because they do not fall within the purview of that method (which can be extended – as Law's notion of a method assemblage indicates – to encompass, for instance, the conceptual and political framing of the research).

This argument for the constructedness of data can be further extended if research is treated as an 'event'. Derived from the work of process philosophers such as Whitehead (1978/1929) and Stengers (e.g., 2005), the event can be regarded as the convergence and 'integration' of very many, very different entities that straddle the human and nonhuman, the cognitive and affective, the social and material, the conscious and unconscious, the macroscopic and microscopic, and the researcher and the researched. Mariam Fraser (2010) argues that, according to these authors, the event should be characterized by the co-becoming of its constituent elements: as they interact in the process of constituting the event, they mutually change. The implication is that, in the morass of transformation, what the research event 'is' is also shifting: the event thus becomes chronically underdetermined, open and unfolding. Thus, while at one level it appears that this event is an 'event of public engagement with science', there are numerous mutual shifts within such an event that might very well render that event 'other'. Participants' many, disparate, emergent reactions – often ignored or 'othered' by the researcher – could be reconfiguring what that event 'is'. Michael (2012a) has (ironically) called these reactions 'misbehaviours' and has argued that they can reconstitute the engagement event into, for example, an occasion for competition or play amongst the participants. Having noted this, the researcher can never be certain of what the research event has become, not least when that researcher herself is likewise embroiled in the process of mutual change. Who is the researcher who has changed in the process of engaging with the 'event of engagement' which itself has changed?

One implication is that if the research event is open, emergent and unfolding toward the 'virtual' or the possible (e.g., DeLanda 2002), then the researcher's task becomes something other than furnishing a more or less accurate depiction of 'that' event. Rather, the researcher's job becomes one of asking – indeed, speculating on – how this event is unfolding and what it is unfolding toward. As Fraser (2010) frames it, this becomes a speculative matter of 'inventive problem making' that, we might say along with Stengers (2010), 'affirms the possible, that actively resists the plausible and the probable targeted by approaches that claim to be neutral' (p57).

In sum, to apply 'event thinking' to a mundane object is to become engaged with the complex multiplicity of emergent relations that comprise that object, while simultaneously acknowledging that what that mundane object 'is' is unfolding in ways that point toward possibilities that afford the prospect of posing 'that' object as an 'inventive problem'. To frame this in the context of the 'public engagement with the mundane', to investigate an everyday technology is both to address, with complexity, the complexity of its composition and to be open to its openness. In the next section, we embark on just such an investigation.

Trying to be complex and open about the complex and open

This section 'speculates' on some of the complex, potential ways in which mundane technologies emerge or, to put this another way, are 'eventuated' (Michael and Rosengarten 2013). The aim here is to illustrate how such mundane technologies – treated as 'eventuations' in which they emerge in their specificity in particular events – can lead to 'inventive problems'. This sets the ground for a later discussion where we begin to think about how we can develop techniques for engagement that facilitate this eventuation, and inventive problem-making for participants.

Let us take two examples of mundane technology: Velcro and wheeled or rolling luggage. In the main, these are not seen as controversial 'in themselves', though as possible members of broader categories – such as products manufactured in the developing world, or potential plastic waste – they might come under critical scrutiny. These technologies are, by and large, seen as efficient, unproblematic tools that are routinely represented as models of convenience (see Michael 2006, 2011). However, a little online exploration and a little reflection readily facilitates a speculative engagement with these artefacts.

Velcro is the brand name for the well-known and ubiquitous 'hook and loop' fastening. Much like 'Hoover' and vacuum cleaners, Velcro (a term concocted by its Swiss inventor, George de Mistral, from velour and crochet, the French for, respectively, velvet and hook) has become synonymous with this form of fastening. One of its key advantages is that it works without the precise alignment required by the zip, button and hole or hook-and-eye fastening systems. Indeed, it has become so iconic of 'easy fastening' that it is applied in other spheres including scientific (nanotubes), corporate (models of work relations and networks) and moral (the loss of an ethos of 'learning through problems') domains. Yet, as Michael (2006) shows, even through the most peremptory online search, its putative convenience is undercut in all sorts of ways – it is eventuated as something other than an unproblematic, efficient technology. Thus, it can get too loose with repeated use, as extraneous material (mud, fluff) gets caught up and accumulates in it; but it can also get too stiff and difficult to unfasten with repeated use. Despite the fact that fastening can occur without alignment, if the hook and loop sides are not accurately aligned, the stiff plastic strips can 'stick out' and rub painfully against the skin (a key example was nappies). Here, we can see how the disjuncture between the promise of a mundane technology and its specific operation opens up the eventuation of Velcro. The sorts of questions that become available might include: What body techniques do we need to develop and put in place in order to enable such technologies to work and to fulfil their promise? What are the means by which we articulate and circulate these technological shortcomings? What constitutes 'expert' advice in this specific domain, what is such 'expertise' grounded in (e.g., technical safety standards versus experience), and what makes it credible?

Michael also reports on the problems that arose when trying to help his two-year-old daughter fasten and remove her cycle helmet: her fine hair would routinely get caught in the Velcro that held in place the cushion pads inside the helmet (these were meant to afford comfort and an adjustable fit). In these cases, it is clear that eventuating Velcro as convenient and efficient again required the development of a choreography of micro-manoeuvres, an inter-corporeal joint action. As this became more practised, so there is a shift from struggle to cooperation, and from a sense of imposition to something like 'mutual care' (in which daughter and father work to avoid each other's distress). Unlike the questions that were posed above (which reflect fairly standard PEST concerns), the latter eventuation of Velcro opens up more inventive questions such as: What sort of heterogeneous relations – let us call it 'technosociality' – does Velcro presuppose and enable? What sort of (dyadic, collective) creativity does Velcro's 'failure' (or, rather, distributed 'success') enable? Do these emergent micro-social and micro-corporeal relations ironically serve to undergird a more complex engagement with Velcro, its producers, its ubiquity and its promises?

Let us consider a second example of a mundane technology, that of wheeled or rolling luggage. This luggage shares with Velcro the air of quiet convenience: in contrast to older forms of luggage, it has incorporated functions that were once assigned to arms, backs and legs (or to porters). Often it is portrayed with a single traveller in an uncluttered space (see Michael 2011), though we know that transport hubs are anything but uncluttered, not least by fellow travellers with their own wheeled luggage. Arguably, in this general setting new body techniques arise. The design of two-wheeled rolling luggage means that it usually trails behind the body rather than at its side. This implies that it cannot be seen directly, without frequent backward glances. Through this movement we can monitor any risks it poses, such as colliding with other travellers' legs and luggage. However, frequent backward glancing means that we are paying less attention to what is ahead of us, raising the risk of getting entangled in others' rolling luggage. In other words, rolling luggage points to a new pattern of mobilizing and distributing attention. Needless to say, this scenario is far too general both culturally (my Australian friends assure me that they would not be worried at all about entangling others) and materially (new four-wheeled luggage demands other sorts of body techniques, not least because such luggage has a tendency to run away on slopes). Yet, it is the specificity of the particular eventuation of rolling luggage that raises interesting problems.

So, let us take the redistribution of attention seriously – a redistribution that 'repairs' these forms of luggage so that they remain 'convenient'. We might now ask whether this is yet another form of neoliberalist 'responsibilization': instead of redesigning the spatialities and temporalities of transport hubs, it is the responsibility of travellers to negotiate these unwieldy space-times. Or we might ask whether the convenience travellers corporeally 'repair' for their luggage reflects a system of travel which seems all too convenient, and yet which is systematically 'repaired' by both self and others (ranging from local transport workers through to those communities who suffer in the extraction, movement and use of fuel). However, echoing the reflection on Velcro, we might also take a more speculative tack and ask whether, in the process of these complex redistributions of attention amongst fellow travellers, there is also a possibility of mutual recognition. In other words, wheeled luggage and inventive bodily practice and coordination provide a socio-material medium for what we might call a momentary (and heterogeneous) sociality – a 'technosociality': the complex interplay of convenience and inconvenience thus enacts a fleeting communitas of humans and nonhumans.

What the examples of Velcro and rolling luggage have hopefully illustrated is that even the most domesticated of technologies can serve as a starting point for critically uncovering related 'controversy', yet they also serve as a basis for speculatively developing 'inventive problems' about their eventuation. Critically, questions can be forged around the marketing of Velcro, or the economics and ecology of transport, and the enrolment of users in managing and dissipating these questions. Here a 'direct' agonistic politics of resistance, protest or participation is implied – how are 'we' to mobilize in order to reform these 'systems' of fastening and transport or mobility? Speculatively, I have pointed to 'inventive problems' that suggest the unforeseen 'pleasures' of the problems that arise with these mundane artefacts. These 'pleasures' – emergent occasions of 'technosociality' – might be momentary, but arguably, as much as they might solve the problems of these mundane technologies, they can potentially serve as a diffuse means by which to build new relationalities, and to do a different, more distributed, amorphous and circuitous 'politics'.

Probing the everyday

If PEST is interested in the issues that collect around a novel or innovative technology (in the making), such technologies are unlikely to be well known to lay publics. As such, a key element in the procedures of PEST is the making of the unfamiliar technology in question, to some degree or other, familiar. This familiarity is not just 'technical' (illuminating the 'workings' of a technology) but also 'political' (elucidating the social risks of the technology). In this way both experts and lay people can begin to appreciate each other's perceptions (e.g., Macnaghten 2010). By contrast, the discussion of Velcro and rolling luggage has attempted to do the opposite insofar as it has taken up what is a standard analytic in the sociology of everyday life, namely making the 'familiar unfamiliar'.

In part, this turning of the familiar into unfamiliar is concerned with becoming sensitive to the extraordinary that accompanies the ordinariness of everyday life. For Highmore (2002), 'the everyday offers itself up as a problem, a contradiction, a paradox: both ordinary and extraordinary, self-evident and opaque, known and unknown, obvious and enigmatic' (p16). For Gardiner (2000), the everyday – marked by the proliferation of sensation – is similarly infused with the unconscious, otherness, exoticism, fantasy and hidden potentiality. The everyday thus additionally possesses the quality of 'marvellousness'. And for Michael (2000) and Bennett (2004), this everyday 'contradictoriness' rests on a view of the mundane as constitutively riven with processes of sociotechnical ordering and disordering. This is, of course, a key observation of early actor–network theory with its analysis of the scripts and affordances of everyday objects such as the door groom (e.g., Akrich and Latour 1992; Latour 1992). Translating all of this into the terminology of the event, the eventuation of the everyday is heterogeneously constituted and what is at stake is attuning ourselves to this heterogeneity.

According to Highmore (2002), accessing this heterogeneity is a complex matter of attempting to represent the everyday that does some sort of justice to its richness (a richness which is multi-sensory) while also acknowledging that it will remain elusive. To this end, avant garde techniques such as montage might be particularly fruitful not least because they can 'bring together' and juxtapose multi-modal materials that span the textual, the visual, the auditory, the tactile, the olfactory and so on. While such techniques cannot convey a complete representation of the heterogeneity of the everyday, they can at least convey a sense of its complexity (see also De Certeau *et al.* 1998).

In the discussion of Velcro and rolling luggage, there was some attempt at practising this juxtaposition of a variety of materials: eventuations of these everyday technologies entailed moving through an account of their (commercial) idealization, their common problematization and their multi-sensory mutual enactment by which they came to 'work' but also by which they evoked particular virtualities (framed as 'technosocialities'). Inevitably, this exercise – and its heterogeneities and virtualities – was 'limited' by the fact that it entailed me partly enacting myself as a particular sort of 'autoethnographic user' of these technologies. Having noted this, autoethnography does not in itself invalidate the foregoing observations – after all, every user is a 'black hole', a Whiteheadian 'actual entity' (see Michael 2012b). The point is that the heterogeneities and virtualities of mundane technology can be further explored through expanding the range of participants, and drawing on methods that multi-modally probe those heterogeneities and virtualities.

'Probe' is not an innocent term here – it connotes the 'cultural probes' that have been a major component of speculative design practice. In what follows, I focus on probes as one means of engaging the mundane. Boehner et al. (2012; also Michael 2012c) argue that 'cultural probes' (henceforth 'probes', not least to connote their multi-sensory or multi-modal operation) diverge from the methods typical of the social sciences. As such, they are developed and operationalized with a number of qualities in mind:

- They are not designed to collect 'data' that can be assessed in terms of their validity, reliability or generalizability so much as to generate 'materials' that are valued for their singularity, oddness and capacity to inspire.
- They are not expected to yield more or less accurate depictions of this or that social phenomenon as tracing, contingently and uncertainly, the way a social situation might unfold or emerge, however idiosyncratic or wayward that might turn out to be.
- They are not supposed to access 'what is'; rather they enable 'what might be'.

While probes are usually part of a design process – they resource the production of speculative prototypes – they can also be put to use in the social sciences as a means of pursuing the speculative project of inventive problem-making.

In developing probes, designers will spend a lot of time considering those likely to use the probes (volunteers may be recruited at random via press advertisements, or snowballed from target communities). Part of this process entails drawing on designers' own and others' relevant experiences, which might entail visits to particular communities as well as detailed discussions about specific settings and events. This will also be supplemented by other sources such as popular cultural and media artefacts, or representations derived from scientific, technological, design and art disciplines. Such materials, which might include statements, photographs and drawings produced by the design team, as well as found texts and images, can be arranged in the form of a workbook. This serves as a common, complex point of reference through which designers can distil a selection of tighter, more focused themes that will inform the design of the probes – themes such as community, future, connection or information. Also important here is an aesthetic that will emerge iteratively with the themes, and a sense of the balance of different probes (drawing on different senses, or accessing different affects). Finally, a central criterion in the making of probes is that they do indeed 'probe'. They need to possess such qualities as novelty, ambiguity, provocativeness and obliqueness. While it is taken for granted that not all probes will work, it is nevertheless hoped that these qualities will enable participants to use the probes in ways that generate responses that are themselves unanticipated, creative, provocative, open and ambiguous.

Concretely, probes are collected together as a probe pack. The probes in the pack might take a range of material forms. Sometimes, they allow for the recording of text. A confessional aerogram encourages people confidentially to admit to practices that waste energy. A plan of a house asks people to translate energy into a sensory experience: what is the smell of energy in the corridor or the sound of energy in the living room? This multi-modally also finds expression in the 'Listening Glass' – a standard drinking glass which is pressed against a domestic wall so that the participant can listen to the building's noises, and make a record of those noises directly on the glass with a marker pen. A probe might also record non-textual material. Simple jotter pads with instructions encourage people to draw in response to specified questions while, for instance, they are talking on the phone. A disposable camera, or an SD card that can be fitted into a participant's own camera, come with directions to photograph such strange sites as the spiritual centre of the home, or the edge of a community.

Typically, probe packs have been sent to individual participants who are given around one month to make use of them. It is made clear that how the probe pack is used is purely up to the volunteers, and it is not expected that all probes within a pack will always be returned. The packs also contain the wherewithal to ensure that completed probes can be sent back to the designers.

Now, obviously enough, these probes have been targeted at individual participants, but there is no reason why they cannot be used in collective participatory events. Thus in the project Sustainability Invention and Energy Demand Reduction: Co-Designing Communities and Practice, a probe workshop was organized to which participants from a variety of groups were invited.[1] These groups included energy communities (communities that had been funded by the UK government to develop energy demand reduction strategies, and implement energy saving or generating measures), government departments and non-governmental organizations (NGOs). Participants were given a number of probe tasks, including one of representing a new collective community that encompassed their very disparate experiences. This was to be depicted on a large map with minimal features (e.g., highly schematic representations of buildings and monuments), with the aid of marker pens and post-it notes and a range of pre-printed stickers designating both obvious (buildings, plants, vehicles, people) and not-so-obvious (guns, large wild mammals, alien space ships, cages) elements. The map also incorporated a timeline and a series of mildly provocative prompt phrases such as: 'sources of shame', 'of value only to residents' and 'easy to get lost'. 'Community' in this collective probe exercise became something altogether more open and unfolding: it was recognized as a construction by the participants; it was marked not by some 'intrinsic' character so much as by certain flows 'through' it (cars, planes); and parklands seemed to be ambiguous sites that combined elements of collective identity, unease (because of degradation) and aspiration (insofar as they signified both 'wilderness' and free social exchange).

So, ideally probes should entail the use of several senses and prompt engagement with the everyday that takes the user outside of the ordinary: they render the familiar unfamiliar. While probes can be regarded as a means of 'accessing' the strange underbelly of the everyday, as it were, they are also, of course, performative – that is, partly constitutive of participants' responses. To reiterate, the corollary is that probes do not simply address 'what is' but serve as speculative media into 'what might be'.

In summary, for Bill Gaver and his collaborators (e.g., Gaver et al. 2004; Sengers and Gaver 2006), the aim is not to derive an overarching, cogent interpretation of the responses engendered through the use of the probes, but to use these as a resource that informs, in oblique, opportunistic and piecemeal ways, the production of a design prototype. For social scientists, the probes can yield 'data' that need not be subject to the usual forms of 'analysis' with their focus on finding patterns – that is, commonalities across the 'data set' that imply underlying structures (such as discourses, configurations of habitus or interests). Instead, treating the analytic phase as another speculative occasion, the data – or, rather, the materials – can be explored for insight into the complexity, heterogeneity and potentiality that makes up the ordinary (in the above case, the community), and affords a hint of how the issue at stake is itself emergent (community is not something that is presupposed, but unfolds in ways which might allow for the making of inventive problems).

Public engagement with the mundane

In this section, we begin to outline some parameters for a 'public engagement with the mundane' specifically as it relates to the problematic of 'science–society' relations. At base, the aim is to draw on a mundane 'technoscientific artefact', imagine how this might be collectively engaged with through the medium of cultural probes, and use this as a means speculatively to address the eventuation of 'science–society' relations.

Public engagement exercises have considered a range of the more exotic biomedical innovations including the generic 'new genetics' (e.g., Kerr et al. 2007), and more specific developments such as nanotechnology (e.g., Delgado et al. 2011; Macnaghten and Guivant 2011) and stem cell research (e.g., Davies 2006; Kotchetkova et al. 2008). In the present case, we will turn to that most mundane of 'biomedical' artefacts, namely what is variously known as the sticking plaster, Elastoplast or Band-Aid. In the discussion that follows, I draw on previous work (Michael 2010) and use this as preliminary 'fieldwork' through which provisionally to thematize the sorts of broad issues that might arise around these artefacts and which will inform the design of probes.

The band-aid can be considered in terms of the histories of 'its' invention and development, but also in relation to the (still relatively sparse) commentaries on its status as an artefact of popular (as well as medical) culture. On this latter score, the band-aid features in rhetoric (a 'sticking plaster' solution), as a cinematic signifier of trivial wounds (innumerable cartoons) and as a proxy for other attachments (e.g., electrode wires, or even digital plasters that serve as remote sensors of body data – see Beaver et al. 2009). It can also, of course, be understood as a commodity that takes particular material semiotic forms. For instance, band-aids can be packaged and marketed as convenient rolls for bespoke sizes, or sets comprising a fixed variety of shapes and sizes; adult oriented or child friendly; appropriate for different skin colours; and oriented to a range of wounds and their healing and aftermath (e.g., accelerating recovery, decreasing scarring).

From these various materials it is possible to derive a series of themes. Here are three.

1. *Size and shape.* Why do band-aids come in the sizes and shapes that they do? What assumptions are made about the range and distributions of everyday wounds? How are these derived? Is there a maximum and minimum sized band-aid? What is signified by an oversized or undersized band-aid? Are other uses implied? What might these be? How do the properties of band-aids (e.g., stickiness, flexibility) suggest other functions?
2. *Scarring and biography.* What does a band-aid say about its wearer? Certainly they are 'injured', but are they, additionally, incompetent, accident prone and unlucky? Band-aids take a range of different forms – plastic, fabric, transparent, spray on and hydro-colloidal. They can signify different concerns – an intent to cut costs, a desire to shun plastic, a determination to protect the wound from germs and a wish to minimize scarring. What sort of identity is enacted through these? How might scarring, or its avoidance, enable particular biographical narratives?
3. *Specialization and expertise.* Band-aids are designed for particular sorts of wounds – what might be called 'domestic' or 'everyday'. But these are slippery categories. How shallow must a wound be for it to remain 'domestic' (as opposed to in need of professional medical assistance)? Under what circumstances does a 'domestic' wound outstrip a band-aid, and require the aid of friends, doctors or even social workers or the police?

From these themes a number of tentative probe exercises can be concocted:

- Participants can place band-aids (or even scars) on photos of each other; the band-aid wearer then has to fabricate a relevant biographical narrative (say out of a pre-existing selection of events).
- Digitize band-aids – ask participants what sorts of information band-aids might transmit and to whom.
- Ask participants to decorate band-aids, turning them into jewellery for specified special occasions.
- Ask participants to invent band-aids that work on psychological, emotional and social wounds, as well as physical ones.
- Ask participants to find uses for very large, very small or oddly shaped band-aids.

It should be obvious enough that these suggested probe exercises need a lot of refinement (a designer's sensibility would certainly be of help in this respect) – at the moment, they remain too hidebound to the object of the band-aid. Nevertheless, embryonic as they are, these outline probes have hints of playfulness, ambiguity and obliqueness that might prove critically and speculatively productive.

At one level, the implementation of such probes, as an alternative to the standard PEST methodologies, might open up a new substantive terrains of analysis of those mundane technologies that co-habit the everyday with us, that shape, in subtle and cumulative ways, everyday relations and actions. No longer will such mundane technologies remain under the radar in terms of what counts as 'controversial' and 'innovative', but be elevated to an appropriate topic for PEST analysis. On a less grandiose scale, in relation to band-aids, probes might raise certain critical issues about, for example, the procedures by which wounds are categorized, modelled and technologized through manufacturers' research, production and marketing assemblages. How do these 'external' – what in PEST might be called 'expert' – processes impact upon everyday relations to the body and its exigencies? How might potential users feed into the processes of product development and thereby, potentially at least, affect the sorts of models of user and body that underpin those processes?

However, more processually, these probe exercises might facilitate a speculative engagement in which the band-aid becomes something other than the partial object, and medium, of a domesticated medical gaze. On a grand scale, the speculative use of probes in PEST points to complex, fragmentary and idiosyncratic relations with mundane technologies – relations that have the potential to resource inventive problem-making. For instance, even in the foregoing admittedly provisional suggestions for probe exercises around band-aids, we have seen that the band-aid serves in the simultaneous enactment of the legal, the moral, the aesthetic, the economic and the autobiographical (to name but the most obvious). The complexification hopefully wrought by the probe exercises eventuates the band-aid as a complex, unfolding 'actual entity' that opens out toward unforeseen virtualities and inventive problems. I've resisted the temptation of suggesting – second-guessing – examples of what these inventive problems might be. Suffice it to say that they might incorporate intriguing admixtures of, say, the aesthetic and the political that invite us, as analysts, to re-imagine the parameters of PEST, and, indeed, the public and its politics (see below).

Conclusion

This chapter has presented an argument for a version of 'public engagement with science and technology' that de-privileges, if not eschews, controversy. Instead of those technoscientific innovations or projects that are actually or prospectively subject to disputation, it has been suggested that mundane artefacts that quietly inhabit everyday life with us can be the 'topic' of a reconfigured PEST analysis. The ubiquitous integration of these entities into the routines of everyday life means that they are necessarily involved in the partial constitution of 'publics'. But further, as 'actual entities', they are comprised of a nexus of heterogeneous relations, and are eventuated in different ways on different occasions.

The implication is that, ordinary though they might seem, mundane artefacts can be used to probe the complex, shifting, 'subterranean' dimensions of science–society relations. A means of accessing this role of mundane artefacts was derived from speculative design and its use of 'probes'. An attempt was made to outline how probes might be designed and mobilized in the context of a PEST exercise focusing on the band-aid.

In any case, speculative design's probes comprise only one route – other techniques might draw on fine art or performing art traditions. The main aim remains the same nevertheless – rendering the familiar unfamiliar so that participants can begin to engage with these entities anew, and, hopefully, emerge with inventive questions about, very broadly speaking, science–society relations (though, of course, there is no guarantee that questions will be directed in this way or, indeed, will be directed at all).

To develop this proposal for a 'mundane PEST' is not to dismiss, or even diminish, 'controversy-oriented PEST' (though, as noted, there are certainly serious issues with the latter). Rather, it is to expand the purview of science–society relations, to pose questions about what additional, unforeseen 'sites' of politics there might be and, above all, to pursue speculatively the specific inventive problems that might emerge with technoscience (mundane or otherwise).

In this respect, we can draw on Gay Hawkins's (2011; see also Hawkins *et al.* 2015) notion of 'infrapublics'. Following Marres (e.g., 2007), Hawkins acknowledges that publics emerge with 'their issues', though in developing a notion of infrapublics she is particularly interested in the *potential* for such emergence. As such, the notion of infrapublics 'challenges the reification of publics' and addresses the 'processual nature of public formation' (Hawkins 2011, 551). More specifically, infrapublics reflect the 'infrasensible' dimensions of 'a' public's responses and reactions. These entail 'affective and visceral registers of subjectivity and intersubjectivity' – registers 'that operate below the culturally organised logics of conscious thinking, feeling and judgement' (Hawkins 2011, 542). Here, then, before a public (with its issues) emerges, there is a pre-public – an infrapublic – characterized by a miasma of tacit, unarticulated affects, visceralities and pre-cognitions – the infrasensible. There is, according to this perspective, a sort of virtuality – a not-as-yet – out of which unforeseen publics and their political responses might emerge, congealing out of infrapublics and infrasensibilities.

The speculative methodologies that have been proposed in this chapter are precisely about accessing – and given their performativity, enacting – infrapublics and infrasensibilities. As such, this chapter concerns the way in which PEST (in whatever forms it might eventually take) might address not only publics and their politics, but also infrapublics and, by extension, their 'infrapolitics' – the virtual, diffuse, amorphous, immanent and circuitous issues that have yet to find 'expression' in voice and action as 'inventive problems'.

Note

1 For details see www.ecdc.ac.uk. The author was heavily involved in this project, initially as the principal investigator and then, after his move to Australia, as an adviser and co-author. On this score, the author would like to acknowledge the RCUK which funded this project (ES/1007318/1) and the following colleagues: Andy Boucher, Bill

Gaver, Tobie Kerridge, Liliana Ovalle, Matthew Plummer-Fernandez and Alex Wilkie. Thanks are also due to all the participants who gave so generously of their time.

References

Akrich, M. and Latour, B. 1992. A summary of a convenient vocabulary for the semiotics of human and nonhuman assemblies. In: W. E. Bijker and J. Law (eds) *Shaping Technology/ Building Society*. Cambridge, MA: MIT Press, pp259–263.

Arksey, H. 1998. *RSI and the Experts: The Construction of Medical Knowledge*. London: UCL Press.

Beaver, J., Kerridge, T. and Pennington, S. 2009. *Material Beliefs*. London: Goldsmiths, University of London, www.gold.ac.uk/media/49materialbeliefs-book.pdf (accessed 10 December 2013).

Bennett, T. 2004. The invention of the modern cultural fact: A critique of the critique of everyday life. In: E. Silva and T. Bennett (eds) *Contemporary Culture and Everyday Life*. Durham, UK: Sociology Press, pp21–36.

Boehner, K., Gaver, W. and Boucher, A. 2012. Probes. In: C. Lury and N. Wakeford (eds) *Inventive Methods: The Happening of the Social*. London: Routledge, pp185–201.

Button, G. 1993. The curious case of the vanishing technology. In: G. Button (ed.) *Technology in Working Order: Studies in Work, Interaction and Technology*. London: Routledge, pp10–28.

Callon, M. and Law, J. 1995. Agency and the hybrid collectif, *The South Atlantic Quarterly* 94: 481–507.

Callon, M., Lascoumes, P. and Barthe, Y. 2001 *Acting in an Uncertain World: An Essay on Technical Democracy*. Cambridge, MA: MIT Press.

Chilvers, J. 2008. Deliberating competence: Theoretical and practitioner perspectives on effective participatory appraisal practice, *Science, Technology & Human Values* 33(2): 155–185.

Davies, G. 2006. The sacred and the profane: Biotechnology, rationality, and public debate, *Environment and Planning A* 38(3): 423–443.

De Certeau, M., Giard, L. and Matol, P. 1998. *The Practice of Everyday Life: Living and Cooking*, Vol 2. Minneapolis: University of Minnesota Press.

DeLanda, M. 2002. *Intensive Science and Virtual Philosophy*. London: Continuum.

Delgado, A., Kjølberg, K. L. and Wickson, F. 2011. Public engagement coming of age: From theory to practice in STS encounters with nanotechnology, *Public Understanding of Science* 20(6): 826–845.

Epstein, S. 2000. Democracy, expertise and AIDS treatment activism. In: D. L. Kleinman (ed.) *Science, Technology and Democracy*. Albany, NY: State University of New York Press, pp15–32.

Fraser, M. 2010. Facts, ethics and event. In: C. Bruun Jensen and K. Rödje (eds) *Deleuzian Intersections in Science, Technology and Anthropology*. New York: Berghahn Press, pp57–82.

Gabrys, J., Hawkins, G. and Michael, M. 2013. From materiality to plasticity. In: J. Gabrys, G. Hawkins and M. Michael (eds) *Accumulation: The Material Politics of Plastic*. London: Routledge, pp1–14.

Gardiner, M. 2000. *Critiques of Everyday Life*. London and New York: Routledge.

Gaver, W., Boucher, A., Pennington, S. and Walker, B. 2004. Cultural probes and the value of uncertainty, *Interactions* 11(5): 53–56.

Gregory, J. and Miller, S. 1998. *Science in Public: Communication, Culture and Credibility*. New York: Plenum.

Hagendijk, R. and Irwin, A. 2006. Public deliberation and governance: Engaging with science and technology in contemporary Europe, *Minerva* 44(2): 167–184.
Haraway, D. 1994. A game of cat's cradle: Science studies, feminist theory, cultural studies, *Configurations* 2: 59–71.
Hawkins, G. 2006. *The Ethics of Waste*. Lanham, Maryland: Rowman & Littlefield.
Hawkins, G. 2011. Packaging water: Plastic bottles as market and public devices, *Economy and Society* 40(4): 534–552.
Hawkins, G., Potter, E. and Race, K. 2015. *Plastic Water*. Cambridge, MA: MIT Press.
Highmore, B. 2002. *Everyday Life and Cultural Theory: An Introduction*. London: Routledge.
Irwin, A. and Michael, M. 2003. *Science, Social Theory and Public Knowledge*. Maidenhead, Berkshire: Open University Press/McGraw-Hill.
Irwin, A., Elgaard Jensen, T. and Jones, K. E. 2013. The good, the bad and the perfect: Criticizing engagement practice, *Social Studies of Science* 43(1): 119–136.
Kerr, A., Cunningham-Burley, S. and Tutton, R. 2007. Shifting subject positions: Experts and lay people in public dialogue, *Social Studies of Science* 37(3): 385–411.
Kotchetkova, I., Evans, R. and Langer, S. 2008. Articulating contextualized knowledge: Focus groups and/as public participation?, *Science as Culture* 17(1): 71–84.
Latour, B. 1992. Where are the missing masses? A sociology of a few mundane artifacts. In: W. E. Bijker and J. Law (eds) *Shaping Technology/Building Society*. Cambridge, MA: MIT Press, pp225–258.
Latour, B. 2005. *Reassembling the Social: An Introduction to Actor-Network-Theory*. Oxford: Clarendon Press.
Law, J. 2004. *After Method: Mess in Social Science Research*. London: Routledge.
Lefebvre, H. 1947. *Critique of Everyday Life*. London: Verso.
Lezaun, J. and Soneryd, L. 2007. Consulting citizens: Technologies of elicitation and the mobility of publics, *Public Understanding of Science* 16: 279–297.
Macnaghten, P. 2010. Researching technoscientific concerns in-the-making: Narrative structures, public responses and emerging nanotechnologies, *Environment & Planning A* 42: 23–37.
Macnaghten, P. and Guivant, J. S. 2011. Converging citizens? Nanotechnology and the political imaginary of public engagement in Brazil and the United Kingdom, *Public Understanding of Science* 20(2): 207–220.
Marres, N. 2007. The issues deserve more credit: Pragmatist contributions to the study of public involvement in controversy, *Social Studies of Science* 37(5): 759–781.
Michael, M. 2000. *Reconnecting Culture, Technology and Nature: From Society to Heterogeneity*. London: Routledge.
Michael, M. 2006. *Technoscience and Everyday Life*. Maidenhead, Berkshire: Open University Press/McGraw-Hill.
Michael, M. 2010. Sticking plasters and standardization. In: V. Higgins and W. Larner (eds) *Calculating the Social: Standards and the Re-Configuration of Governing*. Basingstoke: Palgrave, pp131–148.
Michael, M. 2011. Affecting the technoscientific body: Stem cells, wheeled-luggage and emotion, *Tecnoscienza: Italian Journal of Science and Technology Studies* 2(1): 53–63.
Michael, M. 2012a. 'What are we busy doing?': Engaging the idiot, *Science, Technology and Human Values* 37(5): 528–554.
Michael, M. 2012b. Anecdote. In: C. Lury and N. Wakeford (eds) *Inventive Methods: The Happening of the Social*. London: Routledge, pp25–35.
Michael, M. 2012c. Toward an idiotic methodology: De-signing the object of sociology, *The Sociological Review* 60(S1): 166–183.

Michael, M. and Brown, N. 2005. Scientific citizenships: Self-representations of xenotransplantation's publics, *Science as Culture* 14(1): 39–57.

Michael, M. and Rosengarten, M. 2013. *Innovation and Biomedicine: Ethics, Evidence and Expectation in HIV*. Basingstoke: Palgrave.

Sengers, P. and Gaver, W. 2006. Staying open to interpretation: Engaging multiple meanings in design and evaluation. Proceedings of the 6th Conference on Designing Interactive Systems, University Park, PA, US, pp99–108. New York: ACM Press.

Shove, E. 2003. *Comfort, Cleanliness and Convenience*. Oxford: Berg.

Stengers, I. 2005. The cosmopolitical proposal. In: B. Latour and P. Weibel (eds) *Making Things Public*. Cambridge, MA: MIT Press, pp994–1003.

Stengers, I. 2010. *Cosmopolitics I*. Minneapolis: University of Minnesota Press.

Whitehead, A. N. 1978/1929. *Process and Reality: An Essay in Cosmology*. New York: The Free Press.

Wynne, B. 1991. Knowledges in context, *Science, Technology & Human Values* 16: 111–121.

Wynne, B. 1992. Misunderstood misunderstanding: Social identities and public uptake of science, *Public Understanding of Science* 1: 281–304.

Wynne, B. 1996. May the sheep safely graze? A reflexive view of the expert–lay divide. In: S. Lash, B. Szerszynski and B. Wynne (eds) *Risk, Environment and Modernity*. London: SAGE, pp44–83.

Wynne, B. 2005. Risk as globalizing 'democratic' discourse? Framing subjects and citizens. In: M. Leach, I. Scoones and B. Wynne (eds) *Science and Citizens: Globalization and the Challenge of Engagement*. London: Zed Books, pp66–82.

5
GHOSTS OF THE MACHINE

Publics, meanings and social science in a time of expert dogma and denial

Brian Wynne

Introduction

The uncomfortable sense of an endemic but intensifying crisis, not only for Western liberal democracies but globally, over the relations between science and democracy, is difficult to pin down precisely, but equally difficult to ignore. This sense of mounting disorientation and crisis may be manifesting itself in diverse forms and arenas, but in virtually all of these there are recurrent themes. For example, there is confusion and ambiguity as to what is meant as object of the proliferating references to 'science' as public authority, often for claims which overextend science's proper authority and transform it into normative 'findings' or 'demands' supposedly from nature. There is also the increasing experience of science's subordination to neoliberal political-economic controls, with erosion of what were at least significantly independent influences over scientific cultures, and suppression of healthy scientific dissent when commercial interests are threatened by evidence. These dynamics relate to an unacknowledged process of mutual, emergent co-production (Jasanoff 2004) of policy, commercial and scientific – human and 'natural' – orders, in which each can pass the buck to the other in terms of accountability for problematic and sometimes controversial commitments.

A further recurrent issue in these troubling developments is that of a missing – or elusive and in some ill-defined way problematic – public. I have always argued (compare Wynne 1991, or Wynne 1993, with Wynne 2014) against the grain of mainstream social science,[1] not only of science and policy that we will never understand publics and their science-related attitudes and concerns adequately, until we also problematize the meaning of the object 'science' or 'risk' (for example) which we assume to be their own problem-object. This omission and ensuing asymmetric perspective has consequences, to which I return later. The essence of this asymmetry is that whenever problems are experienced in interactions between publics and science, it is only ever imaginable that publics need somehow to be brought to order; but neither science nor the political economy which shapes it are questioned as to whether it may be they who are responsible for public problems, thus needing significant change.

In the rest of this chapter I summarize some of the key elements of difference which my approach, more influenced by a combination of anthropology and interpretive sociology, and with a distinctly relational and emergent ontology, offers to that mainstream social scientific research deployed for gaining insight into public encounters with, experiences of and responses to the variable interventions into social life which science brings – or is *promised* to bring. Some of the most significant of these interventions are not even recognized as such, and their normative dimensions are thus rendered unacknowledged, unaccountable and undemocratic. This includes the subtle ways in which imaginations of publics, as untested or even unquestioned presumptions by those making social interventions in the name of science, shift from being potentially open-minded and ready-to-learn experimental tests, with potential self-correction by their authors, instead into effectively blind dictatorial imposition.

It does not seem to be recognized by scientists involved as policy authorities, despite many years of it being pointed out to them (e.g., Irwin and Wynne 1996) just how intensely provocative this relentless insult is to the publics subjected to it. Thus as Welsh and Wynne (2013) discussed, there prevails an acute contradiction between, on the one hand, the post-2000 'enlightenment' of public engagement and dialogue with science, thus 'listening' to publics in a two-way fashion, and, on the other, the coincident treatment of publics as threats to sacrosanct principles of modern 'science-founded, democratic' order. I address what this contradiction means for our notions of 'the public' and for how we research this mysteriously evasive but essential collective. I also illustrate the unconventional, distinctly marginal epistemic and ontological research and policy-impact stance I wish to defend and advance. In doing this I use examples from contemporary issues involving science and technology – issues in which publics are naturally involved if often implicitly and indirectly, as well as in always-emergent forms. I suggest that the systematic neglect, even denial, of this crucially important relational and emergent quality of 'the public' or 'publics' has also correspondingly and harmfully rigidified our science into overly deterministic and singular, normatively weighted knowledge claims which are locked into existing dominant forms by contingent political and economic institutional channelling forces (as has been illustrated and explained for agriculture by Vanloqueren and Baret 2009). To stand a chance of developing a more socially robust and respected, as well as more vibrant and innovative, science, these need to be opened out and rendered more plural, reflexive and provisional, and more accountable and negotiable. Similar beneficial effects would then also ensue for the science which we wish to cultivate and make more valuable and effective in society.

In essence, I suggest that the excessive concerns on the part of state authorities (or meta-state, like the European Commission), advisory science and related institutions to produce, control and domesticate their desired imagined publics is increasingly animated by the transnational forces of neoliberal intervention and control. This is not to say that these anti-democratic cultures of science-in-public and corporate science, and their corresponding imaginaries and attempted material formations of 'the public', did not exist until neoliberalist commodification also incorporated knowledge in the late twentieth century. However, those pre-existent unrecognized insecurities over what they were demanding of publics intensified and transformed markedly under neoliberal globalization and commodification of scientific knowledge, as both mode of production and increasingly contentious mode of political authority.

Moreover I argue that this political-economic culture of science and public together is enacted in identifiable ways through 'science' as public authority in an attempted disciplining idiom which citizens typically oppose, and attempt to evade without necessarily overt opposition (Scott 1990, 1998). This attempted enforcement of control, through the medium of science – and in the false name of democracy, responsive science and public engagement – only produces more and more unruly ghosts in the machines of contemporary politics and policy. Making peace with our ghosts, and setting them free, would be a good ambition, not only for this chapter, but more importantly for the enlightened transformation of our disastrously failing existing political – and scientific – culture.

Public engagement as experts–publics dialogue: elephants in the room

Once upon a time, so it was thought, science itself felt no need to refer to any public, even if various versions of this abstract 'actor' could be identified as tacit imaginaries in scientific discourses (Barnes and Shapin 1979; Shapin and Schaffer 1985). Yet empirically speaking, 'the public' might be understood to be an ever-silent (but maybe quietly brooding) collective – what Laclau (2005) called the 'empty signifier' of 'the' (missing – yet essential) *public*. By coining this term Laclau was clarifying the enigmatic quality of a singular 'public' which can never be identified in practice. Yet, he noted, this 'ghost' in the various machineries[2] which have adorned democratic politics of the modern world is – if we wish to uphold democratic ambitions and practices – an *essential* singular; how can 'the public interest', or 'a public mandate', exist for elected representatives to uphold without implying an ultimately singular 'public' as its democratic collective subject? This can be described as one of the 'necessary fictions' (Ezrahi 2012; see also Wynne 1982, 2010) of democratic political imaginaries.

Nowadays we experience a vastly confusing clamour of public reference, as the shift from public understanding of science to public engagement or dialogue with science produced a global industry of public representation for scientific, policy, commercial and even civil society NGO clients. This developed as scientists and scientific advisers to policy, increasingly anxious about their always fragile public legitimacy, became more concerned to engage with those on whom they began to realize science ultimately depends for a licence to operate. This broad (but perhaps superficial) shift in policy culture away from the false 'public deficit model' idea[3] toward an embrace of conspicuous acts of public accountability was represented in the shift from 'public understanding of science' to 'public engagement or dialogue with science'. A central implication of this was the requirement for scientific institutions and practitioners to unpick their conventional understandings of the noble responsibility to inform and to better understand their publics as part of this.

This new relationship required that scientific bodies listen to, as well as lecture, their publics, and impressive investments were made to pursue this. It remains a question, however, whether listening can ever *hear* and *understand* such novel voices, if it does not at the same time also put its own commitments and 'givens' into question, as they 'listen' and engage. This has remained a problem. Genuine readiness to call into question one's own assumptions as part of those engagement processes is not something which the huge investment in a global industry of 'understanding public reactions to science' using standardized social science methods is able to do very easily, especially when those exercises are conducted for policy and scientific clients at some remove and which are themselves under intense political-economic pressures to compete for global commercial investment and support those private agendas. Instead the implied need for greater 'institutional reflexivity' (Wynne 1993) and for 'putting their own assumptions into dialogue with others' (Wynne 2006) are written out. It therefore remains perversely true that only 'the public' is problematized, and existing institutional cultures of science in policy remain taken for granted as true authority. The crux of the problem here is that the implicit (sometimes unapologetically explicit) normative commitments within the 'science' of policy authority remain untouched and unquestioned, even though this lack of reflexivity on the part of scientifically authorized policy institutions is the very problem which publics are typically experiencing, yet finding no acknowledgement of it. Thus 'public dialogue' as the great aspirant-progressive move of the turn of the century has been largely traduced by the dogmatic insistence that there is no need to reflexively problematize dominant institutional scientific and science-framed policy cultures, only to problematize various versions of 'the public', including NGOs and the media.[4]

Since Habermas's heyday this has been known as modernity's legitimation crisis (Habermas 1975), in which the legitimation of the modern political arrangements of the state, science and commerce by reference to rationalist assertions of public security through scientific forms of control were increasingly contradicted by the facts of mounting public experiences of the opposite. Even this uncompromising diagnosis of critical theory may be something of an understatement nowadays, when it can be argued that, 40 years later, the normal conditions of strenuous containment of everyday life-world alienation, anger and mistrust between publics and governments or 'expert authorities' has become so brittle and stretched by repeated denials and false patch-up pretences that they are simply no longer sustainable. This is a part of what Barry (2001) calls the question of 'the space of government': what shifting space(s) – political, cultural, institutional, technoscientific, economic – are open for government which might therefore be brought under a democratic gaze, or imagined democratic possibility (Ezrahi 2012)? Or, as Latour (2004) put it, in his own distinctive way, how do we bring 'Nature' into democratic discourse-practices when we have for too long under modernity been busy instead trying to naturalize Society? These are versions of what STS scholars beyond Latour have described as a constitutional condition (Jasanoff 2011) which is historically shifting, but (like the famous, or infamous, centuries-old democratic British constitution) informal and unspecifiable; flexible and adaptive; open to interpretive (re)negotiation under changing conditions; and also – its downside – evasive of accountability and of deliberate, democratic redesign.

Social science understandings and enactments of public sentiment in relation to science also fall into this ill-charted and more fundamental problematic of constitutional conditions we have to live by as if given, yet which we also need to speak of, problematize and transform. As I tried to articulate in a retrospective (Wynne 2014) on nearly 30 years of developments since the science-initiated and science-centred 'Public Understanding of Science' movement of the late 1980s, a major shift in this constitutional condition in many modern developed world contexts has been the simultaneous emergence of the late twentieth-century eruption and institutionalization of the participation in science movement – but perversely, along with a continuing intensification and proliferation of a sharply contrary force, namely the 'hermeneutic imperialism' of science (Wynne 2014, 62). Thus I argued that while science's public policy role is said to be providing the best available scientific knowledge to inform policy, in practice this has been transformed into new conditions, in which science is allowed to provide and impose (as if revealed from nature) the *meaning* of public issues involving science – for example, by falsely calling them scientific or risk issues, and thus excluding other legitimate concerns and questions, including (but not only) about scientific knowledge's framing itself. An unholy alliance can then be seen between science and politics, where political actors hide behind this scientistic definition of the public issue, and proudly assert that they are only following science in their normative commitments – this scientific advice having already been framed and selectively directed by unacknowledged political shaping of the terms of reference of the 'independent' scientific advisory process (Jasanoff 2004).

Thus, in Stirling's (2005) apt terms, one form of opening up was accompanied – or perhaps more accurately was already undermined – by its ever-present opposite, namely a subtle but powerfully less obvious, pre-emptive closing down. By this I mean a fundamental form of scientism – of the presumed and power-imposed science-centred definition of public concerns and public meanings (the accepted meaning of public issues, especially those involving science, for 'policy' to manage). Thus when, as occurs increasingly frequently, science is given authority to pronounce on public issues involving science, scientific culture's parochial meanings become by default the politically authorized 'democratic' version of *public* meanings and concerns. By this means, more authentic public voices, calling for example in their own terms for greater institutional (including scientific) reflexivity and accountability – such as over what science may not know or is not allowed to ask – are silenced.

This underlying hermeneutic imperialism conducted through science but in concert with political interest, thus silently reshaping and reducing 'the space of government', has seamlessly and subtly grown. Paradoxically this has occurred just as its apparent participatory opposite, the insatiable growth in *invited*[5] public engagement with what is called 'science', has also grown. Thus the whole panjandrum of enlargement of participatory and citizen engagement with science has been celebrated as a transformative opening up of democratic space, while this is less obviously but more powerfully closed down, by a deeper historical process of the colonization of public meanings.

The same syndrome plays out when public unease or concern is manifested over some technoscientific innovation process. The authorities address those concerns, typically about policy and scientific denial of lack of predictive control over the consequences of such innovations, by simply repeating reassurances that scientific risk assessment shows that no harm will result. This not only fails to address the public concerns, which are typically not about *identified risks*, but about unacknowledged scientific ignorance of possible consequences, which risk assessment does not address. It also exacerbates those public concerns by in effect engaging in denial, as well as in suppression of questions about alternative innovation trajectories. Those public expressions of concern are then falsely defined as due to misunderstanding of the science of risk, in classic deficit model form. This deeply entrenched dogmatic institutional policy culture of scientism is reinforced by the unaccountable neoliberal global shifts in the political economy of science. These have also recruited the state as agent of this scientistic definition of public issues and concerns, and as regulator of those publics in compliance with the deregulatory neoliberal scientistic vision.

Perhaps this intransigent reduction of 'the space of government' and of the publics to be governed can be seen as a remedial means of attempting to retain social order, and control, in the face of rising democratic dissent and public alienation, and of incipient overspill of the ('invited') public engagement enterprise into disorder. Parallel insecurities articulated through the state and related institutions in financial and economic contexts (such as the state's fear of publics who may refuse indiscriminate innovation and deregulation) only intensify those pressures to effect public compliance, as 'innovation' of any sort and the deregulation of state relations with science, has become a dominant and defining mantra. Either way, one cannot ignore the changing political economy of science and of the state under neoliberal globalization in trying to understand 'publics', science and democracy.

The scientism as hermeneutic imperialism which has developed as attempted but false cover for a neoliberal private corporate takeover of previously democratically accountable public functions has been allowed to expand into almost every arena of public and private life. The discursive practices, relations and contradictions which this move engenders – such as the imaginaries and definitions of publics and their concerns (Felt and Wynne 2007) – manifestly provoke yet further public disaffection, even if this is typically under-articulated. Of course, this is always wide open to misinterpretation.

This is where a question arises which has accompanied the 20 years or more of research and policy attention to public relations with science as public authority, but which has never been properly addressed. This question concerns the appropriate roles of social science research in this overall domain of public engagement with, experiences and understandings of, and responses to, 'science', in various unclearly defined forms – including unclear or contradictory definitions from science. In 2006, as an 'Afterword' to the Demos pamphlet entitled *Governing at the Nanoscale* (Kearnes *et al.* 2006), I attempted in a very attenuated and rather singular register to address this issue, since we had adopted a rather non-mainstream social science approach, typical of the Lancaster University Centre for the Study of Environmental Change (CSEC), in the ESRC research project which generated, *inter alia*, the Demos pamphlet.

In this chapter, therefore, I update but also elaborate upon that 'Afterword', but not to advocate any singular approach to understanding 'publics and their sciences'. Instead I attempt to assist in establishing a more balanced plurality of such approaches and gain better recognition for an interpretive-relational approach which, despite periodic recognition and even celebration by social scientists as well as policy and scientific actors, is still more celebrated in its neglect and marginality than in its uptake. Mainstream social scientific approaches, whatever their methodological sophistication, cannot be said to have achieved any remote success in engendering more socially robust or socially respected science – indeed, reviewing the last 20 years' experience suggests quite the opposite.

A key ontological difference between these and our approach at CSEC (e.g., Wynne *et al.* 1993; Grove-White *et al.* 1997; Marris *et al.* 2001; Wynne 2001), and since then elsewhere (Macnaghten and Carro-Ripalda 2015), may be that our approach recognizes emergence, and relationality, as fundamental continual human-ontological conditions. Taking this reality seriously disqualifies social science from constructing research on publics as solely extractive (of their attitudes, or understandings) because it cannot avoid being involved in interventions which may change the world it is studying. If by trying to 'understand' publics one is thus professionally involved in 'bringing publics into being' in some way, one also has to accept responsibility and be accountable for that. There are at least two first steps here:

1   Attempting to be reflexive about one's own professional practices by recognizing that our sociological findings are always interpretations of respondents' experiences and concerns. Those findings are therefore hypothetical and tentative representations of those publics, and as such can be peer reviewed not only by social scientific peers, but also by publics themselves. Thus, we have always attempted, where possible, to test and correct these tentative accounts by feeding them back, as questions, not convinced assertions, to the people whom we studied, and asking whether they recognize themselves in them (and if not, then why not?). This further interaction often provides greater insight

and understanding about those publics, including their own routine contradiction of dogmatically entrenched institutional myths, such as that publics are incapable of handling 'scientific uncertainty' or 'risk' – which seems to be an alibi for politicians to require that scientific advisers provide empty reassurances rather than truths about what is not controlled and understood (Stirling 2010); and the depth of unacknowledged ambiguities and inconsistencies in what is represented to them as 'science' by those same official 'scientific authorities'. This reveals a toxic domain of evaded and denied responsibility on the part of such authorities, as they persistently decline to attempt to learn or enforce greater 'institutional reflexivity' (Wynne 1993) and public accountability amongst the institutional web of scientific, policy and commercial powers, and instead place responsibility for the ensuing – and relentlessly persistent – mistrust onto 'the public'.

2 Consistent with the foregoing point, to recognize the basically collective-experimental dynamic of these processes, where all actors might become aware of their own prior assumptions about 'the other(s)' in their interactions, and hold these in continual experimental test, revising them and learning a more robust, honest and effective science-informed (but not science-framed) public policy culture in the process. This also requires trying to have scientific representatives or bodies become more accountable and reflexive in their own worlds, since it is this evident lack of reflexivity, and ensuing manifestations of scientific denial, or dishonesty, which is a major problem for publics in their encounters with 'science'. For example, a common experience I have had in talking and publishing about such research is that I am acting as interpreter of typical public concerns over technoscientific innovations and innovation-processes, and communicating these to their authors, and to the promoters and government regulatory actors of those technologies, as well as to sociological research-peers. This also involves sometimes contradicting some myths about those publics.

Much social science research funding does not recognize a need for this interpretive-relational research approach rooted in a different ontology and a different ethic, which is in its very being normatively imbued, in that it cannot deny its own interventionist qualities. However, this does not mean that it is or should be interventionist in the sense of aiming to influence its publics to advocate or oppose whichever technology or other commitment is in question. It is about encouraging collective *self-engendered* and mutually interactive change, in which new 'others' are recognized by each actor-group, in their world. This recognizes the sociological role of *encouraging a public or publics into being*, in a Deweyian sense (Marres 2005; Rommetveit and Wynne forthcoming); but crucially also, of helping to bring a more modest, responsible, informed and properly publicly accountable policy and scientific culture into being.

Thus, there are at least three elephants in the room of two-way public engagement or dialogue with science. First, there is the lack of problematization of the supposed dialogue-partner, 'science', and the influence of its increasing appropriation by the neoliberal capitalist of science in shaping what it is that people experience as the basic body-language of science, even while it is celebrated as increasingly democratic and citizen-participatory. Second, there is the reluctance on the part of mainstream social science of public attitudes to science, risk and related topics, to problematize science as well as publics, and the consequent questions about different, perhaps more symmetrical sociological methods which can address both science and publics, and their multi-layered interactions. Third, there are the forms of normativity and how to handle them constructively which are left unstated and unaddressed in all research approaches to this overall problematique.

Co-production of the state, science and their publics

It is an essential element of the co-production STS paradigm (Jasanoff 2004) that the scientistic idiom of hermeneutic imperialism has not been only a form of scientific institutional power-grab, as reaction to endemic and pervasive scientific anxieties about public mistrust, indifference or outright opposition to 'science' (more accurately, this is not public opposition to science, but to various normative programmes of social and political intervention promoted *in the name of* science – a distinction that scientific and other institutional voices seem unable to recognize). It has also been a function of the capitulation by political representatives and their officials, in face of admitted threats and demands upon all nation-states (and even to 'meta-states' such as the European Union (EU), with the US-promoted Transatlantic Trade and Investment Pact, the World Trade Organization (WTO) and other global 'free-trade' policemen), to comply with the deregulatory and privatization demands of neoliberal competitive knowledge-economy globalization. The fundamental political responsibility to understand and uphold public needs and concerns in face of such un-localizable threats, insecurities and oppressions has been redefined as a responsibility to redefine 'the public good' as the competitive subordination of public services and welfare to orchestrated 'free-market' forces of privatization in order to compete for international investment and thus 'jobs' and consumption. In these conditions, the state has not so much been rolled back as transformed into a regulator of its own publics, to have them comply with those extremist neoliberal demands, articulated as the non-negotiable requirement of 'science' and innovation (Welsh and Wynne 2013; de Saille 2015; Marris 2015).

Political representatives ought to be joining with citizens to ask what is this intended to achieve? What can science say, and also what it *cannot*, about its own purposes and benefits (to whom, for what, with what opportunity costs)? And what alternatives can be developed, for whichever innovation is at issue for social regulation? Instead, they cravenly bow in worship to this false image of science and its reduced definition of democratic politics, which also embodies a particular corresponding deficit model of publics as only opposing the dominant commercial innovation agendas if they are ignorant of science or, worse, anti-science. Perversely, but understandably, scientific bodies also assume and perform this same normative imaginary, and thus also continue to promote false and repeatedly falsified deficit model publics and false deterministic models of science and innovation, on behalf of whichever innovation trajectory has been promoted by neoliberal commerce in the name of science – 'TINA! There Is No Alternative'.

Science itself has thus been increasingly incorporated in multiple ways as the agent of these political-economic forces, by being defined as the prime factor of production for innovation and competitiveness, controlled by the big knowledge-corporations; but also as the basic agent of reassurance for publics, whose concerns about these science-led developments are insistently defined as exclusively about 'security' and 'risk' (and sometimes, 'consumer benefit' as defined by the industry), thus denying and deleting the more significant public concerns about what kind of society this larger political-economic programme is leading to. Science is therefore integrated into this political project as the key justificatory agent of its sustained political (self?-) deceit and denial, including the particular publics it seeks to cultivate. Although it should not need to be said that science is internally varied, there are nevertheless hierarchies and accommodations with patrons, funders and political-commercial powers; and there is little evidence that science as a whole objects to this unacknowledged and historically novel political role – which is far more than its conventional innocent description as the necessary *informant* of 'science-informed' policy. In light of this use – by scientists, as well as politicians – of the name of science to justify and defend favoured (or seen as inevitable) policy prescriptions, it cannot therefore be deemed surprising or illegitimate if science becomes the focus of overt public complaint and mistrust, when there are public concerns about this subordination of an institution which is supposed to keep wanton power in check. Yet, leading scientific figures such as the president of the London Royal Society (Nurse 2011; see also Wynne 2014) effectively deny any such co-responsibility for this systematic and profound false construction of 'the public' and their prevailing meanings and concerns, or for the corresponding public refusals to trust science. We should also note (and here I have to accept some self-correction, from the critical meaning of the title of Irwin and Wynne, 1996, which was about scientific bodies – *Misunderstanding Science?*) that this is more than only *misunderstanding* of those publics. It is also the more pro-active and normatively interventionist *mis-construction* of them.

While diverse public protest at such emergent 'constitutional' conditions has been a significant factor in animating the development of the 'public engagement with science' front, Welsh and Wynne (2013) have described this apparent opening as being perversely accompanied by a tighter *closure*, around the selection of *which* particular meanings and concerns will be recognized and accepted into those invited spaces, leaving the uninvited to be defined as potential threats to 'security' and 'public order'. Nor can it be claimed that the invited are invariably happy with their representation in consequent official accounts of such processes, and that this distinction of publics represents some categorical distinction of intrinsic types. There is ample evidence of post-hoc participant disaffection from experience of such invited engagement exercises (Chilvers 2013), where often there is no feedback or debriefing apart from the published final report, and no interest in testing or developing provisional interpretations of public concerns and meanings, and ambivalences, by the commercial consultant orchestrators. That there is a much greater complexity, flexibility, ambivalence and public reflexivity in dealing with (science-mediated) authority is deleted, partly because ministers and other political clients of such artificial exercises in pseudo-democratization of 'science' do not normally want to have to address these. Science, including significant bodies of social science, thus ends up, deliberately or not, reinforcing the political representatives' denial of public capacities and complexities, including their relational, reflexive and emergent qualities as social subjects, which is an essential, *sine-qua-non* dimension of a democratic politics. If they recognize the problem at all, the scientists excuse themselves by lamenting (off-the-record, in my experience) the politicians' unwillingness to read or listen to anything more than the 'half-page' account of 'the public', while the politicians then legitimate their simplistic misrepresentations of public concerns and needs by reference to 'the scientific experts', and by assertions that the public issue, whichever it is, 'is a scientific issue' only. This is a co-production process laying the bricks of a dishonest, self-contradictory constitutional architecture built on sand, which no one actually designed or voted for.

Thus neoliberal deregulatory and free-market corporate ideologues and their professional communications contractors (including media corporations themselves) have been ceded a colossal and diverse uncharted space of the politics of science, technology and innovation, by weak political actors who have become accustomed to a craven idolatry to and exaggeration of science and its powers as a false and ultimately self-dissolving buffer against hearing the public they supposedly represent. This space has been redefined and reduced from one that should be vigorous, informed, contentious and inclusive, with deliberation over the proper collective human purposes and priorities of all innovation and related scientific and technical investments (thus also over intended and aspired-to social needs and benefits), into the instrumental assessment and control of (selectively defined) risks and insecurities. It is not surprising that some (Swyngedouw 2010; Lotringer and Marazzi 2007) have deemed this a 'post-political' realm, with little debate or expectation as to whether, and if so how, a more genuine politics can be brought into being to replace the fake, mediatized, and fundamentally scientized 'democratic politics' which now prevails, along with relentless growth in public alienation and withdrawal.

Erasing complex publics, reinforcing scientific and policy illiteracy

I have argued here that organized or invited participatory practices, including in science, are always to an important degree also exercises in imagining their publics, and attempting to bring into material being the kinds of public they desire (McNeil and Haran 2013). In the 'Afterword' referred to before (Wynne 2006), I also tried to reflect upon how our own form of interpretive social science research and intervention is not immune to these processes, which have a normative and not only descriptive dimension, and thus even more strongly imply the need to account for themselves and make themselves open to reasoned challenge. In earlier work, attempting ethnographically to interpret and understand public experiences of and responses to 'science' (Wynne 1992a) – as in the Cumbrian sheep-farmers' responses to the scientific research, advice and interventions in the post-Chernobyl radioactive fallout period after 1986 – I recognized that our research-team's interpretations of farmers' (and scientists') concerns, understandings and attitudes, though partly peer-reviewed by research peers, should also be tested with the people with whom we had done that ethnographic research. In addition to the usual academic and policy outputs, we therefore also organized to present our findings at several public meetings of those farmers, as well as to send our reports on findings (which we had submitted for example to a 1988 Parliamentary Select Committee on *The Government Response to the Chernobyl Accident Radioactive Fallout in Britain*) to farmers' representatives and individuals with whom we had done the fieldwork. Effectively, we were saying, in introduction: 'We have represented (our understanding of) *you* in these findings. Can you please tell us whether you recognize yourselves in them? Are they accurate, and what corrections or changes, or additions, would you wish us to make?' This form of 'extended peer-review' by people who were experts in their domain, just not scientific ones, was a very productive, if also an inevitably approximated, extra process, and it resulted in further insights into the ways of thinking and practice of those farmers and their families.

However, from these interactions we also realized what is a familiar insight from ethnographic fieldwork, which had also been going on more subtly through the primary fieldwork itself, which is that our sociological questioning of this 'public' had been an intervention already, in that it had caused our interlocutors, the farmers, to think about their own situation, their relations with the scientists and policy interveners in their lives, in new ways. Thus the public we were interacting with were not just responding to questions, then acting as before; they were also developing and changing as individuals and as a collective public, not only in response to the experts handling the fallout, and to each other, but to us as well. Relationality rules, regardless of intentions!

This point about 'the' public being an unstable point of reference has sometimes been used to justify ignoring them – with the complaint: 'They are incapable of answering our survey questions precisely, or consistently!' This is in play also when public ambivalence in response to social attitude surveys are coded as 'don't know'. Yet ambivalent responses may be the most interesting and informative, as well as informed, responses (Jupp and Irwin 1989). These so-called 'instabilities' are just the very complexities of real, intelligent and mature social actors (and their interactions which professionalized consultancy-packaged engagement processes) which risk being erased from policy and scientific discourse and understanding, unless they are recognized for the different ontological perspective and being from which they are articulated. The various means of their rationalization and dismissal as somehow spurious or 'unreal', even if they are noticed in the first place, are devices which may make life more tidy, convenient and 'efficient', even 'scientific'; but they are also the very taken-for-granted presumptions and framings which disable policy, science and politics from hearing and addressing what and who is in the arena, and thus from actually dealing with the situation as it prevails. The lack of hearing or understanding here indicates the problem emphasized earlier, which is the lack of reflexivity in the institutional culture of science and policy which governs the salient public and private commercial and scientific processes, but also in the mainstream social scientific research cultures which assume that public attitudes and experiences can be adequately plumbed by methods which reflect such non-relational ontologies. These processes, including the research and surveying of attitudes, understandings, meanings and concerns, are always fundamentally relational and provisional. This is true whether or not a significant actor returns to the scene of engagement and encounter – and it is true for typical publics, who treat such encounters in a naturally relational way, even if the other party (a social researcher, or scientists in a public engagement process) does not.[6] Silence, or lack of recognition of the other on the part of scientists and other 'authorities', speaks volumes to ordinary people; and if they are themselves silenced and excluded, by presumptive simplifications of their own complexities, and misrepresenting or just ignoring a concern they have, then their own autonomous capacities and processes of making collective meanings beyond official or social scientific representations continue unabated. Typical publics and their own autonomous social networks can and do – incessantly: they are practised at it – read unwritten institutional bodylanguages (Wynne 1992a), whether or not they read correctly, in terms of what their 'other' meant to communicate.

James Scott (1990) has described how public groups can live as autonomous human cultures under the political radar of surveillance from the authoritarian regimes 'governing' them. They do this by using various ingenious collective practices developed as buffers against detection, and allowing multiple collective identities to be manifested – in Scott's cases, one apparently compliant and obedient, the other free and self-reliant, with improvised and intelligently managed shifts as appropriate between one and the other. Likewise in public engagement contexts where public reactions to authority and its comfort-blanket of 'science' are anxiously monitored and examined, often to be smothered with misconceived and self-defeating communications aimed at control, silent publics may signify positive assent to whatever policy or commitment is the question. They may also signify withdrawal, and alienation, and even hostility to dominant programmes and plans. A more difficult and typical stance is silence – which may reflect ambivalence, or reluctant quiescence, misread as acceptance.

Thus in the 1980s when I was dealing with nuclear industry leaders and scientists over growing opposition to nuclear power, their dominant reaction to evidence of public concern or worse was the only partly rhetorical question: 'Why has the public shifted so absolutely from their previous full acceptance of us, to unrelenting opposition?' The usual collectively self-generated answer was to blame NGOs (with the continuing model of the public being only capable of being led by the nose, whoever succeeds in doing the leading). My response was to point out that just because the vast majority of the typical UK public was silent over the two or three decades while nuclear stations were being built and extravagantly celebrated by the industry and government, this told us nothing about their possible enthusiasm, or perhaps total abhorrence or ambivalence, for nuclear technology. The most typical representations of British public attitudes in relation to nuclear technology in those days were those imagined and expressed by nuclear industry scientists themselves (e.g., Wynne 2011, 21). The elementary but crucial point that citizens have their own autonomous modes of making collective meaning, and also sense, and monitor continually what came to be called 'the opportunity-structures' of politics, and act accordingly, was totally missed (Welsh 2000). Democracy, as Arendt (2005) has noted, requires that these autonomous forms of public meaning, not only public opinions or 'choices', are recognized and respected – which does not mean giving them automatic sovereignty.

The contingent experimental character of NGO campaigning over issues involving science, like GMOs, also came into focus when the House of Lords (2000) Science and Technology Select Committee Report on *Science and Society* was published. As a special adviser to the committee's inquiry, I attended the press conference, but as normal, I was expected to be seen and not heard, so had to remain silent alongside the chair, Patrick Jenkin. One press question posed to Jenkin was based on the false premise that environmental NGOs campaigning against GMOs are anti-science, and asked why their illegitimate anti-science views were being so successfully transmitted to the British public. Jenkin accepted the premise, and added the further one that NGOs only ever managed to select successful campaigns (the Greenpeace victory over *Brent Spar* was still in warm memory from its highly controversial success in having many European citizens rapidly mobilize *en masse* to boycott Shell petrol, thus causing Shell to abandon its marine dumping plan, and to assuage Greenpeace and public opinion by towing it to a Norwegian fjord to await a recycled future). Such NGOs were thought to be in command of some mysterious evil genius when it came to communicating their so-called anti-science, which neither science nor industry nor government could for some reason remotely match and counter. Thus the myth was perpetuated that 'the public is anti-science', and both *Brent Spar* disposal on the North Atlantic sea-bed and GMOs (which were then both matters of intense media coverage) were 'scientific' issues. I was unable to point out a contradictory understanding of my own experience of working and talking with NGOs, which is that they adopt a careful but experimental approach to campaigns, and whether they gain 'take-off' with the media, and publics. Some succeed; but some fail as soon as they leave the NGO front door, and no one ever sees them again. Both the *Brent Spar* and GMOs campaigns succeeded in taking off, even though the *Brent Spar* success especially was a total surprise to Chris Rose, then the campaigns director for Greenpeace UK. The general point is that NGOs have no magic, but are necessarily experimental, and when they do gain success, it is because the campaign resonates significantly with public concerns out there in society, which then of course amplify the campaign.

It went unappreciated by the typical media practitioner (and by my Select Committee chair) that this is a highly contingent and real-time experimental process, however it might appear in typical public–political exchanges. Collective expression of public values and priorities is basically similar – except that they are not campaigning. When we conducted research at CSEC (Wynne *et al.* 1993) on attitudes in West Cumbria around the Sellafield nuclear complex, particularly into the then-mooted deep repository for radioactive wastes, ordinary members of the public there explained how they did not trust Greenpeace any more than they trusted the industry or the government, as they saw the latter 'in bed' with the industry. The reason they supported Greenpeace in this case was because they were the ones, as a voluntarist civil society organization, with no formal powers of inspection, and no guaranteed resources, who were regulating the industry, and not the government which was supposed to be doing so (from continuing monitoring, Greenpeace had recently found an illicit discharge of radioactivity from the site to the North Sea, and went on to win a legal case to prosecute the operators BNFL).

This illustrates how typical public meanings are collectively and autonomously – and in endlessly emergent, thus never completed form – composed and articulated. Only close-quarters interpretive, exploratory and itself open and contingent, improvisory (and, yes, fallible) social research was able to identify these forms of public meaning, relational practice and concern. Mainstream social science attitude surveys, or public engagement methods, however diligently conducted, are unable to achieve this. They perform a different function. Large-scale questionnaire surveys are fine for simple questions, and can code and gain representative standing for the answers to such simple attitudes questions. However, the issues with which we are involved here, and with which 'the' public is involved, are not of this kind. Nor are the typical publics as simple as to hold attitudes, concerns and meanings which such methods have to require of their respondents in order to ask the kinds of simple questions which can be coded into meaningful quantitative outputs. The same with public engagement processes, which are one-off events (even if multi-stage), whereas publics have continuing and emergent developing experiences and relations with those intervention programmes they are then expected to engage with but only in the artificially packaged and restricted manner of a formal participatory exercise. Those exercises may be better than nothing. Some may even extract useful insights. But their degree of penetration into the complexities and dynamics of public experiences, relations, meanings and attitudes should not be exaggerated, as they invariably are.

Public as 'empty signifier': ghost of the scientized neoliberal machine

As I explained earlier, Laclau coined the term 'empty signifier' to describe the paradoxical concept of 'the public' which, as he reminded us, is an essential, *required* term if we want, practically, to try and develop democratic politics in our society.

Yet, he continued, when we empirically attempt to identify this public, it/they disappears, like a ghost. We know something/someone is there, but when we try to touch it, or feel it, or demonstrate it, frustration and confusion ensue. Thus there is an important sense in which we have to communicate in poetic and multivalent register, when we speak of 'the' public. This also requires some degree of self-reflective sensibility on the part of the communicator, or dialoguer. It helps if we turn explicitly empiricist, and pluralize our ghost into 'publics', which we can then try to differentiate meaningfully, by definitions and classifications which more or less capture identifiable social groups, but copious ambiguities and imprecisions always remain. It only partly resolves the problem to differentiate 'publics' if the non-relational ontology and un-reflexive ethic remain intact, and 'listeners' are still, as is typical, not putting their own assumptions into question, as well as those of their public interlocutors. Laclau's theoretical concept of 'the public' could be considered as a case of Gallie's earlier, more general idea of the 'essentially contested concept' – like 'democracy', the true stable version of which has never been experienced or demonstrated, but endlessly argued over and aspired to, even died for, as something decidedly real.

With the empirical turn to publics, we might think that there is some solid ground underlying each definition of an empirical public we have composed. But we should remember that the same complexities and properties of relationality and emergence still prevail, even as we become less ambitious in scope. Dewey's (1927) way of conceiving publics has become influential over the last decade or so (e.g., Marres 2007), following Latour's (2004) materialist reading of issues and publics. In this, the latter come into being as a particular identifiable collective thanks to material issue-formation arising from the overspill of prior actions into unintended consequences for different citizens, which gives rise to collective identity-formation as affected parties with mobilized collective demands for remedial action. Indeed, this is Dewey's nutshell version of how a state is developed – in order to attend to such emergent public demands. Each such 'public', like those identified by Dewey-inspired contemporary scholars such as Latour and Marres, though in principle identifiable by the processes they describe, can change in form and scale, *not only* according to material 'overspill' consequences and their particular social distributions, *but also, beyond this materialist account*, according to voluntaristic self-definition as an 'affected party' by anyone who chooses to associate ethically or politically with the cause of those who may have been materially affected. This boundary between a 'directly affected' and a voluntaristic 'affected' public is ill defined and fluid (and this provides copious work for legal experts in defining rights of redress, or of protest).

At the same time, there is ample scope for even directly affected citizens to avoid overt protest and to be silent, but still to hold (and in their own life worlds share with networked others) acute but unmobilized concerns about things they appear to accept. This was the case with the 1980s widespread UK (and evidently elsewhere[7]) public reticence about existing concerns over nuclear power described earlier, and which reticence was mistaken by authorities and nuclear promoters to be active public support. This point lends substance to the conceptual understanding of publics as ghosts, appearing or disappearing apparently whimsically, to unreflexively self-referential expert authorities complacent about their apparently willing compliance, and anxious about controlling them once this false representation is contradicted. *Modest* witness seems to be a prerequisite for institutional self-enlightenment, and more effective understanding of 'the public'.

I am suggesting that no matter how small scale or even individual we go, there is good reason to accept that 'publics' and not only 'the public' are also always emergent, in the making and a function of relational conditions, and, *inter alia*, thus not the essentialist individuals of classic liberal-democratic social science. In addition, they can be understood as following what I once called, only half in jest, the inverse relationship between power and reflexivity (Wynne 1993). In this view, all else equal (which it is not), typical publics are more reflexive than, for example, scientists and other authorities precisely because, as powerless and vulnerable actors, they have to reflect continually upon their situation, including their relations, insofar as they can discern them, with those in power. As I have also explained elsewhere, this rational preoccupation with such chronic dependencies is why objective risk remains an unavoidably social-relational issue, about trust, and not just a scientific one. Even the best scientific risk assessment can never guarantee that future consequences are all known – thus publics are *always* dependent upon those institutional actors who will be responsible for handling those surprises which escaped scientific prediction. The institutional inability to be reflexive about this endemic responsibility, and the effective denial of possible surprise by the pretence that risk assessment deals with scientific unknowns (Wynne 1992b; Stirling 1998), which typical citizens instinctively know to be untrue, generates a festering sore of incipient public mistrust, which could only be healed by a transformation of those entrenched institutional cultures of science-framed policy. Thus, a more careful investigation of those ghostly publics leads us to understand their problems with the deaf machinery of the science–policy–commerce mutually ordered institutional cultures of contemporary governance.

Conclusions: escape routes from chronic apocalypse?

In keeping with my observations thus far, I now attempt to point towards possible ways of escape from the unviable, unacceptable and, despite its clamorous scientific self-justifications, deeply unreflective and unthinking trajectory which seems to intensify the prevailing sense of insecurity and crisis. This has largely failed to generate public assent, even while having invested hugely and impressively in apparently democratic public engagement initiatives of many kinds. Of course, to some this is precisely why we now have this impending crisis, or crisis-mentality, as public deliberation hampers and misdirects crucial innovations for economic viability. The first essential condition for turning this in a better direction, one which could gradually gain self-propelling as well as collectively self-defining directions and forms, would be to recognize and address critically the basic forces of erosion of democracy which underlie, and delimit, the apparently progressive moves over the last 20 or so years to develop and strengthen public engagements with science and innovation.

One of the key visions of such moves was to shift primary political and scientific attention 'upstream' in such life-cycles, to engender inclusive deliberation less about downstream risks and consequences alone, and more about the unchallenged imaginaries, purposes, priorities and aims of scientific research and innovation. This was meant to induce or enforce institutional reflexivity, and public accountability, over these issues, practices and commitments (Wynne 1993) rather than the entrenched presumption and rigid dogma about indiscriminate innovation and deregulation, both ruled by a politically orchestrated, so untrustworthy 'science', which prevailed and still prevails with only patchy exceptions today.

This difficult normative challenge would need a politically courageous requirement to develop *institutional reflexivity*:

- in the sense of critical, open and public reflection upon profound yet elementary intellectual failures, in not recognizing the normative political choices which frame and thus restrict what is claimed to be independent regulatory risk assessment science, and not recognizing how risk assessment as a scientific public policy discourse of authority is crippled in this role by the pretence that it can and does also encompass not only scientific uncertainties but also scientific ignorance and contingency; and
- about those scientific and policy institutions' corresponding imaginaries of their publics, as unable to handle 'uncertainty' ('so we must keep it from them'), and craving zero-risk.

The abstract recognition that scientific knowledge is always provisional sits in direct contradiction to the identity which science is given in risk assessment, which is invariably made to imply that there are only marginal uncertainties in given variables, and no unknowns as to what the salient variables are, and that all significant consequences have been captured as predicted. Scientifically defined policy authorities such as EFSA writhe endlessly in a tortuous riddle of ambiguity here, under procedures for risk assessment which cannot seriously be called scientific, except by dogmatic declaration which betrays democratic principles.

This failure of risk assessment[8] has probably been if anything exacerbated by the visibly anti-democratic force, if not intention, behind what I called a deeper counter-revolution of the hermeneutic imperialism of science, undermining and corrupting the public engagement, public participation and citizen science fronts. This is explained earlier, but in summary here, it can be described as the gradual extension of one of science's key public roles, to inform public policy and deliberations with the best available scientific knowledge into the fundamentally different and democratically treacherous – but unremarked – role of defining the *meaning* of public issues involving technoscience. As Hannah Arendt (2005) put it, if public meanings come to be imported or imposed from somewhere outside of the collective arena of inclusive mutual negotiation of those public meanings, concerns and priorities – and this includes pre-emptive imposition from science, or from private corporate global actors using their economic power over science as their discursive instrument – then democracy is lost.

I argue that this process of exogenous imposition of public meanings by science as influenced hugely by those global neoliberal agents has been proceeding as a negative force that has deeply undermined the public engagement movement which, in turn, has not remotely been able to counter it. Indeed, civil society movements that have mobilized to articulate their own meanings and concerns that diverge from instrumental science-defined ones have nevertheless been pilloried, and even treated as threats, by the scientistic response to such divergence. This has rendered publics silent but unimpressed whilst remaining, contrary to the conventional wisdom, *interested* onlookers. Thus I argue that the deeply inadequate imaginations of what a collective public might be, and enacted by social science as well as science as policy authority, can only ever be transformed if that dominant idiom of science is also transformed, along with the behaviourist imaginary of its publics-as-consumers which neoliberal market capitalism demands. One of the several further innovations which this condition needs for its overthrow lies undeveloped in the social sciences of risk, public attitudes, public engagement and governance. Critical to this innovation is a better integration (or failing integration at least a peaceful coexistence and mutual recognition) with humanities-inspired and ontologically relational/emergent social research with, or on, those publics in their increasingly proliferating and dense encounters with current and forthcoming technosciences, in both their public authority forms and their material and imaginary (i.e., promissory) technological-social forms.

While current science-framed policy and politics is bent on rendering democratic politics *non*-political, and restrictively reframing and mistranslating public voices into corresponding science-framed meanings, as Marres (2005) points out, following Dewey, an authentic democratic politics which would encompass technoscientific research and development (R&D) and innovation and its embedded global commercial interests is waiting to be built, if grounded public experiences and concerns can be more sensitively understood and articulated – by social research, but also in policy and scientific as well as political institutions. Such humanities insights – with their emphasis on meanings, relationality and interpretive-reflective realities – are there, of course; but the closer these come to encounters with science-framed policy processes and actors, the more they become marginalized by the scientistic mainstream. Meeting these challenges will be painfully slow, powerfully resisted and misunderstood; and uncertain in outcome. Risk, imagination and trust, of other than the usual social science kinds, are likely to be at a premium. There is a lot more work of analysis, collective reflection and of strategically informed collective experimental learning to be done. The public resources and potential are there, awaiting real engagement.

Notes

1 There is a vast and sprawling body of social scientific work I combine as 'mainstream' here, such as: Slovic (2000); Bauer and Gaskell (2002); Pidgeon et al. (2003); Renn and Roco (2006); Horlick-Jones et al. (2007).
2 I borrow and adapt here from Arthur Koestler (1967), *The Ghost in the Machine*, who in turn borrowed this phrase from the philosopher Gilbert Ryle, analysing the mind–body dualism problem.
3 As I have described elsewhere (see Wynne 1991), the 'deficit model' is characterized by the assumption that public opposition or questioning of scientifically authorized policy commitments – whether the claimed safety and value of the MMR triple vaccine for children, or the same for GM crops, or synthetic biology (and so on) – is due solely to moral or intellectual public deficits. A corollary of this conception of public knowledge deficits is that public concerns about science and technology further the assumption that they need not be taken seriously and their concerns investigated.
4 This remains true even when the more benign but naively misconceived opposite practice is adopted of asking the public 'what it expects' of science (usually meaning: not 'what behaviour in relation to publics and policy or private interests?', but 'what products?').
5 This distinction between the invited and uninvited publics of 'public engagement' (Wynne 2007) opens the further question of silenced voices, and excluded (non-) publics, or, as de Saille (2015) has usefully elaborated, 'disinvited publics'. This is treated later.
6 An example of this is the remark to me (in November 2001) of an eminent UK public scientist, then director of the UK government Food Standards Agency, after reading the five-country EU project report (Marris et al. 2001), which I had led from 1998 to 2001, on public attitudes to GM crops and foods in Europe. He thanked me for what he generously described as the depth and subtlety of our research and findings, but then explained: 'But as far as I am concerned, *if they buy it, they like it!*' When I countered that in effect this was unrestrained behaviourism for public policy, and on scientific authority too, he gave no evidence that this was something to be reconsidered.
7 When I acted as amateur representative for objectors at the unprecedentedly big, five-months-long 1997 Windscale nuclear public inquiry, media publicity for my role provoked an unexpected stream of letters from previously unknown UK but also international citizens expressing their support.
8 I mean this in both senses: scientific risk assessment has been failed by policy, by requiring it to be so unaccountably politically framed as 'science'; but also, risk assessment has failed by not challenging its policy institutional clients, in EFSA's case DG-SANCO of the European Commission, to allow it to speak of these framing limitations, and other contingencies, such as possibly significant scientific unknowns, which always characterize scientific knowledge used to represent practical situations such as those which deliver real risks. These failings of risk assessment apply for existing technologies, even before we come to ultra-ambitious and promissory emergent technologies such as synthetic biology, or nano-bio technologies, where the technological agent giving rise to causal consequences cannot even be precisely scientifically defined.

References

Arendt, H. 2005. *The Promise of Politics*. New York: Schocken Books.
Barnes, B. and Shapin, S., eds. 1979. *Natural Order*. London and Beverly Hills, CA: SAGE.
Barry, A. 2001. *Political Machines: Governing a Technological Society*. London: Athlone Press.
Bauer, M. W. and Gaskell, G., eds. 2002. *Biotechnology: The Making of a Global Controversy*. London and New York: Cambridge University Press.
Chilvers, J. 2013. Reflexive engagement? Actors, learning, and reflexivity in public dialogue on science and technology, *Science Communication* 35: 283–310.

de Saille, S. 2015. Dis-inviting the unruly public, *Science as Culture* 24(1): 99–107.
Dewey, J. 1927. *The Public and its Problems*. Athens, OH: Shallow Press.
Ezrahi, Y. 2012. *Imagined Democracies: Necessary Political Fictions*. Cambridge and New York: Cambridge University Press.
Felt, U. and Wynne, B. 2007. *Taking European Knowledge Society Seriously*. Brussels: European Commission.
Grove-White, R., Macnaghten, P., Mayer, S. and Wynne, B. 1997. *Uncertain World: Genetically Modified Organisms, Food and Public Attitudes in Britain*. Lancaster: CSEC, Lancaster University.
Habermas, J. 1975. *Legitimation Crisis*. New York: Beacon Press.
Horlick-Jones, T., Walls, J., Rowe, G., Pidgeon, N., Poortinga, W., Murdock, G. and O'Riordan, T. 2007. *The GM Debate: Risk, Politics and Public Engagement*. London and New York: Routledge.
House of Lords. 2000. *Science and Society*. London: The Stationery Office.
Irwin, A. and Wynne, B. 1996. *Misunderstanding Science? The Public Reconstruction of Science and Technology*. London and New York: Cambridge University Press; paperback edition, 2003.
Jasanoff, S., ed. 2004. *States of Knowledge*. London and New York: Routledge.
Jasanoff, S., ed. 2011. *Reframing Rights: Bioconstitutionalism in the Genetic Age*. Cambridge and New York: Cambridge University Press.
Jupp, A. and Irwin, A. 1989. Emergency response and the provision of public information under CIMAH: a case study, *Disaster Management* 4(1): 33–37.
Kearnes, M., Macnaghten, P. and Wilsdon, J. 2006. *Governing at the Nanoscale*. London: Demos.
Koestler, A. 1967. *The Ghost in the Machine*. London: Hutchinson.
Laclau, E. 2005. *On Populist Reason*. Beccles: Verso.
Latour, B. 2004. *Politics of Nature: How to Bring the Sciences into Democracy?* Cambridge, MA: Harvard University Press.
Lotringer, S. and Marazzi, C. 2007. Autonomia: Post-political politics, *Semiotext* 1: 22–31.
Macnaghten, P. and Carro-Ripalda, S. 2015. *Governing Agricultural Sustainability: Global Lessons from GM Crops*. Abingdon: Earthscan-Routledge.
Marres, N. 2005. Issues spark a public into being: A key but often forgotten point of the Lippmann–Dewey debate. In: B. Latour and P. Weibel (eds) *Making Things Public: Atmospheres of Democracy*. Cambridge, MA: MIT Press, and Karlsruhe: ZKM, Centre for Art and Media, pp208–217.
Marres, N. 2007. The issues deserve more credit: Pragmatist contributions to the study of public involvement in controversy, *Social Studies of Science* 37(5): 759–780.
Marris, C. 2015. The construction of imaginaries of the public as a threat to synthetic biology, *Science as Culture* 24(1): 83–98.
Marris, C., Wynne, B., Simmons, P. and Weldon, S. 2001. *Public Attitudes to Agricultural Biotechnologies in Europe, PABE*. Final report to D-G Research, European Commission, Brussels.
McNeil, M. and Haran, J. 2013. Publics of bioscience, *Science as Culture* 22(4): 433–451.
Nurse, P. 2011. Science under attack, *BBC Horizon* [television programme], 22 January, www.bbc.co.uk/programmes/p00ddylw (accessed 17 August 2013).
Pidgeon, N., Kasperson, R. and Slovic, P., eds. 2003. *The Social Amplification of Risk*. London and New York: Cambridge University Press.
Renn, O. and Roco, M. 2006. *Nanotechnology Risk Governance*. Geneva: International Risk Governance Council.
Rommetveit, K. and Wynne, B. forthcoming. Imagining public issues in the technosciences, *Public Understanding of Science*.

Scott, J. C. 1990. *Domination and the Arts of Resistance: Hidden Transcripts.* New Haven, CT: Yale University Press.
Scott, J. C. 1998. *Seeing Like a State.* New Haven, CT: Yale University Press.
Shapin, S. and Schaffer, S. 1985. *Leviathan and the Air-Pump: Hobbes, Boyle, and the Experimental Life.* Princeton, NJ: Princeton University Press.
Slovic, P. 2000. *The Perception of Risk.* London and New York: Earthscan.
Stirling, A. 1998. Risk at a turning point? *Journal of Risk Research* 1(2): 97–109.
Stirling, A. 2005. Opening up or closing down: Analysis, participation and power in the social appraisal of technology. In: M. Leach, I. Scoones and B. Wynne (eds) *Science and Citizens: Globalization and the Challenge of Engagement.* London: Zed Books, pp218–331.
Stirling, A. 2010. Keep it complex, *Nature* 468: 1029–1031.
Swyngedouw, E. 2010. Apocalypse forever? Post-political populism and the spectre of climate change, *Theory, Culture & Society* 27(2–3): 213–232.
Vanloqueren, G. and Baret, P. 2009. How agricultural research systems shape a technological regime that develops genetic engineering but locks out agroecological innovations, *Research Policy* 38: 971–983.
Welsh, I. 2000. *Mobilising Modernity: The Nuclear Moment.* London and New York: Routledge.
Welsh, I. and Wynne, B. 2013. Science, scientism and imaginaries of publics in the UK: Passive objects, incipient threats, *Science as Culture* 22(4): 540–566.
Wynne, B. 1982, 2010. *Rationality and Ritual: The Windscale Inquiry and Nuclear Decision-Making in Britain.* Chalfont St. Giles, Bucks, UK: British Society for the History of Science. Republished, with new Introduction, as *Rationality and Ritual: Participation and Exclusion in Nuclear Decision-Making.* London and New York: Earthscan.
Wynne, B. 1991. Knowledges in context, *Science, Technology & Human Values* 11: 1–19.
Wynne, B. 1992a. Misunderstood misunderstandings: Social identities and public uptake of science, *Public Understanding of Science* 1(3): 281–304.
Wynne, B. 1992b. Uncertainty and environmental learning: Reconceiving science and policy in the preventive paradigm, *Global Environmental Change* 2(2): 111–137.
Wynne, B. 1993. Public uptake of science: A case for institutional reflexivity, *Public Understanding of Science* 2(4): 321–337.
Wynne, B. 2001. Creating public alienation: Expert cultures of risk and ethics on GMOs, *Science as Culture* 10(4): 445–481.
Wynne, B. 2006. Afterword. In: M. Kearnes, P. Macnaghten and J. Wilsdon (eds) *Governing at the Nanoscale.* London: Demos, pp70–85.
Wynne, B. 2007. Public participation in science and technology: Performing and obscuring a political–conceptual category mistake, *East Asian Science, Technology and Society: An International Journal* 1(1): 99–110.
Wynne, B. 2011. 'Introduction' to second edition, *Rationality and Ritual: Participation and Exclusion in Nuclear Decision-making.* London and New York: Earthscan.
Wynne, B. 2014. Further disorientation in the hall of mirrors, *Public Understanding of Science* 23: 60–70.
Wynne, B., Waterton, C. and Grove-White, R. 1993. *Public Perceptions and the Nuclear Industry in West Cumbria.* Lancaster: Centre for the Study of Environmental Change (CSEC), Lancaster University.

PART II
Making participation

6

STATE EXPERIMENTS WITH PUBLIC PARTICIPATION

French nanotechnology, Congolese deforestation and the search for national publics

Véra Ehrenstein and Brice Laurent

Introduction

In certain areas of public life, public participation has become a quasi-mandatory exercise for governmental action both in the so-called developed and developing worlds. Authorizing new kinds of consumer products (e.g., GMOs), deciding about nuclear waste, building transportation infrastructures or preserving natural environments now comprise participatory initiatives. This empirical phenomenon has been accompanied by social scientific studies, which in turn have contributed to shape 'public participation' as a specific domain of scholarly interest and political intervention (Callon *et al.* 2009; Gourgues 2012).

In the public participation literature, the state is often present as a negative reference, the locus of activities from which interesting (in both analytical and political terms) initiatives distance themselves. The usual tendency is to oppose 'participatory' politics and 'representative' democracy. This casts a contrast between deliberative, decentralized and inclusive initiatives, on the one hand; and electoral mechanisms and authoritative monitoring inscribed in the traditional functioning of the state, on the other. The analytical exploration of this opposition takes various forms. One can identify a normative perspective that describes participatory instruments as ways of contravening the constraints imposed by state government. This perspective comprises works praising the spontaneous and vibrant political awareness of civil societies (Bayart *et al.* 1992; Scott 1998), and describing the political value of deliberation, which, contrary to electoral systems, does not presuppose that individual will can be unproblematically assessed (Manin *et al.* 1987). But the opposition between traditional government and participatory activities also appears in a critical guise. Such perspectives analyse the very devices that claim to be 'more democratic' as micro-processes, through which political power is exercised in concealed manners. The critical perspective sees participatory mechanisms as a ruse creating obstacles toward a 'radical democracy' (Laclau and Mouffe 2001).

When expressed in the Foucauldian language of governmentality, it focuses on the micro-politics at play within devices meant to construct 'deliberating' citizens or 'empowered' communities (Cruikshank 1999; Li 2007). For both the normative thinkers and the critics, 'real participation' is beyond what the state can do, and happens elsewhere.

This chapter seeks to rethink the question of the state in participation studies. It examines two attempts at articulating public participation and governmental interventions in domains related to the public administration of environmental issues: a national debate about nanotechnology conducted in France in 2009–2010, and participatory initiatives addressing national forest loss in the Congo in 2011–2012. By contrast to the perspectives outlined above, the chapter adopts an agonistic perspective about what 'public', 'participation' and 'the state' mean. Its objective is to account for situations where participation and governmental action are mutually at stake, considering that the perimeter of the state (Mitchell 1991a) and the modalities of its intervention are not to be taken for granted but should be treated as outcomes of processes and analysed accordingly.

Through its focus on a French national public debate on nanotechnology and participatory initiatives in the Congo dealing with deforestation, this chapter demonstrates the relevance of approaching public participation as a vehicle for the empirical and theoretical investigation of the state. Despite their differences, these situations – a developed country confronted with the production and potential consumption of a new technology and a developing country seeking to participate in an international environmental regulation – illustrate the increasing use of *ad hoc* participatory interventions to deal with issues framed as national. The first section of the chapter explains why we propose to apprehend them as 'state experiments with public participation'. The chapter then analyses a first situation, in which French governmental agencies with different technical expertise must coordinate in order to address an elusive national public, and a second situation, in which Congolese civil servants, expatriate consultants and members of the so-called civil society must temporarily speak in one voice. The analysis of nanotechnology in France and deforestation in the Congo display two different political constructs. But they both highlight moments when the state describes itself as a unified and delimitated political entity in front of an audience. In the two cases, the (difficult) implementation of public participation at national levels contributes to the possibility of a self-description that can be qualified as a 'state demonstration' (Linhardt 2012), expected to make the democratic quality of the state visible.

Experimenting with public participation

Public participation does not happen in an institutional void, but is often, in one way or another, related to governmental initiatives. Whereas the literature on 'dialogic democracy' in the late 1990s approached participatory mechanisms through stylized empirical examples in order to present them as potential models to be replicated (Callon *et al.* 2009), our purpose here is to examine public participation as part of broader political operations and to focus on processes whose participatory qualities are *a priori* more ambiguous. Situations where issues constructed as national are at stake appear particularly interesting in that respect.

Consider, for instance, the case of the British national dialogue on genetically modified organisms, GM Nation?, held in 2003 and meant to be a response to growing controversies about biotechnology. This governmental initiative has been widely commented on, by science and technology studies (STS) scholars (Irwin 2006) and students of public participation interested in organizational or evaluation matters (Rowe et al. 2005). This now abundant literature treats GM Nation? as a paradigmatic site for the study of public participation. But the initiative was also a process through which the very existence of a national public was problematized, as well as the modalities of governmental intervention. The British government sought to gather the opinion of a national, GM-attentive public. In fact, and as STS scholars have argued, the mechanisms drawn on to make this new entity speak actively performed a national public opinion about GMOs. By setting several meetings in various places, with selected and non-selected participants, the organizers of GM Nation? ended up producing many kinds of stakeholders (such as consumer groups or environmental movements), of interest for the British government and private actors (Jasanoff 2005, 129; Reynolds and Szerszynski 2006). The intervention aimed at constructing a common national position after years of controversy. It framed GMO issues as a general topic of concern for public opinion, whereas they could have been dealt with through localized regulatory puzzles targeting different products. The participatory initiative here was initiated by the state, albeit in a way that did not directly commit the government to follow its outcomes. It was an attempt by the state to demonstrate that it acts in a democratically acceptable manner on an issue framed as a concern for its population. This example invites us to simultaneously problematize the identity of publics who are expected to take part in nation-wide participatory mechanisms, the definition of the issues to be dealt with by these government-led mechanisms, and, eventually, the characteristics of the state engaged in public participation.

In order to further develop the analytical reflection about the connections between public participation and the state, we suggest focusing on 'state experiments with public participation'. State experiments with public participation, as we have labelled them, are empirical situations in which governmental institutions conduct experiments with public participation. But state experiments with public participation are also situations in which the modalities of state action are experimented with. As we argue, the notion allows us to identify and examine settings and moments where the ability to configure the autonomous action of the state on a nation-wide issue is questioned.

A first step in the development of this approach is to clarify what we intend to capture with the vocabulary of the 'experiment'. The term has been widely used in STS, first to describe scientific activity and display the mutual construction of scientific knowledge and social order (Shapin and Schaffer 1985; Latour 1993; Jasanoff 2004), and then to extend the realm of STS analysis to political initiatives where the nature of the objects and issues at stake are redefined and where the form of democratic activities are questioned (Callon *et al.* 2009). Participatory initiatives, for that matter, have been analysed as experiments, whether laboratory experiments undertaken in closed, controlled conditions (Lezaun 2011; Bogner 2012) or open experiences within which the ontological characteristics of mundane things and of the people interacting with them are put to trial (Marres 2012). From this body of work, we take a notion of experiment that helps us to grasp situations characterized by public demonstrations performed through sociotechnical instrumentations, and meant to explore social and/or technical uncertainties.

The language of experiments points to three related dimensions, namely the apparatus used to conduct participatory initiatives; the recombination of the identities and issues at stake; and the demonstrations conducted in order to assess the validity of the undertaken initiatives. The experiments we analyse lead to 'state demonstrations' (Linhardt 2012). The constitution of a public for managing issues framed as national concerns is the expected experimental outcome, and the modalities of state intervention are questioned in front of various witnesses, including the sociologist. As shown by the two examples described in the following sections – French nanotechnology and Congolese deforestation – state experiments with public participation offer an analytical opportunity for characterizing the state in situations where explicit public participation is required but state power still matters.

The French state on trial: experimenting with a national public debate on nanotechnology

Making nanotechnology governable

The French public bodies in charge of science policy made nanotechnology a priority in the early 2000s. Public funding was attributed for research, itself conducted by private companies and major public research bodies such as the Commissariat à l'Energie Atomique (CEA), the historic research institution for nuclear energy which recently diversified in other technoscientific areas. In France as in other places, nanotechnology was considered as a source of potential public concerns, at environmental and/or ethical levels. A particularity of the French situation is that the opposition to nanotechnology in France has been extremely vocal. Nanotechnology has been targeted by activists claiming that nanotechnology developments would further increase human control over nature and the human body (by developing personalized medicine, introducing new chemicals in the environment, and offering paths for greater control thanks to application in electronics). Activists, gathered within a group called PMO (Pièces et Main d'Oeuvre), have been particularly active in Grenoble, in south-east France, a major hub for nanotechnology research, particularly at CEA (Laurent 2007).

In this context, the national consultation on environmental regulation that the newly elected president Nicolas Sarkozy launched in 2007 discussed nanotechnology under some pressure. Civil society organizations that were present at the time (mostly France Nature Environnement, a federation of French environmental organizations) called for a 'national debate' on nanotechnology and for the traceability of nanotechnology objects. These propositions were eventually included in two laws, through which the French government committed itself to organize such a national debate, and introduce a mandatory declaration of nanotechnology substances.

The dual concern for the national position within the 'global nanorace' (Hullmann 2006) and for the 'responsible development' of nanotechnology is not specific to France. The French case has two interesting particularities, though. In France, many of the participatory activities about nanotechnology are directly state sponsored and connected to public bodies. In other countries, the boundary between public and private initiatives related to 'public engagement in nanotechnology' has been more diffused. For instance, the notion of 'upstream public engagement' took shape in the United Kingdom within a collection of think tanks, consulting firms and research laboratories (Wilsdon and Willis 2004), themselves part of a diverse landscape of public participation with a wide range of connections with public bodies (Chilvers 2010). Second, France has been active within international arenas as well as through national initiatives in the redefinition of risk regulation processes for nanotechnology, and regularly argues in front of European institutions for the need to reshape the regulation of chemicals in order to take the risks of nanomaterials into account.

When faced with its self-commitment to organize a national debate on nanotechnology, the French government chose to ask a public agency called the National Commission for Public Debate (CNDP) to organize it, with the mission of collecting the opinion of the French public about nanotechnology development and its issues, and using it to 'enlighten public decision-making' (*éclairer la décision publique*). Created in the mid-1990s, CNDP has developed an expertise in the organization of public debates. Its usual procedure consists of a series of public meetings during which stakeholders are invited to speak and answer questions from the audience. The meetings are complemented by written contributions submitted by all interested parties. A procedure inscribed in the legal regulation of industrial activities makes the organization of CNDP *débat public* mandatory for industrial projects reaching a certain amount of investment. In 2002, CNDP became an Independent Administrative Authority (*Autorité Administrative Indépendante*), meaning that it received additional autonomy from the government itself. It then became competent for the organization of so-called *débats d'option générale*, that is, debates in which the question being discussed is a broad policy decision. The national debate about nanotechology was organized under this legal provision. Before that, CNDP had organized only two such *débats d'option* on national public policy issues that were also connected to local development projects.[1]

Commissioning CNDP to organize the national debate about nanotechnology was a prudent choice by the members of the public administration in charge. They hoped to benefit from the ability of CNDP to act as a neutral organizer of public debates held in controversial situations. The Independent Administrative Authority statute is only one component of a series of practices through which the commission may demonstrate its independence during the public debates it organizes. The loose standardization of the procedure, defined in very broad terms in the law and adapted to every new case according to principles and practical advices gradually defined by CNDP and circulated to organizing teams, and the composition of the teams themselves mostly made of retired civil servants, have allowed CNDP to find ways of answering recurrent questions about whether or not it is indeed independent from the parties involved.

For the French government, this expertise was particularly valuable in a situation where the preferred modes of public action were far from certain, and where a device such as the CNDP public debate was considered to be potentially useful, if applied to complex large-scale science-related issues. At a time when CNDP had become an illustration of 'public participation *à la française*', as French political scientists named it (Revel *et al*. 2007), using the public debate procedure was meant both to test the participatory device on larger issues than local infrastructure projects, and to experiment with potential new ways of crafting the national technology policy.

Making national publics speak

The CNDP debate on nanotechnology was conceived as a replication of a procedure well adapted to local infrastructure projects. This replication translated into the division of the discussions into sub-themes, related to the local places where the meetings were held. For instance, the applications of nanotechnology in cosmetics were discussed at Orléans, where the cosmetic company L'Oréal owns a production factory. A similar approach was adopted in other geographical sites. Overall, the division of the national debate into localized sub-issues can be understood as a way of grasping the variety of nanotechnology as a science policy programme, as well as an attempt to attract local concerned publics in the hope that the collection would constitute the national public the nanotechnology debate looked for.

Yet it was difficult for the organizers to maintain the representation of nanotechnology in various sub-issues, as participants systematically displace the original framing of the discussions. The processes through which public discussions escape predefined framing are familiar to public participation specialists, including to those who have studied the CNDP procedure (Fourniau 2007). The particularity of the CNDP debate on nanotechnology was that the pervasive uncertainty about nanotechnology objects and programmes prevented any easy location of the concerns expected to be discussed. Participants regularly stumbled over whether or not particular industrial sectors used nanomaterials, and witnessed the impossibility of the public administration to identify the circulation of these potentially hazardous substances within French territory. For instance, lengthy discussions occurred during a public meeting about whether or not the food industry used nanomaterials. NGO representatives claimed it was the case, while the industry representatives argued for the opposite. Officials from different ministries publicly expressed their disagreements, and eventually a senior official from the ministry of health concluded that 'all depended on the chosen definition'.[2]

The identification of the concerned publics that the public debate hoped to attract was as difficult as that of nanotechnology objects. Among the expected concerned publics, civil society organizations were explicitly targeted by the organizers, who were keen to encourage them to submit written contributions and participate in the meetings – even if most of them were poorly acquainted with nanotechnology. But the most active participants in the public meetings were PMO activists strongly opposing nanotechnology and the participatory initiative itself. The activists considered nanotechnology as a global programme pertaining to the control of humans and nature for economic interests. Opposing the participatory objective of the CNDP debate, the anti-nanotechnology activists interrupted several meetings. They proposed a counter-model of citizen participation based on a critical distance to nanotechnology. They grounded this counter-model on what they label 'critical inquiry' – that is, a combination of critical descriptions of technological development programmes and public interventions targeting participatory initiatives.[3] The attempt at constituting a national public and a manageable national problem through the division of the debate into many local debates was subverted by PMO. It became an opportunity to realize the 'converging fights' (an expression regularly used by the activists) by bringing together all anti-technology movements in the various places where the debate circulated: nanotechnology was not a collection of sub-issues any more, but a national programme that comprised the debate itself.

Eventually, the activists' interventions made it necessary to adapt the public meeting format so that this uncontrollable public could be isolated. Organizers made extensive use of Internet exchanges and claimed that the online public was 'more representative' than the physical one, by virtue of its larger size. They attempted to divide the meeting rooms in two parts, one truly public and another one closed, where only invited participants sat, and where discussions were held – phone communications from the first to the second room remained possible, in a somewhat symbolic gesture meant to maintain the possibility of a public dialogue. Eventually, the last public meetings were all but cancelled, and replaced with closed ones held in locations that were only revealed once the would-be participant had duly registered. After the initial opposition from PMO and the subsequent adaptation of the device, most of the civil society organizations withdrew from participation, while only the representative of France Nature Environnement (the same federation of French environmental organizations that had argued that a national debate was necessary) participated in all the discussions, including the final secluded ones.

Displaying the French state in action

The French government reacted to the national debate on nanotechnology two and a half years after it ended. The document it released in 2012 explained that the debate had demonstrated that the state should adapt its strategies toward nanotechnology:

> [The conclusions of the debate] imply that the state ought to propose and put in place instruments that could:
> - take health and environmental risk issues of nanomaterials into account, as well as the social and ethical issues related to nanotechnology;
> - facilitate the integration of these issues in industrial strategies and technological diffusion.[4]

Through this conclusion, the French government recognized the potential risks and problems associated with nanotechnology and attempted to make industries aware of their responsibilities. No sign of the opposition appeared, and the debate became a demonstration that new regulatory categories for nanotechnology needed to be created, and that new devices for dealing with uncertainties about the publics and objects of nanotechnology were necessary. This final move referred to the initiatives in which the same civil servants active in the national debate were involved. In 2011, France was the first country to introduce a mandatory declaration of 'substances at a nanoparticulate state' in the public regulation. Through an innovative legal intervention on a still widely uncertain technological domain, the new regulatory category of 'substances at a nanoparticulate state' granted a legal existence to nanotechnology objects. For the civil servants in charge, this was also a way of demonstrating at the European level that a new regulation for nanomaterials was possible (Laurent 2013).

In its final declaration answering the conclusions of the CNDP debate, the French government spoke in one voice, in the name of the state. The state appeared as a consistent entity, made of diverse, yet well-organized, components (i.e., the public agencies in charge of implementing the mandatory declarations; CNDP that organized the public debate; and the government itself). This can be seen as an answer to the multiple requests, during the debate itself, for 'the state' to act voluntarily in front of the uncertainties related to nanotechnology.[5]

This final move is a reconstruction of the state as a single actor, able to speak to an external and internal witnesses in one voice. But the debate had made this configuration much more complicated to sustain, as it exposed the diversity of the 'state'. Thus, the preparation of the document presenting the position of the French government on nanotechnology was complex. Views of the issue at stake were different across the ministries involved, which caused difficulties for the coordinator of the preliminary report. He was an adviser to the minister of the environment,[6] in charge of ensuring the internal consistency of a document made up of contributions written in different ministries. Contributors from the ministry of research stressed the crucial role of national science policy programmes in harnessing the scientific and economic potential of nanotechnology. At the ministry of health, the civil servants in charge considered that their duty was to 'prevent another health crisis'[7] and that the national debate should explore the relevance of such policy tools as risk-benefit analysis or labelling for the public management of nanomaterials. This diversity was reflected in the team constituted across various ministries, in charge of monitoring the debate and its further development, and which intervened during the public meetings mostly to restate the (unclear) objectives of the debate, and the involvement of the government in the public management of nanotechnology.

But the state was also present during the national debate as a stakeholder among many others, through the contributions of public agencies working on health, environmental and occupational risks, and public research institutions such as the Commissariat pour l'Energie Atomique (CEA) or the Centre National de la Recherche Scientifique (CNRS). All these public bodies submitted written contributions to the debate, alongside private companies and civil society organizations. This attracted criticisms, as some of the civil society organizations withdrew from the debate while claiming that the role of these institutions was improperly drawn: they considered that CEA, being the main research operator and filling up the ranks of the public administration for science policy issue at the ministry of research, should not have been a stakeholder among many others, but the operator whose projects are debated.[8] The critics had a particular version of the CNDP debate in mind, namely the debates organized about local infrastructure projects. In these latter cases, of which CNDP had become expert, the company in charge of the project commissions the debate and presents its project in front of all stakeholders, who can voice their opinions and concerns. The critics of the nanotechnology debate hoped to restore this configuration by drawing separations among the public bodies involved. For them, the state was to be the neutral organizer and the neutral regulator of nanotechnology, the development of which was undertaken by CEA, a public research body that resembled the private companies developing infrastructures (as in the case of local public debates). What was questioned at this point was the ability to distribute the roles and responsibilities of various public bodies so that the state could conduct nanotechnology policy in a participatory manner. This distribution was criticized in a wholly different way by the most radical activists, for whom the very possibility of distributing state interventions was not possible. For PMO, organizing public debates, regulating nanotechnology or fostering its development were all related to the making of nanotechnology as a global programme – precisely what was targeted by their critique. Through the gaze of PMO, the state could not act for the interest of the national public since it made both scientific development and participatory practices components of a threatening technological programme.

Throughout all these discussions, the debate questioned the possibility of the state to speak as a single entity acting on a problematic topic framed as a national issue. The response to the conclusions of the debate by the French government marked an attempt to reconstruct the consistency of the state as an actor integrating in its missions the public management of social and technical uncertainty, for the benefits of the national community and within the external witnessing of European institutions.

Extending the expertise of the French state

With the CNDP debate, the state experimented with public participation as much as the state was experimented with. The picture that emerges from this event is that of a state acting from multiple places, which attempts to reinvent its way of dealing with technological issues. Yet as it appears through the nanotechnology-related initiatives, this reinvention implies that the state redraws the perimeter of its activities, bases them on the centralized works of the public administration, while also reorganizing the various centres from which it intervenes in technology policy.

This movement is inscribed into, as much as it contributes to shape, the trajectory of a powerful state, expected to guarantee the neutrality of administrative expertise and prone to integrate new concerns in this very expertise. Political scientists have described how the French state managed to integrate environmental issues related to industrial activities into the centralized public administration of industry (Lascoumes 1994). Others have analysed the response to health crises in the 1990s and showed that the French state created health agencies meant to ensure the neutrality of its technical expertise while also taking participatory concerns into account (Benamouzig and Besançon 2005). These works display a state constantly attempting to integrate new components in a centralized expertise that grounds the legitimacy of its intervention. This powerful state is able to act through an expertise owned by various government components, public agencies and research organizations, brought together for the sake of the development and control of technology. Nanotechnology in France, as a public programme of technological development expected to be managed and controlled by the public administration, takes place within this definition of the state. Accordingly, the CNDP debate appears as a particularly explicit moment of trial for the French state, as it attempts to make both the conduct of public participation and the public management of technological uncertainty part of its roles and responsibilities, and a component of the expertise on which its interventions rely. In the meantime, the CNDP debate also displays the multiple centres from which industrial development is conducted and regulated, whether they are various ministries with different priorities, 'independent administrative authorities' organizing public debates, public research bodies, or health agencies providing technical expertise. This state experiment is a lens into the French state, for which industrial development is a crucial matter in defining its own identity (cf. Hecht 1998), and is dealt with in a plurality of inter-connected centres of administrative expert interventions.

The preparation of the Congolese state: making civil society participate in decision-making about national deforestation

Our second case provides a contrast in which the nature of state intervention is radically different. It focuses on a series of participatory initiatives undertaken in the Democratic Republic of the Congo in the context of international climate policies.[9]

Since 2005, climate change negotiations have been investigating how to create economic incentives that would encourage developing countries to reduce their deforestation, considering the carbon dioxide (CO_2) emissions attributed to the phenomenon. The international debate on 'REDD+' – 'Reducing Emissions from Deforestation and forest Degradation in developing countries' – has stimulated a set of interventions across the world. Supported by public funds from Northern governments, international organizations have launched so-called 'readiness' or 'preparation' programmes that intend to prepare developing countries to participate in the future incentive-based global policy. Since 2009, the Congo has benefited from this financial and technical assistance. Because negotiators decided to frame the reduction of forest loss as a problem in need for state actions, preparation programmes target forestry administrations and expect national commitments to the issue. This section considers one of the devices tested within the Congolese preparation for REDD+ and expected to demonstrate such a commitment: the Thematic Coordination Groups (TCGs) initiated in 2011. These groups are participatory settings gathering civil servants, NGO representatives and members of private companies in order to explore various policy options that could reduce national deforestation (e.g., energy alternatives to charcoal, sustainable agriculture, community-based forest management). While numerous countries are undertaking readiness processes, the Congo is an interesting case because of both its reputation as an undemocratic state and its strong dependence on foreign aid.[10] Therefore, the attempt to display participation through the creation of the TCGs led to situations where the distinctions between national and international, and 'civil society' and the state, are questioned by the actors themselves.

Preparing a country for REDD+

In order to understand the experimental character of the Thematic Coordination Groups and how this participatory initiative connects to governmental action in the Congo, it is necessary to situate it within the preparation process for REDD+. Since the end of the war in 2002, the Congo has been subjected to many aid interventions (Trefon 2011), such as the construction of infrastructures or the implementation of health or education policies targeting the Congolese population. The preparation for REDD+ is quite different from these interventions, which resemble the processes through which colonial or post-colonial states have emerged (Mitchell 1991b, 2002). It was deliberately tentative (Ehrenstein 2013) and its first and main achievement was institutional. A *Coordination Nationale* (National Coordination) in charge of preparation activities and belonging to the ministry of environment was constituted. This temporary organization comprised both Congolese and non-Congolese people. All of its members were working with the status of consultants hired on an annual basis by the two funding programmes (one managed by the World Bank, the other by the United Nations Development Programme), and most of them, whether national or expatriate, were previously employed in other development projects in the Congo or elsewhere. With limited resources, the *Coordination Nationale* sought to make tropical deforestation a national concern. Its members attempted to identify development projects that were already undertaken with international resources and that could be slightly modified in order to become, as they put it, 'REDD+ governmental actions'.[11] For instance, the conception of a large programme funded by the World Bank, which was meant to improve transportation infrastructures for the revival of the agricultural sector, was seen as an opportunity to include provisions for non-deforestation. The state's capacity that the preparation for REDD+ intended to harness for dealing with forest loss was mainly constituted by aid interventions.

As an experimenter, the *Coordination Nationale* performed demonstrations addressed to a variety of audiences. Among them stand the funding agencies, but also international NGOs based in Europe or North America and connected to Congolese organizations or individuals.[12] The NGOs and their partners were attentive to ensure that the preparation for REDD+ was in line with criteria like the participation of local communities and the respect of indigenous communities' rights.[13] This close monitoring had been stimulated by past development projects in the forestry sector (e.g., the reform of the forestry legislation) that were criticized for benefiting logging companies at the expense of people whose livelihoods depend on forestry resources. In 2009, such surveillance targeted the early steps of the preparation process. At that time, the minister of environment and the newborn *Coordination Nationale* decided to delegate to the consulting firm McKinsey & Company the definition of possible national options for decreasing forest loss. Greenpeace publicly condemned the collaboration. The NGO considered that McKinsey's recommendations would benefit the private sector by planning to provide incentives for the regulation of the forestry industry, while blaming the agricultural practices of local communities as one of the main causes of deforestation (Greenpeace 2011). According to the critics, the report was serving private interests and not the national one. The creation of the TCGs, which would meet regularly over a relatively long period of time, was an outcome of the controversy. The groups were expected to convince these external witnesses that the ministry regarded the formulation of policy options for REDD+ as an inclusive and progressive process that involved, among other stakeholders, civil society organizations speaking on behalf of local communities.

Representing civil society

For the *Coordination Nationale*, the demonstrative role of the TCGs (i.e., displaying the participatory nature of the preparation process) was clear. Less so was the content of the initiative, or the material and human resources it required. Who should and could participate in the TCGs? And how? Looking at the answers experimented by the team of consultants is a way of not taking for granted the reality of the Congolese society, but to decipher how it is actively performed by participatory devices. This is an important analytical task in a context where the means of political representation are saturated with development projects, whether they are related to international organizations like the World Bank or to international NGOs such as Greenpeace attentive to the participatory nature of aid and speaking for local communities.

The TCGs were expected to represent the diversity of the public concerned by deforestation by involving spokespersons of the so-called civil society, representatives from private companies and public administration officials. The use of the term 'civil society', which stems from the political sciences, is worthy of analytical attention in this context. The notion was used in the 1990s by development agencies to rethink their actions and promote associative ways of dealing with public problems and services instead of supporting the public sector (Ferguson 2006). Today in the Congo, what is referred to as civil society is a set of numerous small NGOs specialized in the different sectors targeted by international cooperation – sanitation, health, education, etc. (Trefon 2004). For instance, one of the organizations focusing on forestry issues and the preparation for REDD+ was the Réseau Ressources Naturelles (RRN; 'Natural Resource Network'). This Kinshasa-based team of about 15 people represented more than 150 environmental small organizations disseminated on the Congolese territory interacting closely with people directly affected by forestry policies. It was funded by the European NGO Rainforest Foundation Norway, particularly active in the promotion of community-based forestry management and the defence of indigenous rights. 'Consulting civil society', as the *Coordination Nationale* says when speaking in front of international NGOs or governmental donors, means that a few meetings have been held in Kinshasa with members of organizations such as the RRN. Congolese civil society is as much an outcome of the relationships between international and national organizations as it is a collection of Kinshasa-based individuals mobilizing for the sake of environmental concerns, understood in the broadest sense of the term, given that campaigns for forests are also campaigns for the protection of human rights and against economic inequalities.

In this complex landscape, an analytical account that would describe, in a Deweyan manner, the formation of concerned groups related to the emergence of new issues not dealt with by existing institutions (cf. Marres 2007) would fail to grasp the specificity of the process. The highly equipped and tentative dynamics that might constitute national publics in the Congo is well illustrated by the functioning of the TCGs. In Kinshasa, participation in workshops and meetings related to projects receiving foreign aid is paid. Participants usually receive lump sums for travel costs, as well as daily allowances known as *per diem* (about US$10 per person).[14] They are also provided with snacks and soda ('sweets', *sucrés*, is the colloquial term). Compared to the average salary, these benefits are significant amounts of money and sources of revenue in their own rights for participants who may attend up to six workshops a month. In Kinshasa, civil society and civil servants are in 'the race for workshops', as an expatriate expert involved in the reform of the forestry sector puts it.[15] Therefore, receiving 'sweets' and *per diem* was a fair expectation according to the participants in the TCGs, but a budgetary challenge for the organizers.[16] The practical and moral issue of the payment of allowances was directly connected to the construction of a convincing channel for the civil society to speak. During discussions among members of the *Coordination Nationale*, differentiating the economic interest to participate from the concerns (whether environmental, economic or political) the participants were expected to represent proved impossible and finally irrelevant: indeed, being a member of a TCG was considered as a kind of consultancy work that should be rewarded.

If the *Coordination Nationale* agreed on turning civil society representatives, ministerial officials and private-sector actors into mission-paid consultants (like themselves), its members disagreed on the work to expect from the TCGs. For some of them, the groups aimed at collectively exploring how different policy options could be translated into concrete interventions for reducing deforestation, which could be implemented by applying for additional aid assistance and funds. But for other consultants of the *Coordination Nationale*, the TCGs were a unidirectional channel, through which they would explain the options the ministry had chosen to this assembly of spokespersons, and for them to then provide the information to the rest of the Congolese population. With the TCGs, the preparation process for REDD+ was oscillating between two configurations: trying to set up a dialogic forum or establishing a mere information channel.

Speaking for the Congo

While some participants in the TCGs were members of the civil society, others were civil servants from various administrative bodies, and these were particularly reluctant to contribute to the participatory initiative. Whereas the *Coordination Nationale* considered their participation as part of their mission given that, according to one expatriate consultant, 'the state asks them to',[17] the ability of its international staff to represent 'the state' was challenged. The civil servants were unwilling to consider the demand formulated by the *Coordination Nationale* as a governmental one. The role of the TCGs as state-sponsored initiatives was questioned through the mixed status of the *Coordination Nationale*. The latter was formally a part of the ministry of environment, under the authority of one of its senior officials as well as the minister himself. But its staff were dependent on and accountable to international organizations – United Nations and World Bank programmes – that provided their salary.

That the *Coordination Nationale* acted as a national public body and owed its existence to international policy is of no surprise in the development world (see examples in Ferguson 2006). For our analysis, it is more interesting to note that the preparation process and the experiment with the TCGs challenged its capacity to talk on behalf of the Congolese state. This issue was raised even more straightforwardly in international settings. Climate change policies such as REDD+ are designed during regular international negotiations. During these meetings, side events are organized to present publications or interventions related to climate issues. In December 2011 at the negotiation session in Durban, South Africa, the preparatory activities conducted in the Congo were displayed as those of a country speaking in one voice.[18] What mattered was to represent a state-party of the United Nations, in front of an international community composed of aid organizations, governmental donors and international NGOs. This took the form of a careful choice of speakers, complemented by various rhetorical strategies. In South Africa, the work done by the *Coordination Nationale*, such as the creation of the TCGs, was presented by the Congolese members of the team only (and not the expatriate ones) and by a few representatives of Congolese NGOs (including RRN) who acknowledged the participatory qualities of the preparation process.[19] The side event concluded with two speeches. In the first, a senior official from the World Bank congratulated 'the Congo' for its achievements. In the second, the spokesperson of a UN funding agency stated that this example of preparation for REDD+ was led by national will. He used a colloquial expression within international cooperation – 'country ownership'[20] – implying that the Congo was considered the main actor of the process, and not a passive beneficiary of foreign assistance.

A particular situation is configured during public events such as this one in Durban. It is characterized by the self-description of actions undertaken by a united state that, albeit in need of external resources, is able to harness them for the sake of a national interest. It relies on a staging during which these actions are displayed in front of witnesses positioned as external to the country. The event reconstructs the Congo as an autonomous political actor, worthy of international attention, partly given its experiments with public participation. That such a reconstruction does not account for the many practical difficulties encountered by the preparation process is not what matters here. At stake is the possibility of making the Congo speak as a member of an international community, able to deal with nation-wide issues in the right way because it involves civil society. Accordingly, one of the outcomes of the TCGs experiment was to make such a description persuasive.

While the demonstration in South Africa seemed to convince its audience, the national unity was soon challenged. One year after the Durban performance, RRN and other Congolese NGOs publicly voiced their dissatisfaction with the preparation for REDD+. In a letter addressed to the World Bank, they explained that the *Coordination Nationale* was asking the TCGs to rapidly provide recommendations, even though they felt that the work done by the groups was not completed and feared that, once they provided the outputs, they will not be further involved in designing concrete interventions (Groupe de Travail Climat REDD 2012). They also highlighted the fact that the TCGs never received what they described as 'needed operating costs' – that is, *per diem* and sweets. By circulating this complaint within international institutions, the civil society distanced itself from the preparation process and its actors – Congolese and expatriate consultants, the ministry of environment and the funders. For a short moment in December 2011, it had been possible to construe the Congo as a unified political entity by separating national actors and external witnesses. One year later, Congolese NGOs drew another separation that opposed civil society (be it national or international) and actors directly involved in ministerial initiatives (again, national and international).

Making the Congo an actor of the development world

One can identify issues similar to the case of the French nanotechnology debate: How to speak in the name of the state? How to constitute publics within participatory settings linked to governmental intervention? How to frame a national problem for the state to deal with? Here, the Congo experiments with the TCGs as much as the Congolese state is experimented with through the participatory initiative. Such a state experiment with public participation questions the possibility of constituting an entity able to play a role in international arenas and to conform to moral imperatives such as 'public participation'. It is an analytical opportunity to see the Congolese state in action, where it has to produce a self-description, to manage a public administration problem, and to organize channels of representation under the gaze of witnesses like the World Bank and international NGOs that are deeply involved in the practical conduct of the experiment and in the constitution of the political entity called the Congo.[21]

Analysing state experiments with public participation is useful to characterize the Congolese state. It has often been described as failed (Reno 1999; Trefon 2009) or more recently as fragile (Karsenty and Ongolo 2012). These qualifying terms tend to see Congolese leaders as unwilling to act for the general interest, and the administration as unable to implement political decisions. Indeed, one cannot pretend that everything is fine and democratic there. Yet, the Congolese state might also be characterized by highlighting the permanent displacement of its perimeter. Consultants working in the *Coordination Nationale*, Congolese organizations speaking for civil society and backed by international NGOs or civil servants hoping to receive economic benefits additional to their official salary all act at some point in the name of the Congo, but they all also manifest uncertainty about where exactly the Congolese state begins and ends. The study of participatory devices tested in the preparation for REDD+ shows that the elusiveness of the boundaries of this political entity is a modality of its action within international settings.

Conclusion

The two cases examined in this chapter have described state experiments with public participation. They constitute interesting opportunities to reflect upon participation and state action in (at least) three ways. First, our analysis apprehends the use of public participation by governmental actors within broader state interventions. Second, state experiments question the channels of representation when the issues to be dealt with, such as technological development or environmental concerns, are constructed as national, partly through the participatory devices themselves but also because they relate to the integration of the countries in international programmes and regulations. Third, these experiments end with state demonstrations during which external witnesses are created to observe where the state is and what it does.

The chapter focused on situations that cannot be described in terms of emerging concerned groups willing and able to participate. In the French case, it proved impossible to create a physical arena where pacified discussion could happen given that one of the most concerned and active stakeholders stood against any participation in the debate, which was considered as another extension of technocracy. In the Congolese case, it proved necessary to include in the organization of participatory meetings the distribution of monetary rewards to both ministerial officials and representatives of civil society, whose attitude toward the initiative was quite ambivalent, praising its participatory qualities and then denouncing its flaws in front of the same audience. The rooms hosting the French national debate and the bank notes required for the functioning of the Congolese TCGs can be seen as examples of the instrumentation through which public participation is produced, but also through which two different forms of state action and state power are displayed. Separating the invited speakers of the debate on nanotechnology from the rest of the public, including anti-technology activists, says something about the French state, where a powerful technoscientific expertise is at the core of decision-making. Using *per diem* to indifferently incite civil servants and environmental associations to participate in regular meetings about policy-options to reduce deforestation says something about the Congolese state, where the public–private demarcation, which characterizes an ideal Western modern state, is irrelevant for a resource-less public administration.

One of the main analytical interests of state experiments is to offer empirical entry points for the exploration of different kinds of political organizations called states. The example of nanotechnology displays the centrality of expertise in the constitution of France. As the French state attempts to govern the social and technical uncertainties of nanotechnology, the variety of administrative resources and expertise centres requires coordination. The CNDP debate then appears as a moment when the polycentric organization of the administration is made explicit, publicly questioned and then re-stabilized under the guise of the single state able to publicly engage in technological development in front of internal (French citizens) and external (European institutions) audiences. By contrast, the constitution of the Congolese state as a political actor in international arenas stems from processes that intertwined aid projects and ministerial initiatives dealing with deforestation. The example of the preparation for REDD+ shows the blurring of the divide between public and private interests, in addition to casting light on the problematic fluctuation of the frontier between what is considered as Congolese and what is not. The participatory tool tested by expatriate consultants on behalf of the ministry of environment comprises trials for the elusive localization of the Congolese state.

But our analysis of state experiments with public participation has also emphasized similar mechanisms. It showed that framing issues as national concerns may occur through *ad hoc* participatory devices that supplement the more traditional representation processes. We accounted for operations through which states are delimited as entities in front of an audience (citizens or partners). In both cases, experimenting with alternative forms of public participation contributes to the possibility of such a self-description. In the Congo, it enabled the *Coordination Nationale* to show to external partners that the whole country was united in the preparation for REDD+, more than if only a couple of ministerial officials would have appeared on stage. In France, the experiment sought to create a national public, and required coordination among public bodies for the state to address this audience's concerns. What is at stake in the two situations we analysed is the possibility of constituting the state as a sovereign political power while demonstrating the democratic quality of its interventions.

Identifying state experiments helps the sociological analysis of the state. In the two cases the configurations in which participatory initiatives aim to construct publics for state intervention and public concerns for the state to act on are not pre-given. The state experiments analysed in the chapter contributed to create these configurations and maintain their stability. As a result, they display to the sociologist what the French and the Congolese states are made of and how they are expected to act.

We introduced this chapter by highlighting the proliferation of explicit public participation initiatives in state interventions. This calls for examining further the connections between these experiments and other kinds of state experiments. For instance, the analysis of the CNDP debate on nanotechnology can be pursued with that of the regulatory innovation that the mandatory declaration of nanotechnology introduced, and the description of the TCGs as a state experiment suggests the need to account for the other initiatives undertaken under the preparation process. But one can also rephrase the specificity of the participatory domain (both as a characteristic of collective undertakings and as an area of scholarly investigation) by considering that it is itself an outcome of state experiments to analyse. The opposition that the national debate on nanotechnology faced might well make the French government reluctant to commission other large-scale debates to the CNDP. In the Congo, whether or not the TCGs are indeed participatory will be known once the *Coordination Nationale* manages to convince its external witnesses – which include NGOs with practical experience and a normative perspective on public participation – that it is indeed the case. Framed this way, arguing for a specific scholarly domain devoted to public participation risks predefining what is best understood as outcomes of state experiments, namely areas of collective action known as 'participatory'. Rather, this chapter attempted to show that an empirical theory of the state grounded in the analysis of state experiments would allow us to locate the multiplicity of empirical sites where state power is problematized, and to identify the operations that make participatory initiatives specific components of state intervention.

Notes

1 One was related to transportation policy in south-east France, the other to nuclear waste policy choices related to a zone selected for burying them.
2 This quote is an excerpt from the transcription of the meeting (translated from French).
3 The critical descriptions are presented in short articles distributed in various meetings, and published online on the organization's website (www.piecesetmaindoeuvre.com).

4 *Réponse aux conclusions du Débat Public*, translated from French.
5 All the exchanges during the public meetings were transcribed and published online by the services of CNDP. One can easily locate the occurrence of 'the state' in this body of text. 'The state' is the entity that is almost always called for by participants in the public meetings when it comes to envisioning potential actions to manage nanotechnology and its concerns. We do not attempt to use this evidence to draw conclusions about the importance of an abstract notion of 'the state' for the French public. Yet this helps to characterize the situation that was constituted through the CNDP public debate, namely a public display of the uncertain ability of the state to deal with an issue, nanotechnology, that was itself uncertain.
6 He was interviewed by one of the authors in October 2009.
7 This expression was used by a civil servant from the ministry of health during an interview with one of the authors. The interviewee was then referring to a health crisis that happened in France in the 1990s.
8 These arguments were publicly stated in *Le Monde* ('Nanotechnologies, osez mettre en débat les finalités', *Le Monde*, 18 February 2010).
9 This section is based on doctoral research conducted by one of the authors (Ehrenstein) from 2009 to 2014. It comprised ethnographic fieldwork within the Congolese ministry of environment (March–April 2011) and at the Durban negotiation session on climate change in December 2011.
10 In 2013, the World Bank's initiative included more than 30 countries (Forest Carbon Partnership Facility 2013).
11 This expression was used by an international consultant during a weekly team meeting attended by one of the authors in spring 2011.
12 See the description of 'transnational advocacy networks' by Keck and Sikkink (1998).
13 Participation is also required by the World Bank for this programme (Forest Carbon Partnership Facility 2008).
14 For a similar account of the conduct of clinical trials in Kenya see Geissler (2011).
15 This expression was used by the regional coordinator for Central Africa of the projects funded by the World Resources Institute (WRI) on forestry issues during an interview in spring 2011.
16 Whereas the usual practice in development projects is to organize a few meetings that gather a maximum of 100 people, the TCGs were designed to assemble almost 300 people, which would meet within groups of 15 on a regular basis for two years.
17 The expression was used during a meeting of TCGs organized within the *Coordination Nationale*, which one of the authors attended in spring 2011.
18 One of the authors attended the session and the side event.
19 NGO representatives participated as members of the country delegation, whereas there is a distinct NGO status (independent from any country) within the climate negotiations.
20 'Ownership' has become a watchword for the World Bank and other aid organizations since 2000. The objective of such an orientation in development assistance is to guarantee that governments own the projects funded (and fostered) through international cooperation (Anders 2010).
21 Other countries are often more explicit regarding the external control they face. For instance, Brazil refused the introduction of constraining participatory requests during the negotiations conducted at the 2011 session of the Conference of the Parties.

References

Anders, G. 2010. *In the Shadow of Good Governance: An Ethnography of Civil Service Reform in Africa*. Leiden and Boston, MA: Brill.
Bayart, J.-F., Mbembe, A. and Toulabor, C. M. 1992. *Le Politique par le bas en Afrique noire. Contributions à une problématique de la démocratie*. Paris: Karthala.

Benamouzig, D. and Besançon, J. 2005. Administrer un monde incertain: les nouvelles bureaucraties techniques. Le cas des agences sanitaires en France, *Sociologie du travail* 47(3): 301–322.

Bogner, A. 2012. The paradox of participation experiments, *Science, Technology & Human Values* 37(5): 506–527.

Callon, M., Lascoumes, P. and Barthe, Y. 2009. *Acting in an Uncertain World*. Cambridge, MA: MIT Press.

Chilvers, J. 2010. *Sustainable Participation: Mapping Out and Reflecting on the Field of Public Dialogue on Science and Technology*. Harwell: Sciencewise Expert Resource Centre.

Cruikshank, B. 1999. *The Will to Empower: Democratic Citizens and Other Subjects*. Ithaca, NY: Cornell University Press.

Ehrenstein, V. 2013. Les professionnels de la préparation. Aider la République Démocratique du Congo à réduire sa déforestation: Programme REDD+, *Sociologies pratiques* 2: 91–104.

Ferguson, J. 2006. *Global Shadows: Africa in the Neoliberal World Order*. Durham, NC: Duke University Press.

Forest Carbon Partnership Facility 2008. *Forest Carbon Partnership Facility Information Memorandum*. Washington, DC: Carbon Finance Unit, the World Bank.

Forest Carbon Partnership Facility 2013. *Forest Carbon Partnership Facility 2013 Annual Report*. Washington, DC: Carbon Finance Unit, the World Bank.

Fourniau, J.-M. 2007. L'expérience démocratique des 'citoyens en tant que riverains' dans les conflits d'aménagement, *Revue européenne des sciences sociales* 136: 149–179.

Geissler, P. W. 2011. 'Transport for where?' Reflections on the problem of value and time à propos an awkward practice in medical research, *Journal of Cultural Economy* 4(1): 45–64.

Gourgues, G. 2012. Avant-propos: penser la participation publique comme une politique de l'offre, une hypothèse heuristique, *Quaderni* 3: 5–12.

Greenpeace 2011. *Bad Influence: How McKinsey-Inspired Plans Lead to Rainforest Destruction*. Amsterdam: Greenpeace International.

Groupe de Travail Climat REDD 2012. *Mémorandum de la société civile environnementale sur le processus REDD en R.D. Congo*. Kinshasa.

Hecht, G. 1998. *The Radiance of France: Nuclear Power and National Identity after World War 2*. Cambridge, MA: MIT Press.

Hullmann, A. 2006. Who is winning the global nanorace? *Nature Nanotechnology* 1(2): 81–83.

Irwin, A. 2006. The politics of talk coming to terms with the 'new' scientific governance, *Social Studies of Science* 36(2): 299–320.

Jasanoff, S., ed. 2004. *States of Knowledges: The Coproduction of Science and Social Order*. London: Routledge.

Jasanoff, S. 2005. *Designs on Nature: Science and Democracy in Europe and the United States*. Princeton, NJ: Princeton University Press.

Karsenty, A. and Ongolo, S. 2012. Can 'fragile states' decide to reduce their deforestation? The inappropriate use of the theory of incentives with respect to the REDD mechanism, *Forest Policy and Economics* 18: 38–45.

Keck, M. E. and Sikkink, K. 1998. *Activists beyond Borders: Advocacy Networks in International Politics*. Ithaca, NY: Cornell University Press.

Laclau, E. and Mouffe, C. 2001. *Hegemony and Socialist Strategy: Towards a Radical Democratic Politics*. London: Verso.

Lascoumes, P. 1994. *L'éco-pouvoir. Environnements et politiques*. Paris: La Découverte.

Latour, B. 1993. *The Pasteurization of France*. Cambridge, MA: Harvard University Press.

Latour, B. 2007. Diverging convergences: Competing meanings of nanotechnology and converging technologies in a local context, *Innovation* 20(4): 343–357.

Latour, B. 2013. Les espaces politiques des substances chimiques, *Revue d'anthropologie des connaissances* 7(1): 195–221.

Lezaun, J. 2011. Offshore democracy: Launch and landfall of a socio-technical experiment, *Economy and Society* 40(4): 553–581.

Li, T. M. 2007. *The Will to Improve: Governmentality, Development, and the Practice of Politics*. Durham, NC: Duke University Press.

Linhardt, D. 2012. Épreuves d'État. Une variation sur la définition wébérienne de l'État, *Quaderni* 2: 5–22.

Manin, B., Stein, E. and Mansbridge, J. 1987. On legitimacy and political deliberation, *Political Theory* 15(3): 338–368.

Marres, N. 2007. The issues deserve more credit: Pragmatist contributions to the study of public involvement in controversy, *Social Studies of Science* 37(5): 759–780.

Marres, N. 2012. *Material Participation: Technology, the Environment and Everyday Publics*. London: Palgrave Macmillan.

Mitchell, T. 1991a. The limits of the state: Beyond statist approaches and their critics, *American Political Science Review* 85(1): 77–96.

Mitchell, T. 1991b. *Colonising Egypt*. Berkeley: University of California Press.

Mitchell, T. 2002. *Rule of Experts: Egypt, Techno-Politics, Modernity*. Berkeley: University of California Press.

Reno, W. 1999. *Warlord Politics and African States*. Boulder, CO: Lynne Rienner Publishers.

Revel, M., Blatrix, C., Blondiaux, L., Fourniaux, J.-M., Hériard-Dubreuil, B. and Lefevre, R. 2007. *Le débat public: une expérience française de démocratie participative*. Paris: La Découverte.

Reynolds, L. and Szerszynski, B. 2006. Representing GM Nation?, *PATH Conference: Proceedings*, www.macaulay.ac.uk/pathconference (accessed 4 September 2015).

Rowe, G., Horlick-Jones, T., Walls, J. and Pidgeon, N. 2005. Difficulties in evaluating public engagement initiatives: Reflections on an evaluation of the UK GM Nation? public debate about transgenic crops, *Public Understanding of Science* 14(4): 331–352.

Scott, J. C. 1998. *Seeing Like a State: How Certain Schemes to Improve the Human Condition Have Failed*. New Haven, CT: Yale University Press.

Shapin, S. and Schaffer, S. 1985. *Leviathan and the Air-Pump*. Princeton, NJ: Princeton University Press.

Trefon, T. 2004. *Reinventing Order in the Congo: How People Respond to State Failure in Kinshasa*. London: Zed Books.

Trefon, T. 2009. Public service provision in a failed state: Looking beyond predation in the Democratic Republic of Congo, *Review of African Political Economy* 36(119): 9–21.

Trefon, T. 2011. *Congo Masquerade: The Political Culture of Aid Inefficiency and Reform Failure*. London: Zed Books.

Wilsdon, J. and Willis, R. 2004. *See-Through Science: Why Public Engagement Needs to Move Upstream*. London: Demos.

7
TECHNOLOGIES OF PARTICIPATION AND THE MAKING OF TECHNOLOGIZED FUTURES

Linda Soneryd

It can be quite frustrating that the hope for democratization in science and technology development is still being raised by scholars in those fields. It is now clear that such a hope, which Langdon Winner expressed in his 1977 book *Autonomous Technology*, has not been fulfilled. But even so, 'the notion of participatory technology has continued to attract the attention of a wide range of scholars and practitioners' (Brown 2007, 327). A number of social scientists, some of whom themselves were previously involved in developing and evaluating public engagement with science and technology, now claim that we need to approach these exercises in more critical ways (Levidow 1998; Rayner 2003; Irwin 2006; Chilvers and Evans 2009). Despite this 'critical' turn, expressions of this hope still linger, even though it is now often formulated in more careful terms, or hidden in the embedded premises of studies.

It is not a lack of signs of the potential realization of a democratized science that makes this lingering hope so frustrating. Rather, it is the type of expectations that are raised – the wish to attain such abstract principles and ideals – that makes some of the existing analyses of public engagement exercises futile. The hope itself is often not subject to critical reflection – that is, there is a lack of critical analysis of *the effects* of continued beliefs in democratized science in this abstract way.

To some extent, this is related to understanding democracy and public engagement in what Marres calls '"issue-less" terms'. Marres defines 'issue-less' as 'implying that democratization refers mainly to the project (or hope?) of making policy-making more inclusive and accountable' (Marres 2007, 763). But also, I argue, it has to do with the lack of attention to *the effects* of institutions within science and technology studies (STS). That is, they do not recognize or critically approach the fact that democracy *is* widely accepted in issue-less terms. Continued belief in such abstract principles, like democracy (as well as rationality, efficiency and so on), has spread transnationally through blueprints and standards, creating similarities across a range of organizations, governments and private companies.

STS lacks a concept that can capture *institutions* (institutions in the sense of taken-for-granted beliefs) and how they can produce similarities among organizations across cultural and national boundaries. This lack poses a challenge to STS scholars when approaching 'technologies of participation' (i.e., relatively stabilized configurations of expertise and blueprints for engaging publics). The question is, thus, how can we account for both the production of similarity and its implications when technologies of participation circulate across various sites, while acknowledging the localized effects of participatory experiments?

In order to approach this question, this chapter will combine some insights from STS approaches to technologies of participation with new institutionalist approaches to the transnational spread of ideas and standardized forms of governance. The relevance of combining these theoretical approaches will be discussed in relation to two empirical examples of formalized public deliberation exercises. The first aimed to demonstrate and test how participation and transparency can be implemented in nuclear waste management programmes. The second gathered citizens in order to generate visions that could be relevant for future areas of development in science and technology and provide a set of policy options to future European framework programmes.

The rationale for discussing these two cases is based on the fact that they are both examples of designs that have not remained completely localized, experiential or informal; they have, at least to some extent, travelled and linked multiple sites and actors together. A commonality between the cases is that they are both connected to research projects funded by the European Commission. In some ways, the cases have many of the characteristics of what Bogner (2012) calls 'lab participation': they are highly choreographed by experts in public participation; they are organized in the context of a research project and funded by a third party; and they are very well documented, ending up in research reports that can be used to circulate the results of the deliberations. Differences between the cases, and divergences from Bogner's account, make them interesting to discuss in relation to the question of how travel and the local context changes the effects of technologies of participation. Bogner's account of lab participation suggests that technologies of participation can cut off what happens within the lab (e.g., the particular, organized, participatory event) from participatory practices, science and politics outside the lab. This is not what happens in these two cases; rather, the deliberations that take place within the events are implicated in all sorts of relations with politics and science, albeit very differently between the two cases.

The formats for stimulating public deliberation, as well as the issues which citizens are expected to deliberate upon, differ in the two cases. Despite these differences, the official presentations of these events are in many respects very similar: they both appear to have achieved what they set out to do. Through this work of documenting, evaluating and reporting, formal designs for public deliberation can be reassembled and set off to travel to new sites and problem areas. In addition to the documentation of 'good examples' and 'best practices' that can show what makes a particular design work well, there are also powerful organizational carriers of the designs, and the ideas inscribed in these designs, including these organizations' visions about technological innovation, and the roles that are granted citizens in such processes.

Taking the reflexivity of organizations into account, and trying to understand why changes (that mainly produce similarities across organizations) occur without being transformative (in a sense that it truly changes practice), is crucial for understanding the uptake of public engagement exercises by government bodies across a range of policy fields and political cultures. This is a rather different approach than speaking about the spread and uptake of public engagement exercises as 'partial progress' (Irwin 2014, 73) or continued hopes that learning and constructive critique can make governing actors reflexive and thereby 'more responsive, responsible, and accountable' (Chilvers 2013, 285). It is different because it sees the *reflexivity* of organizations and continued belief in progress as part of the problem and an object of study, rather than a desired state or development.

If we want to understand how expertise and designs for public deliberation both form a variety of relatively stable configurations, and how these scripts are translated and transformed, there are many valuable insights to gather from scholars of STS and their recent interest in technologies of participation. The discussion can be advanced when it comes to understanding the interaction between the design and implementation of formalized public engagement procedures, as well as processes of circulation, and who the prime 'movers' are in these endeavours. For this purpose, it is suggested that some of the valuable insights in STS can be combined with insights from organizational studies and new institutional theory. The next section will make a brief excursion into some current STS approaches to public engagement with science and technology and identify some issues for which it would be beneficial to combine STS and new institutional theory. The sections after that will use the two aforementioned empirical cases to illustrate how we can approach these issues in order to understand the circulation of technologies of participation and their effects. Finally, the chapter ends with a summary and discussion.

Combining STS and new institutionalist theory approaches to technologies of participation

After some decades of promoting public deliberation and the inclusion of laypeople in the governance of science and technology, STS scholars witnessed a turn towards public deliberation in policy circles. While it was new that policy actors recognized the importance of taking laypeople's views into consideration, scientific governance in many other respects remained the same. When the need to take public engagement seriously became 'a standard part of the policy repertoire' (Irwin 2006, 300), this led to calls from STS scholars to approach public engagement in more critical ways.

The call for more critical approaches generated a number of studies and publications that, each in unique ways, analyse how publics are 'constructed' or 'made' through the formats and methods for public engagement (Irwin 2001; Goven 2006; Braun and Schultz 2010), and the tendency that new designs for public deliberation primarily seek to engage 'pure publics' or those citizens that have no prior engagement on an issue (Lezaun and Soneryd 2007; Bogner 2012). Formalized methods for public deliberation include specific objectives, ways of inviting groups of the public, as well as guidance and facilitation, which frame issues of concern and forms of interaction that only allow deliberation to take place in very limited ways. Along these lines, it has been argued that 'the public' 'is never immediately given but inevitably the outcome of processes of naming and framing, staging, selection and priority setting, attribution, interpellation, categorisation and classification' (Braun and Schultz 2010, 406).

A focus on formalized designs, it has been argued, gives too much power to these designs; this focus lends only a very poor understanding of the agency and active sense-making of citizens, and how publics come into being (Marres 2007; Felt and Fochler 2010). Rather than being a direct reflection of how policy elites imagine publics, citizenship is made in the interaction between how people are interpreted as citizens, and the way in which they themselves make sense of this in relation to their own histories and visions of the future (Jasanoff and Kim 2009; Felt and Fochler 2010, 2011).

In order to account for both the design of deliberative events, and how participants engage with that design, Felt and Fochler (2010) used the well-known insight from STS scholars that technologies are never transferred in a linear sense but are the result of translations, and always involving processes of both inscription and description (Akrich 1992). On the one hand, 'designers imagine and try to frame an object's context of use and its user's behaviour by inscribing their vision of the world in the artefact's design, thus pre-scribing certain forms of use', and, on the other, 'users and participants might have very different ideas about the technology, the world inscribed in it, and their attributed roles. They might struggle with, attempt to shift, or even reject the script – hence de-scribing the technology' (Felt and Fochler 2010, 220).

This suggested approach for studying technologies of participation certainly has its advantages when it comes to understanding how citizens can resist and transform the roles and identities ascribed to them. But in order to account for both the production of similarity when technologies of participation circulate across various sites, and the localized effects of participatory experiments, it is not very helpful since it gives prevalence to fluidity of meanings rather than stabilized boundaries, and to difference, rather than homogenization and reproduction of power structures.

The following sections discuss three aspects for which it could be beneficial to combine an STS approach with some insights from new institutional theory. The first aspect has to do with the travel and circulation of technologies of participation, a question for which STS scholars, and in particular Actor Network Theory (ANT) approaches, tend to give more focus on how people and things are combined in new ways – that is, more focus on localized effects than on the production of similarities.[1] From a new institutionalist perspective, it is the reflexivity of organizations that make them attentive to organizational trends, and inclined to adopt similar organizational recipes and/or policy instruments. This does not happen through some mysterious force, even though new institutionalists often speak about 'institutional forces', or through evolutionary democratic progress, but through powerful organizational carriers. For example, the European Commission (EC), the World Trade Organization (WTO) and the Organisation for Economic Co-operation and Development (OECD) are such powerful organizational carriers that provide means and scripts for good governance.

The second aspect concerns the question of *what* it is that travels. If we choose to focus on particular designs and formats for public engagement as standardized configurations that travel from place to place and become reconstituted and replicated at multiple sites, we are, to be honest, left with very few examples that actually do so. Apart from the much-touted consensus conference, there are not many standardized mechanisms that have travelled worldwide. Most examples of public deliberation exercises remain localized and one-off. The stabilized technologies that are often in focus for the ANT researcher are simply not present. Most cases of public engagement exercises are rather characterized by more or less stable configurations between organizational carriers and institutionalized values.

Finally, a third aspect has to do with the importance for organizations to engage in self-presentations (cf. Strathern 2006). Increased interdependencies in a transnational world, as well as increased focus on audit, ranking and other evaluation practices, have certainly increased the importance for nation-states and local governments, as well as companies and NGOs, to engage in activities of good self-presentations. New institutionalists could also add that there is an urge for organizations to present themselves as *modern* organizations, as the idea of 'the modern actor' (i.e., an entity with clear boundaries and preferences). This highly institutionalized myth affects how both individuals and organizations tend to think of and present themselves (Meyer and Jepperson 2000; Sundström *et al.* 2010). This last aspect is most of the time neglected by ANT researchers, since the focus is more on the fluidity and impurity of the boundaries of modernity, than on the efforts with which actors try to uphold them.

In what follows these three aspects will be discussed in turn, and explored in relation to the two empirical cases focused on in this chapter. These two cases of public deliberation exercises are examples of blueprints for good governance that, at least to some extent, have travelled to other sites. The first case concerns the involvement of stakeholders in order to increase transparency and improve dialogue on nuclear waste management in the Czech Republic. The case is an example of testing expertise and, to some extent, replicates a particular format for stakeholder dialogue on nuclear waste management. A relatively stable configuration consisting of Swedish government bodies and consultants had, from the early 1990s and onwards, organized several public deliberation exercises on nuclear waste in Sweden (Elam *et al.* 2010). Through this, a model for communicating with and engaging stakeholders emerged, which also made a few excursions outside the nuclear waste area – for example, into the issue of mobile telephony (see Lezaun and Soneryd 2007; Soneryd 2007, 2008). The efforts to replicate this model for communication in the Czech Republic was part of a European research project called Arenas for Risk Governance (ARGONA) funded by the European Commission. The main deliberation exercise was organized as a public hearing in the Czech Republic in 2009. The following analysis is based on documentation from the project, the evaluation report from the hearing and interviews with the key actors involved in the organization of this event (who had also organized similar deliberative events in connection to nuclear waste management in Sweden).

The second case was much broader in terms of the issues that were under discussion and was organized through a number of citizen panels in the seven countries (Denmark, Belgium, Malta, Hungary, Finland, Bulgaria and Austria). During the first citizen consultation, the selected panel, composed of 25 citizens randomly selected from each country, spent a weekend developing 'visions for the future' that would provide input for the European Commission's framework programme for research. The format, and especially the procedure for selecting citizens and putting together the citizen panels, resembled the ways and methods in which the Danish Board of Technology previously had been working with its famous consensus conference. This exercise was also part of an EC-funded research project called Citizen Visions on Science, Technology and Innovation (CIVISTI). The first citizen consultations took place in 2009, and the following analysis is based on documentation from the project, interviews with the key organizers at the Danish Board of Technology (DBT), and observation of the first consultation with the Danish citizen panel.

The travel and uptake of formalized public engagement procedures

The first aspect that is a challenge to STS scholars when approaching the widespread use of formalized procedures for public engagement with technoscientific issues has to do with how such formats travel and circulate across national and cultural boundaries. This is a problem partly because STS scholars, and in particular ANT approaches, tend to give attention to how people and things are combined in new ways – that is, more focus on reconfigurations, translations and localized effects than on imitation, replication and the production of similarities.

As described earlier, some STS scholars have either tended to give too much power to how publics are invited, categorized and framed by policy actors, or analysed how citizens themselves make sense of their participation, and how they also sometimes reject the narrow roles ascribed to them. But when publics are increasingly invited by governments to deliberate upon technoscientific matters, how can we account for processes of translation as 'a process of replication or imitation *and* differentiation at the same time' (Barry 2013, 415)? For this purpose, studies by new institutionalist scholars on how management ideas travel and shape business practices can be useful (cf. Czarniawska-Joerges and Sevón 1996; Sahlin-Andersson and Engwall 2002). From a new institutionalist point of view, citizens participating in deliberative events would not be seen as the primary users. Rather, the focus would be on the organizations that decide to use a particular configuration of expertise and design for public deliberation: government bodies, private corporations or foundations.

Globalization, and shifting patterns of governance, indicates that blueprints, standards and management ideas migrate across politico-geographic borders, and impact upon existing regimes, as they form part of an increasingly transnational regulatory order (Djelic and Sahlin-Andersson 2006). The circulation of 'best practice models' simultaneously leads to homogenization and reproduction of dominant views and power structures, as well as increase in diversity, as standards and blueprints are always translated and given new meaning when they are adopted and implemented in different institutional and local contexts. As technologies of participation are most often 'non-stabilized technologies', the innovators/designers are likely to be present (cf. Akrich 1992, 211) and often able to adopt a pragmatic situated view of the user, adjust the design according to its needs, and 'package' them in ways that make them attractive to potential users.

What makes a particular design or organizational recipe popular is, according to new institutionalists, related to established ideas and norms of what it means to be a 'modern organization'. They have attended to the abstract similarities among organizations placed in very different cultural environments, and explain such similarities by the attractiveness of certain institutionalized, taken-for-granted values such as 'democracy', 'rationality', 'efficiency', 'development' and so on (Røvik 2002). ANT and new institutionalist approaches, in fact, share a joint interest in how ideas become widely accepted as truth, and/or the 'right way' of doing things. However, while for ANT researchers the main assumption is that it is difficult to achieve stability, and that this is always only achieved temporarily, the new institutionalists have ascribed more power to institutions. In her analysis of the similarities and differences between these two approaches, Liv Fries (2009) phrases this as a matter for ANT to ask *how the taken-for-granted emerges*, while for a new institutionalist, the question is rather why similarities are produced and *institutions become the answer*.

If we turn to the two cases of deliberation exercises, they were both packaged in the form of promises and objectives of what the exercises were intended to achieve. These promises were primarily directed to the needs and objectives of the governmental bodies. For example, the starting point for the ARGONA project was that participation and transparency are key elements in effective risk governance. The needs that this project promised to meet were primarily formulated from the point of view of implementers of nuclear waste programmes. It was also important to tell the story of achievements concerning stakeholder involvement that had been made in Western Europe. It was assumed that new European Union (EU) member countries would want to make these same achievements (ARGONA 2006). The project promised to address 'how effective risk governance can be achieved by concretising how participation and transparency can be implemented', and to develop 'guidelines for the application of novel approaches that will enhance real progress' in nuclear waste management programmes (ARGONA 2014).

One of the explicit aims with CIVISTI was 'to help European decision makers in the process of defining relevant and proactive research agendas'.[2] The objectives that it set out to meet were to 'encourage views on the future, which are new or not generally recognised as policy issues', and to 'represent trends of relevance' to science and technology developments in the future. The project aimed to 'contribute to the expansion of the European foresight capacity' in a 'cost-effective' way, and potentially develop a method that could be used in the future to 'execute citizen consultations across all member states' (CIVISTI 2014).

Both cases involved powerful organizational carriers of the idea that citizen deliberation is needed to improve governance; the European Commission, with its ambivalent but forceful promotion of new forms of citizen involvement, funded both projects (for a view of the EC as an ambivalent promoter of public deliberation, see European Commission 2000, 2001, 2002; Hagendijk 2004; Felt and Wynne 2007; McNeil and Haran 2013). Another important carrier of the ideas and formats for public deliberation was the DBT that, since early 2000, had been developing formats for citizen involvement as pilot projects to be used on a Europe-wide scale (this resulted in the European Awareness Scenario Workshop). Additionally, the Swedish nuclear waste authorities had not only been working with stakeholder involvement in Sweden, but were also involved in projects and networks aiming to transfer their experiences of stakeholder dialogue to other organizations (RISCOM I and II; and the OECD's Nuclear Energy Agency's Forum for Stakeholder Confidence). Both the DBT and the Swedish nuclear waste actors were thus already well known and had a good reputation.

Previous studies of how organizational recipes flow show that they become popular when they are associated with organizations or individuals that are 'widely recognized as authoritative actors and models' (Røvik 2002, 122). Social authorization is made, for example, through stories about the origins of the model as well as stories of contemporary and previously successful uses. The cases contained combinations of social authorization by highlighting what these organizations had achieved previously, and by claims that the designs were universally applicable and not limited to the particular context in which they had been developed. The way in which these organizational carriers packaged ideas into objects that could be attractive to others is also a matter of timing (Røvik 2002, 132). This was relevant especially for the nuclear waste case: the ARGONA project was introduced in the Czech Republic when the negotiations related to geological disposal for radioactive waste had reached a deadlock.

The travel and uptake of these formalized designs can be explained through the way they were packaged, and how they were made attractive to their organizational users. But what happened when they were displaced (e.g., relocated to these new sites)? Are the inscriptions strong enough to have any effects other than just yet another localized experiment that leads to new combinations between people and things?

What travels? Worlds inscribed and displaced

It is perhaps not surprising that government bodies are attracted to promises to improve governance and guidance in how they can be more open, participatory and efficient at the same time. However, if guidelines, ways of conducting public engagement exercises and the roles attributed to participating citizens become decontextualized as they are adapted to the new context each time, what is it, exactly, that has travelled from one site to another? If it is not a standardized design, then what travels? If the expertise and blueprints for public engagement that travel to a new site carry nothing with them – that is, if they do not inscribe ways of understanding the world – then the effects are not reproduced at the new site. Or in Latour's words: 'Without the displacement, the inscription is worthless; without the inscription the displacement is wasted' (1986, 16).

Imaginaries that public elites use to frame publics can be powerful, but they are also time and context specific, intrinsically embedded in the history and practice of particular organizations. For example, the assumption of publics as non-existing entities, as lacking hermeneutic capacities or as threats has emerged in relation to particular times and contexts of scientific governance in the UK, as shown by Welsh and Wynne (2013). So, what is it, then, that is inscribed when public engagement practices are displaced at other contexts and sites?

Publics as valuable entities

Inscribed into the designs of both cases, and also prevalent in the practices they generated, is an imaginary image, an assumption that publics are valuable entities. The use of ready-made designs for public engagement was thus not a direct employment of a particular modality of communicating with publics, but an abstract inscription of publics as potential resources and valuable for the productivity of government.

In the CIVISTI case study, this productive value was connected to the elicitation of citizens' 'positive dreams and wishes for the future'. It was crucial for the organizer that the visions truly reflected the participants' views. However, at the same time, the organizer also wished to exert some effect on participants, and, to some extent, prescribed the content of the visions the participants eventually provided. The visions needed to be formulated in positive terms and in present tense. Participants were to imagine that, and write as if, their dreams had already been fulfilled. Visions were also supposed to reflect personal wishes, as well as societal and European needs. The content of the visions should not be manipulated or framed by the organizer, but it was crucial that the visions involved 'technology'. The organizer expressed some worries about how to meet these requirements, but these were resolved through adopting particular facilitating methods, as discussed at an internal meeting between the organizers and the facilitator (observation, May 2009).

In relation to nuclear waste, the modalities of communication that governments, organizations and companies use to engage with publics have changed over time in many nuclear power countries (cf. Callon *et al.* 2009), and the communication model for engaging publics that later was used in the Czech Republic was shaped at the very breakpoint of this change in Sweden. Welsh and Wynne (2013) describe how citizens were imagined as passive non-entities, illustrated with a picture of a nuclear power station from the 1950s surrounded by a halo-like light, but emptied of people and publics. This modality was also prevalent in Sweden during the 1980s, when the implementer of the nuclear waste programme started with site investigations in the north of Sweden, without the consent of municipal governments, and without informing the population, as if there was no constituency concerned with the enterprise other than the experts involved in the programme. When this caused riots and local protests at the sites, it brought about changes in the strategies of both the implementer (the nuclear waste industry) and Swedish nuclear waste authorities.

A relatively stable configuration of Swedish authorities, the same communication model that was later used in the ARGONA project and a commitment to dialogue emerged at this point. This fostered a particular modality: publics were imagined as a resource for eliciting a wide range of stakeholder viewpoints that could transform nuclear waste management in a broad sense from being closed and opaque to transparent and open for critical exploration of the ongoing industry-led nuclear waste programme. This configuration was based on a separation of the implementer's turn to a voluntary approach (meaning that the implementer's strategy to only work with municipalities in which they were welcome, in combination with the legislated veto-right, gave the municipalities a relatively strong position) and the government authorities' turn to dialogue. This, together with funding for the involved municipalities and NGOs for their engagement in the process, granted a relative autonomy of involved actors.

This configuration was not and could not easily be transferred to the Czech Republic in its entirety. It was inscribed only in two weak forms: in the expectations that the format for dialogue could be universally applicable, and in the assumptions about the involved actors' relative autonomy. The abstract value of public deliberation inscribed in the design for stakeholder involvement was strong enough to generate a range of activities, including the adoption of professional knowledge of formalized public engagement, and pledges that the nuclear waste authority was now truly committed to dialogue (a commitment that lasted only for a short while; see Konopásek *et al.* 2014).

The travel of design elements

There are more concrete things, devices or 'design elements' that are transferred from previous public engagement exercises to the new sites. In the case of nuclear waste there was a list of principles for good dialogue (Andersson and Wene 2008); a reference group was formed; an agreement to follow the principles was signed by all involved participants in the reference group; and a standardized format was employed for the public hearing that was later held in the Czech Republic

(Agreement on Cooperation 2008). All of these elements had also been present, nearly in an identical way, in the public discussions that had been conducted in Sweden (RISCOM I and II).

In the case of citizens' visions for a European research agenda, more elaborate forms for organization were used. The selection of participants was made according to a procedure that was very similar to the consensus conference that the DBT has become famous for developing. A *Cookbook* – a manual describing the planning, preparations and practical organizing – was implemented.[3] In addition, a *Guide for the Facilitator* was employed, which described the venue, equipment (flip charts, notebooks and markers) and notes about what the national organizer should bring up in the welcoming speech, roughly what the facilitator should do and say, and detailed prescriptions of how long the introduction and discussion rounds within sessions should take.[4] Inscribed in the case is thus also a technologized view of democracy: the idea that democratic values can be attained through formalized and partly standardized procedures.

Technologized democracy

In the CIVISTI project description, it is stated that it is possible to gain access to citizens' concerns and expectations 'with the right facilitating methods' (CIVISTI 2014). These facilitating methods are, as discussed above, quite detailed and described in written documents, but also based on years of experience and a 'gut feeling' that takes a long time to develop (interview with DBT). Public deliberation used as a tool for producing visions that can be 'transformed into relevant research agendas' (CIVISTI 2014) is an example of how valuable a resource the elicitation and mobility of citizens' views is for effective governance – values that were inscribed in the design.

The dimension of technologized democracy is more obvious and explicit in the CIVISTI case, although it is also present in the ARGONA case. According to the vision of the organizer, the value of 'gathering the major national stakeholders under the same roof' (ARGONA 2006) was not to reach consensus, or to legitimize decisions that had already been made, but to increase transparency. However, it was a limited version of transparency – as all devices to reach transparency necessarily are limited – that primarily exposed the statements and arguments of all the involved stakeholders, and which included 'the documentation of all discussions and points of view' (ARGONA 2006). The expertise involved was less technical, in comparison to the CIVISTI case, in the sense that less elaborate methods for facilitation were used. Rather, it was based on a type of 'political expertise' – that is, a very context-sensitive expertise (cf. Barry 2013) concerning, for example, the right timing to test the model in Eastern European countries, and to serve as ready-made solutions to the problems government authorities faced in relation to nuclear waste siting.

New institutional studies have attended to an increased homogenization among organizations in terms of *abstract* similarities, and in terms of *managerialization*. This means that popular reforms and organizational recipes that circulate will have very different local effects. Additionally, organizations' strategies for dealing with pressure from their environments (e.g., anything outside the organization itself) will become more and more elaborate. When organizations attend to new problem areas and try to deal with them, they will tend to mainstream and simplify problems so they become manageable (cf. Strathern 2006, 192). Both events were translated and acted upon in ways that were not foreseen by the designers or organizers. These unexpected changes did not cause any serious problems in terms of the abstract value of public engagement, and the practice of evaluating the exercises could be sustained and potentially carried on to subsequent exercises at other sites.

The importance of organizational carriers and organizations' self-descriptions

It can be argued that as long as the use of devices 'do not diverge too radically from those predicted by the designer', the script can be employed as the main key for interpreting these events, according to Akrich (1992, 216). This brings us to the third point, which also relies on insights from new institutionalist studies: because of the urge for organizations to present themselves in ways that show that they are efficient, rational, democratic, etc., they will engage in efforts to purify the events, so they can present themselves as having achieved what they had set out to do (cf. Meyer and Rowan 1977; Meyer and Jepperson 2000; Strathern 2006; Soneryd et al. 2010) or at least so that the hope can be maintained that they will do so in the future (Brunsson 2006). After all, failures can always be ascribed to something outside the organization, and can be attributed to contextual circumstances, rather than to inherent flaws in the design itself. What was achieved, according to the organizers in the two empirical examples, is first of all 'that it happened at all' (i.e., the accomplishment of gathering participants and making them discuss the respective issues) and, second, that it happened 'according to the scripts' (i.e., that the guidelines of the formalized designs were followed strictly enough).

The ARGONA project took place at a point when the process of finding a suitable place for nuclear waste siting had reached a stalemate in the Czech Republic, which had led to a temporary moratorium on these activities. The organizer of the public hearing on nuclear waste management in the Czech Republic expressed the opinion that the nuclear waste authority in the Czech Republic 'possibly did not really know what to do when the moratorium would finish. ARGONA gave them some kind of programme, a direction to go' (interview with ARGONA). Since relations between the government authorities, environmental authorities and municipalities were destroyed and all communication had been stopped by a stalemate in dialogue, it took some time for the organizer to gather a reference group that would consist of representatives of all these actors, and who would be interested in pursuing the public debate. That it was possible at all 'to get all these parties around the same table' (interview with ARGONA) was therefore interpreted as a success.

In a study of nuclear waste management in France, Michel Callon and his co-authors show that, 30 years ago, social concerns could easily be dismissed as irrational public fears, while today 'social concerns' are recognized to be central (Callon et al. 2009, 25). Their conclusion is that this leads to hybrid forums (e.g., discussions between laypeople and technical experts), which will blur the boundary between the 'technical' and 'social'. However, this will also induce new boundaries, as governments and organizers of these public events will be engaged in separating these impure practices from the official presentations of these events. Each application of the communication model necessarily needs to be pragmatic, and adapt to the situation at hand, as the organizer of ARGONA admits, but the principles were nevertheless applied as 'strict as was possible in the Czech Republic' (interview with ARGONA). It was subsequently clear that the efforts to test the communication model in the Czech Republic also worked as a catalyst for a range of other activities among stakeholder groups, who translated the project into their own objectives and interests. This led the organizer to distinguish between what elements belonged to the tested communication model, and potential failures that could be dismissed as mere contextual factors, or misunderstandings of the model.

The abstract productive value of public deliberations was reinforced by the events that were organized in the context of CIVISTI. In the final report of CIVISTI, it was stated, 'a high risk was taken' in the project since 'this kind of methodology had never been tried before'; the fact that the citizens' production of visions worked out at all was therefore conceived as a success. The primary results in this case were not merely related to the actual gathering of citizens in the same room – though this was also seen as an accomplishment, as the project suffered from a drop out of participants. But what the participants produced was, in fact, quite a lot, at least in quantitative terms: 69 citizen visions for the future and 100 recommendations produced by a group of experts and stakeholders. A priority of these recommendations by the citizen panels resulted in a list of 30 recommendations for new and emerging issues for future European science and technology policies (CIVISTI 2011, 24).

Though both cases were conducted in ways which did not appear to be standardized (e.g., contextualized, pragmatic ways), they both entailed strong scripts for how the events should be interpreted. These scripts proved to be valuable, and remained somehow intact, even when they were 'de-scripted' (i.e., translated into other forms of action and other meanings of these actions by the participants). They could potentially work as productive features of government, and in subsequent organized public deliberation exercises. It is not only the abstract value of public engagement – which can be concretized in terms of mobilizing important partners in relation to nuclear waste siting, or in producing visions that can be valorized in relation to European science and technology programmes – that is reinforced through these exercises and reports, but also other values related to ideas of the nation, European citizenship and standards for 'good governance' as well.

The worldwide circulation of ideas – which travel through organizational carriers and blueprints for 'good governance' – reproduce ideas of centre and periphery.

It is from the 'centres' that 'best practices' can be exported to the 'peripheries' that are not as well equipped with democratic traditions and experiences. This might explain how a 'Swedish model' of stakeholder involvement in nuclear waste management found its way to the Czech Republic, even though only fractions of the configuration found its way with ARGONA.

The value of engaging with publics, and the evaluation of such engagement, can be seen as part of the increasing importance of audit practices, in a broad sense (cf. Strathern 2006, 192). Being able to show that government organizations are taking public concerns seriously *is* an example of the increased reflexivity of some organizations. But, obviously, this is another type of reflexivity than what some STS scholars make calls for. Brian Wynne, for example (including Chapter 5 in this volume), makes persistent arguments for reflexive institutions that can recognize more heterogeneous roles, as well as the hermeneutic capacities of the public. Organizations that are attracted to ready-made designs for public engagement, on the contrary, seem to be reflexive in ways that simplify a complex environment into understandable and governable issues, and leads to abstract similarities among organizations. It also leads to the potential tendency that government bodies are increasingly approaching democracy in managerial ways.

Conclusion

The two example designs for public deliberation that have been discussed in this chapter differ quite substantially concerning the basic assumptions they make about who was going to be involved, and what the exercises were going to accomplish. They are also different in terms of how elaborate the designs were, and their respective requirements concerning expertise on facilitation. In what ways can these examples illustrate tendencies towards the production of similarities through the transnational circulation of public engagement exercises, rather than just being examples of singular, isolated events that produce difference and a multitude of possible interpretations? Of course, they are experiments, which, by their nature, spur a range of interpretations, activities and uncertain outcomes. However, they are *also* expressions of new ways of problematizing and relating to publics. Public deliberation – as abstract and productive values for governments – leads to more elaborate organizing and tendencies to technologize issues of public concern. This chapter has argued these two points with the help of new institutionalist scholars and their previous studies on the flow of management ideas.

The focus that new institutionalist studies place on organizational carriers can be said to correspond to the focus on devices, such as computer programs, indexes and technical objects in ANT studies (Fries 2009). Most cases of public engagement exercises cannot be understood through technical objects in this sense, but rather through relatively stable configurations between organizational carriers and institutional values. In the case of ARGONA, the displacement of the design for dialogue was accompanied with a strong script. The organizer had strong *assumptions* about how the design could and should form a part of a configuration of relatively autonomous actors. But, its *in*scription was too weak; it was not enough to fix the distribution of actors that could shape a successful configuration.

In the CIVISTI case, a whole, relatively stable configuration (e.g., the DBT, the EC and a range of standardized elements that could be recognized from previous public deliberation exercises conducted by the DBT) was presented as a new method, and positioned at multiple sites across seven countries. In this case, there was never really any need to form a new configuration. Therefore, it 'worked', but it was relatively detached from the sites in which it was implemented: problems with drop-outs, and suspicions from the organizer that some visions were not really citizens' 'real' visions, certainly say something about the constructed, or detached, character of the citizens' dialogue that actually took place.

What travels and produces effects at multiple sites are abstract similarities across organizations; this means that organizations in different cultural and national contexts tend to be attracted by similar ideas and governance tools (such as adopting voluntary standards and certification procedures, or commissioning public participation professionals). Organizational carriers that manage to spread their ideas transnationally tend to generalize and theorize the ideas they champion, thus making them abstract and universally applicable. Effects beyond de-scripted and localized effects will therefore never be very concrete, and comprise particular characteristics of 'the public' as a valuable constituency. The combination of this general idea, and other institutionalized values held by its organizational carriers, however, also turns this phenomenon into something more specific. Through the study of transnational regulation across a diverse set of regulatory fields, Djelic and Sahlin-Andersson (2006) identified a number of institutional forces that they argue make up a transnational culture to which organizations relate and adapt. Strong tendencies towards scientization, marketization and formal organization, together with an increased understanding of democracy through deliberation, can be conceived of as contradictory. But they can also reinforce each other, and lead to increased managerialization and a technologized view of democracy.

Notes

1 Exceptions to this are, of course, approaches that focus on the performativity of economic theories (Callon *et al.* 2007), and in concepts such as 'immutable mobiles' the idea is that things remain, to some extent, alike when they travel to new sites (Latour 1986).
2 The *Cookbook* (2009) was made available to the author by the Danish Board of Technology.
3 The *Cookbook*.
4 The *Guide for the Facilitator* (2009) was made available to the author by the Danish Board of Technology.

References

Agreement on Cooperation 2008. Annex No. 1 to Vojtechova, Hana 2009. Evaluation, testing and application of participatory approaches in the Czech Republic application of the RISCOM model in the Czech Republic, Arenas for Risk Governance EC FP6/Euratom.

Akrich, M. 1992. The de-scription of technical objects. In: W. E. Bijker and J. Law (eds) *Shaping Technology/Building Society: Studies in Sociotechnical Change*. Hong Kong and Cambridge, MA: MIT Press, pp205–224.

Andersson, K. and Wene, C.-O. 2008. *The ITA Process™: The Institutionalised Transparency and Accountability Process for Clarity in Policy Making*. Karita Research and Wenergy, www.karita.se/publications (accessed 28 November 2014).

ARGONA. 2006. *Arenas for Risk Governance*. Description of Work. Proposal/Contract FP6-036413, Sixth Framework Programme, Euratom Management of Radioactive Waste.

ARGONA. 2014. www.argonaproject.eu/project_summary.php (accessed 7 July 2014).

Barry, A. 2013. The translation zone: Between Actor-Network Theory and international relations, *Millennium – Journal of International Studies* 41(3): 413–429.

Bogner, A. 2012. The paradox of participation experiments, *Science, Technology & Human Values* 37(5): 506–527.

Braun, K. and Schultz, S. 2010. '... a certain amount of engineering involved': Constructing the public in participatory governance arrangements, *Public Understanding of Science* 19(4): 403–419.

Brown, M. B. 2007. Can technologies represent their publics? *Technology in Society* 29: 327–338.

Brunsson, N. 2006. *Mechanisms of Hope: Maintaining the Dream of the Rational Organization*. Fredriksberg: Copenhagen Business School Press.

Callon, M., Lascoumes, P. and Barthe, Y. 2009. *Acting in an Uncertain World: An Essay on Technical Democracy*. Cambridge, MA: MIT Press.

Callon, M., Millo, Y. and Munesia, F., eds. 2007. *Market Devices*. Oxford: Blackwell Publishing.

Chilvers, J. 2013. Reflexive engagement? Actors, learning and reflexivity in public dialogue on science and technology, *Science Communication* 35: 283–310.

Chilvers, J. and Evans, J. 2009. Understanding networks at the science–policy interface, *Geoforum* 40: 355–362.

CIVISTI. 2011. Collaborative project on Blue Sky Research on Emerging Issues Affecting European S&T, Final Project report, www.civisti.org/files/images/Civisti_Final_Report.pdf (accessed 29 November 2014).

CIVISTI. 2014. *Citizen Visions on Science, Technology and Innovation*, www.civisti.org/files/images/CIVISTI_project_description.pdf (accessed 24 April 2014).

Czarniawska-Joerges, B. and Sevón, G., eds. 1996. *Translating Organizational Change*. Berlin: de Gruyter.

Djelic, M.-L. and Sahlin-Andersson, K. 2006. Transnational governance: Institutional dynamics of regulation. In: M. L. Djelic and K. Sahlin-Andersson (eds) *Transnational Governance: Institutional Dynamics of Regulation*. Cambridge: Cambridge University Press.

Elam, M., Soneryd, L. and Sundqvist, G. 2010. Demonstrating nuclear fuel safety – validating new build: The enduring template of Swedish nuclear waste management, *Journal of Integrative Environmental Sciences* 7(3): 197–210.

European Commission. 2000. *Science, Society and the Citizen in Europe*. Brussels: Commission of the European Communities.

European Commission. 2001. *European Governance: A White Paper*. Brussels: Commission of the European Communities.

European Commission. 2002. *Science and Society: Action Plan*. Luxembourg: Commission of the European Communities.

Felt, U. and Fochler, M. 2010. Machineries for making publics: Inscribing and de-scribing publics in public engagement, *Minerva* 48(3): 219–238.

Felt, U. and Fochler, M. 2011. Slim futures and the fat pill: Civic imaginations of innovation and governance in an engagement setting, *Science as Culture* 20(3): 307–328.

Felt, U. and Wynne, B. 2007. *Science and Governance: Taking European Knowledge Society Seriously.* Luxembourg: Office for Official Publications of the European Communities.

Fries, L. 2009. Framtiden för nyinstitutionalism och ANT Gemensamma frågor och nyinstitutionell colonialism. A future in common? Common questions in neo-institutional and Actor-Network Theory, *Nordiske Organisasjonsstudier* 11(3): 45–61.

Goven, J. 2006. Processes of inclusion, cultures of calculation, structures of power scientific citizenship and the Royal Commission on Genetic Modification, *Science, Technology & Human Values* 31(5): 565–598.

Hagendijk, R. P. 2004. The public understanding of science and public participation in regulated worlds, *Minerva* 42(1): 41–59.

Irwin, A. 2001. Constructing the scientific citizen: Science and democracy in the biosciences, *Public Understanding of Science* 10(1): 1–18.

Irwin, A. 2006. The politics of talk: Coming to terms with the 'new' scientific governance, *Social Studies of Science*, 36: 299–320.

Irwin, A. 2014. From deficit to democracy re-visited, *Public Understanding of Science* 23: 71–76.

Jasanoff, S. and Kim, S.-H. 2009. Containing the atom: Sociotechnical imaginaries and nuclear power in the United States and South Korea, *Minerva* 47: 119–146.

Konopásek, Z., Soneryd, L. and Svačina, K. 2014. *Working Paper: Czech Dialogues by Swedish Design,* InSOTEC, Project co-funded by the European Commission under the Seventh Euratom Framework Programme for Nuclear Research & Training Activities, www.insotec.eu/publications/topical-reports (accessed 4 September 2015).

Latour, B. 1986. Visualisation and cognition: Drawing things together. In: H. Kuklick (ed.) *Knowledge and Society: Studies in the Sociology of Culture Past and Present.* Stamford, CT: JAI Press, pp1–40.

Levidow, L. 1998. Democratizing technology – or technologizing democracy? Regulating agricultural biotechnology in Europe, *Technology in Society* 20(2): 211–226.

Lezaun, J. and Soneryd, L. 2007. Consulting citizens: Technologies of elicitation and the mobility of publics, *Public Understanding of Science* 16(3): 279–297.

Marres, N. 2007. The issues deserve more credit: Pragmatist contributions to the study of public involvement in controversy, *Social Studies of Science* 37(5): 759–780.

McNeil, M. and Haran, J. 2013. Publics of bioscience, *Science as Culture* 22(4): 433–451.

Meyer, J. W. and Jepperson, R. L. 2000. The actors of modern society: The cultural construction of social agency, *Sociological Theory* 18(1): 100–120.

Meyer, J. W. and Rowan, B. 1977. Institutionalized organizations: Formal structure as myth and ceremony. In: W. Powell and P. DiMaggio (eds) *The New Institutionalism in Organizational Analysis.* Chicago, IL: University of Chicago Press, pp41–62.

Rayner, S. 2003. Democracy in the age of assessment: Reflections on the roles of expertise and democracy in public-sector decision making, *Science and Public Policy* 30(3): 163–170.

Røvik, K. A. 2002. The secrets of the winners: Management ideas that flow. In: K. Sahlin-Andersson and L. Engwall (eds) *The Expansion of Management Knowledge: Carriers, Flows, Sources.* Stanford, CA: Stanford University Press, pp113–114.

Sahlin-Andersson, K. and Engwall, L. 2002. *The Expansion of Management Knowledge: Carriers, Flows, and Sources.* Stanford, CA: Stanford University Press.

Soneryd, L. 2007. Deliberations over the unknown, the unsensed and the unsayable? Public protests and the 3G development in Sweden, *Science, Technology & Human Values* 32(3): 287–314.

Soneryd, L. 2008. A traumatising transparency exercise on mobile phones and health. In: C. Garsten and M. Lindh de Montoya (eds) *Transparency in a New Global Order: Unveiling Organizational Visions*. Cheltenham: Edward Elgar, pp223–240.

Soneryd, L., Furusten, S. and Sundström, G. 2010. Democratic values and the organizing of actors in governance structures. In: G. Sundström, L. Soneryd and S. Furusten (eds) *Organizing Democracy: The Construction of Agency in Practice*. Cheltenham, England: Edward Elgar, pp131–146.

Strathern, M. 2006. Bullet-proofing: A tale from the United Kingdom. In: R. Annelise (ed.) *Documents: Artefacts of Modern Knowledge*. Ann Arbor: Michigan University Press, pp181–205.

Sundström, G., Soneryd, L. and Furusten, S., eds. 2010. *Organizing Democracy: The Construction of Agency in Practice*. Cheltenham, England: Edward Elgar.

Welsh, I. and Wynne, B. 2013. Science, scientism and imaginaries of publics in the UK: Passive objects, incipient threats, *Science as Culture* 22(4): 540–566.

Winner, L. 1977. *Autonomous Technology: Technics-out-of-Control as a Theme in Political Thought*. Cambridge, MA: MIT Press.

8
PARTICIPATION AS PLEASURE
Citizenship and science communication

Sarah R. Davies

I want to start with two anecdotes, both drawn from a single day in the city of Phoenix, Arizona. On Saturday, 15 October 2011, in the intense heat and sun of the Arizona autumn, I attended Phoenix Maker Faire – an event which was, as the organizers wrote in a flyer for the event, 'a newfangled fair that brings together science, art, craft, and engineering plus music in a fun, energized, and exciting public forum'. Under the baking sun, hundreds of people wandered round an empty lot that had been filled with fire-breathing robots, 3D printers, biodiesel demonstrations, homemade wearable TVs and open-source cars. Aside from the 40 degree Celsius temperatures, I was reminded of nothing so much as an English village fete, albeit one in which tinkering, playing and asking had been brought to the fore. The aesthetic was one of hot-dog eating leisure and firework-watching delight combined with the intensity of the hobbyist.

The 15th was a busy day in Phoenix. On my way home I stumbled across the tail-end of Occupy Phoenix's inaugural gathering: many hundreds of people, still passionate after a day in the sun, straggling up the street after their first general assembly. With its emphasis on critical discussion, use of hand signals and human microphones to enable open and accessible debate, and a determination to circumvent established modes of representative democracy, the Occupy movement is in many ways a direct democrat's dream (see, for instance, the discussion in Taylor 2011). It speaks to notions of citizenship that are communal, intentional and expressly active: it is open, deliberative and consensus-oriented. Maker Faire's hackers and makers, with their small-scale interventions into technological development and their concern with spectacle for its own sake (fire-breathing robots!), seem to operate in a different world.

It is on the contours of this distinction that I want to reflect within this chapter. Those two encounters, taken together, highlight many of the themes I will discuss: pleasure and leisure; lay interventions into technoscience; and how citizenship can be performed. At the heart of my discussion is the question of whether these events were, indeed, operating in different worlds. Was one about leisure (even consumption) and the other about citizenship? Or can we understand these two modes of being in the world as interlinked?

Science communication, public understanding of science and affect

My starting point is an interest in science communication, by which I understand organized processes that seek to engage lay publics with scientific knowledge, but which generally do not seek to directly inform science policy or the practice of research (Davies *et al.* 2009). This will include, for instance, science festivals, the Café Scientifique movement (and related science café and science-in-the-pub events), the majority of museum galleries, exhibitions and events, sci-art projects, university open days and outreach activities, and citizen science projects (where non-scientists may have an opportunity to assist in producing data sources useful to scientific knowledge, and, more rarely, to help guide its direction and priorities). We should also include mass media (TV, radio, popular science books and magazines) and Web 2.0 science within this category; similarly, we can frame making and hacking activities – such as those showcased at Phoenix's Maker Faire – as science communication (though it perhaps becomes a little less clear exactly who is doing the communicating, and to whom). These activities are not frivolous; that is not the right word, but are definitely predicated upon their audiences and participants *wanting* to engage in them. They are leisure activities, and as such are framed around affects such as pleasure, interest, delight and enjoyment. They are thus organized, and experienced, slightly differently to other, more formal, forms of public participation in science, such as consensus conferences, deliberative workshops or focus groups, or citizen juries.

My discussion focuses on science communication for two reasons. First, exactly because of their focus on the pleasures of science as a means of engaging lay publics: this emphasis on the affective, I will argue, offers us resources as we consider how to 'remake', or at least reimagine, public participation with science. And, second, these practices have been relatively understudied within the science and technology studies (STS) literature, which has tended to focus on forms of engagement with more direct links to scientific governance (see, for instance, the discussion in Stilgoe *et al.* 2014). Exploring science communication in more detail can, I suggest, open up the way in which all participation is situated and emotional, as well as offering us opportunities to enlarge our conceptions of scientific citizenship. I will therefore build upon previous work that has, first, emphasized the role of the affective in everyday life and, second, explored the ways in which particular affective publics are constituted through (discussion of) public engagement with science.

Both of these developments can be understood as rooted within a wider 'affective turn' in social research (Gregg and Seigworth 2010). At its most basic, this has involved a calling of attention to features of social life outside of the discursive, and thus to the role of the material and emotional (Wetherell 2012); stronger versions have posited a pre-discursive, autonomous world of affect and are rooted in a radical return to the body and the biological (Massumi 1995; see discussion in Leys 2011). Work that draws on affect theory has varied from histories of emotions, studies of 'everyday affect' and accounts of the political performativities of particular emotions (see Stewart 2007; Ahmed 2008; White 2009). My discussion here, however, is primarily concerned with highlighting the need for attention to the material and emotional within spaces where publics engage with science. As suggested above, I am particularly interested in the role of the positive affects (such as joy, delight, curiosity and pleasure) within these spaces, and in reflecting upon the capacities that such positive affects may engender. As such I have been influenced by the work of Jane Bennett (2001), who has argued that wonder and enchantment may provoke ethical reflections and actions far beyond the context of their immediate elicitation.

This is a somewhat different direction to existing accounts of the positive affects within public engagement with science (see Wynne 2007a; Kearnes and Wynne 2007; Thorpe and Gregory 2010). These have tended to focus on the way in which public 'confidence' has been constituted as both a necessary outcome of engagement and the (sole) correct way for scientific citizenship to be performed; as Thorpe and Gregory (2010) have argued, public engagement has been 'constructed as a technique for producing the public confidence regarded as essential to the stability of the "innovation system"' (p286). This constitution of confident and enthusiastic publics of science is, of course, in line with previous policy iterations of the science–society relationship and in particular the 1980s drive for 'public understanding of science', with its implicit logic that 'to know science is to love it' (Irwin and Wynne 1996; Turney 1998). The danger of emphasizing the pleasures of science communication, then, is that such processes enable delight without critique. My discussion will therefore seek to tread the line between an assessment of the ways in which affective publics are constituted by themselves and by others: I am interested in the unpredictability of affective flows and rhythms – their disturbing ability to effect change in surprising directions, to overflow categorization, or to transcend immediate contexts and purposes (Wetherell 2012). Who is to say what strange stirrings particular pleasures may elicit – what it is, exactly, that wonder at the sheer size of the universe, interest in the intricacies of synthetic biology or delight in the uncanny behaviour of the nano-hummingbird may ultimately result in?

I use a number of literatures, cases and examples to orient this discussion. In particular, I draw on recent (and not so recent) thinking from political and deliberative theory, and on the activities of the hacking and making movement mentioned at the start of the chapter. While I make reference to different kinds of events and activities present within the contemporary landscape of science communication, and to STS's analysis of these, I also use data from an interview study, carried out in 2012, exploring the way in which those who use hackerspaces around the US talk about those spaces and the practices found within them.[1]

Affective publics, affective practices

I want to start, then, by suggesting that science communication activities offer us a valuable resource for our thinking on public participation because, put briefly, they are about enjoyment. They foreground affective engagement with science and technology. They are fun. As such they draw attention to the fact that participation in and with science is not something solely mediated by discourse, or configured through 'reasoned argument'. Instead, public participation is always situated, embodied and emotional.

Of course, to say that science communication activities are straightforwardly 'fun' is misleading: certainly not all of them are, all of the time (boredom is also an affect). But these practices are largely predicated on the need to entertain and attract. However drily educational their aims, nobody starts to plan a science communication activity by saying: we need this to be really dull. (To quote the title of one article on public engagement: 'Oh Yes, Robots! People Like Robots; the Robot People Should do Something'; Wilkinson et al. 2011.) As such they are designed both to showcase the affective dimensions of science – wonder, interest, excitement, awe, delight, curiosity – and to foster these affects within those participating, whether scientist, communicator or citizen (or all three, in the same person). And their affective make-up is, exactly because they need to engage their publics, largely driven by lay demand. Exhibitions such as *Body Worlds* or *Grossology* respond to a carnivalesque desire for spectacle; the pomp of Royal Society lectures (wine reception included!) or the hands-on playfulness of a science fair similarly meet the needs of their respective audiences. We see this most explicitly in grassroots science such as Maker Faires or hackerspaces, which are entirely user-driven. If people are taking pleasure in opening up science in these spaces, in building and making and deconstructing and learning, that is because they want to, because they desire such activities and experiences. While excitement, interest and confidence in science may be outcomes desired by those who fund public engagement, at least some of these affects are expressed by lay publics as they engage with science on their own terms.

Indeed, hackers and makers explicitly make reference to the very tangible pleasures of the technoscience with which they engage. For one hacker in a Boston hackerspace, for instance, a common feature of the group is that they

> all love making things that are interactive. I actually can't think of any project where that's not the case. It's kind of fun … You start without knowing that much. That's why we're here. So it was very interesting. Okay, here's the problem. How do we go at it … I did want to know how electronics works. It's just kind of this mystery, electronics. You make a circuit, things happen. What's really going on? I guess the whole process, everything about the process kind of intrigued me.

For this hacker, Winni, hacking projects arise at an intersection of pleasurable affects: a love of interactivity, fun, interest and intrigue, and the satisfaction of understanding how something previously mysterious works. Technology is something delightful, rather than a chore or education; more than this, it is inherently emotional – something to be 'loved' – rather than affectless and dry. Winni's emphasis on interactivity and discovery as a form of pleasure (which is echoed by other hackers and makers) is itself something which constitutes the participants of hacking: publics are co-created with hacking's projects and spaces, as those who actively seek knowledge, and who want to use it. This form of enjoyment, in this instance, is thus one that is active, and that relates to the pleasures of self-directed engagement and the acquisition of new skills and understanding. There are certainly other kinds of pleasurable affects present in other forms of science communication – we might compare, for instance, the more passive role of many of those (such as myself) who wandered around Phoenix Maker Faire, where the enjoyment derived from the encounter with other people's projects and skills. Similarly, many participants in less interactive science communication activities, such as public lectures, science cafés or exhibitions, report a desire for information and enjoyment of being able to access and absorb this. Wilkinson et al. (2012), for instance, quote one audience member at a science café-style event as saying that they

> go down the [names venue] reasonably regularly, it's a quite entertaining place to go when I've got a couple of hours to kill midweek, evening, it's basically an excuse to exercise my brain outside of the confines of work … and they've got some decent wine and food there as well.
>
> (Wilkinson et al. 2012, 7)

Here participation is clearly framed as leisure. The science café is understood as an experience at the intersection of several different points of enjoyment, including 'entertainment', the satisfaction of 'killing' time, exercising one's brain and being able to consume 'decent wine and food'. The pleasure here is thus somewhat different to the very active involvement, the alert unpicking of particular technologies, that Winni describes as being part of hacking; what links these accounts, however, is a straightforward delight in finding new things out. It is this kind of pleasure, I would suggest, that marks many of the ways that publics experience science communication. While, of course, it is possible to critique the way in which such processes frame their publics (as, for instance, simple consumers of science–food–entertainment), it is important to note that it is participants themselves who articulate the satisfactions of this framework. The publics of science communication thus understand themselves as hedonistic in their pursuit of, and pleasure in, scientific knowledge: it is something they seek out, and enjoy, in and of itself.

The broader point here, beyond outlining specific instantiations of lay pleasure in technoscience, is that science communication is a useful resource in its highlighting of emotion (and not necessarily only pleasurable ones: the London Science Museum's use of sculptures by Anthony Gormley, science cafés' deliberate presentation of controversial science and bio-art projects that grow and display synthetic meat, amongst many other instances, all aim to disturb and unsettle rather than straightforwardly engender enthusiasm or delight). In contrast, formalized processes of participation such as consensus conferences or deliberative workshops have, in both their practice and the way in which they are analysed and assessed, tended to focus on the fundamentals of the exchange of reasoned discourse (and in doing so have drawn on longstanding traditions in deliberative democracy; Bächtiger et al. 2010). Their situatedness, and the inevitable presence (and importance) of bodies, passions and places, can thus be elided through attention to the information flows, discourse and texts associated with them (see Rowe and Frewer 2005; Horlick-Jones et al. 2007). It can be easy to forget that any form of public engagement will be, to quote Matthew Harvey, dramatic and emotional (Harvey 2009).

This emphasis on scholarship and practice is problematic because, as a number of scholars of deliberation have argued, the emphasis within the theory of participation and deliberation on discourse, reasoned argument and consensus-building can itself be undemocratic (Sanders 1997; Young 2001; see discussion in Bächtiger et al. 2010). The Habermasian model of deliberation, featuring 'flat' power structures, participants at all times open to the best argument and the end point of 'rationally motivated consensus' (Cohen 1989, x), presents not only an ideal that is never realizable (Davies et al. 2006) but which effectively functions to disadvantage those participants who are unfamiliar with or unskilled in the norms of universalistic, disinterested argumentation (Sanders 1997). In the context of science, the use of such deliberative ideals to structure participatory processes such as consensus conferences has tended to ensure that deliberation is, as Elam and Bertilsson (2003) have suggested, a 'democratic politics played out on scientists' home turf' (p242). Public participation has thus, very often, been stripped of affects such as outrage, curiosity or interest in the name of ideals of fairness or in the search for 'pure', non-partisan publics (Lezaun and Soneryd 2007). As Iris Marion Young (2001) has argued, there are therefore good reasons, at times, for abandoning these ideals, circumventing deliberative processes and making use of techniques designed not to enable rational debate but to disturb, disrupt and startle.

The implication of these critiques of the deliberative process is a broadening of our notions of how participation should be organized and carried out. It is not only formalized processes, with invited publics and norms of fairness and disinterest, which are valuable, but other, more messy instances of engagement – those where partisan publics intervene, or where protest and activism insert themselves into decision-making (what Brian Wynne, 2007a, 2007b, has called 'uninvited participation'). Similarly, it is not only calm and objective argument that is appropriate, but any other means of persuasion, including storytelling, anecdote, theatre, polemic or the use of music (see Sandercock 1998; Dryzek 2000). Here, then, we start to see some of the ways in which science communication may offer resources for other forms of public engagement and deliberation. If deliberation should go beyond the discursive – if it should incorporate not just reasoned argument about the technical, and its 'implications', but expression of the emotions and materialities implicated in particular technological presents and futures – then participatory instruments should foreground and normalize emotion, rather than suppressing it. Science communication, with its presentation of the material and emotional aspects of science and its ready stimulation of public affects, may offer a set of tools for doing this.

One example of what this might look like can be drawn from the activities of the NISENet (Nanoscale Informal Science Education Network), an NSF-funded US network of science museums and centres devoted to 'fostering public awareness, engagement, and understanding of nanoscale science, engineering, and technology'.[2] One of the many activities the network funds and supports is the development of forum theatre pieces: loosely inspired by Boal's participatory theatre (Babbage 2004), these ten-minute plays present a scenario involving a potential nano-technological application, aiming to raise questions and stimulate discussion. The piece 'Let's Talk about It',[3] for instance, shows two sisters sorting through their dying mother's belongings; she has cancer, and had been investigating an experimental treatment involving nanoparticles. Though the play includes some rather technical details about the potential of nanotechnology both for medical use and environmental harm, it is also unashamedly emotional, evoking the confusion and desperation of illness. The learning outcomes NISENet suggests for it are diffuse, including examining 'both sides of the argument', but beyond these aims the play can function, I would suggest, as a means of introducing a different dimension to technical discussions. Exactly because it foregrounds the emotional, it can be a disturbing piece to watch. Rather than enabling a bloodless debate based on the potential of a particular technology, the ethics of experimental treatments and patient rights and agency, it acts, or could act, to turn our assumptions about deliberation on its head – forcing the question not of what *we* should do about this technology, but, rather: what would *I* do in this situation?

This is, of course, not to suggest that adding drama, art or activism (see Sandercock 1998; Webster 2005; Wehling 2012) into participatory processes is straightforward. Any engagement process, whether it foregrounds reasoned argument or more emotive interactions, will shape its publics and work to other some people and modalities (Law 2004; Marres and Lezaun 2011). ('Let's Talk about It', for instance, clearly speaks to an audience that prioritizes health above all else, and which is able to understand and access experimental medical treatments, as well as to a world in which cancer, and the disruption of family relations, is the key life-changing threat.) Rather, it is to say that public participation always overflows the theoretical and analytical frames we place upon it. Being overt about its multiplicities, aware of the limitations of any one process or analytical lens and even consciously and consistently reflecting on the different possibilities of any particular moment can enable better understanding of the richness of all of the ways that publics engage with science.

The deliberative society

Science communication, then, may offer us resources for opening up other features outside of the narrowly rational and discursive within public participation in science. My second point is that science communication activities and practices also lead us to reflect on the nature of scientific citizenship, and to reconsider models in which there is a strong distinction between 'civic' activities and those that are private, informal or individualistic (Campbell 2005). In other words – and to return to the anecdote with which I started the chapter – I want to suggest that we might expand our notion of (scientific) citizenship in a way that allows us to understand Maker Faires and consumption of science communication as similarly active, though rather different, democratic practices as the direct action of Occupy. I do this by drawing on recent work in deliberative theory that has argued for a move from a focus on deliberative *processes* to one on deliberative *societies*.[4] This entails an analytical shift from the design and analysis of (ideal-type) deliberative processes to the multiple, partial and often flawed sites in which deliberation actually takes place within systems of public talk about particular issues. Here, then, there has been a move to effectively broaden how deliberation is conceptualized, and to acknowledge its presence in multiple fora.

The background to this discussion is the 'deliberative turn' in political and democratic theory (Chambers 2003), and its translation, over the preceding decades, into experiments with participatory and deliberative 'mini-public' formats. Scholars of political theory have been instrumental in both designing and analysing these – the political theorist Mark E. Warren, for instance, was involved in the British Columbia Citizens' Assembly, one important experiment in participatory democracy (Warren and Pearse 2008) – and the pathologies and opportunities of different forms of mini-publics have by now been extensively discussed (Carpini *et al.* 2004). Recently, however, this *format*-oriented approach (which has been mirrored in STS, with its interest in one-off deliberative processes; see Powell and Colin, 2009, for one discussion of this) has been criticized (Parkinson 2006; Parkinsonand Mansbridge 2012, 176). A number of deliberative democracy theorists have suggested that the discipline's analytical emphasis should move towards deliberative *systems* rather than processes. For example, Mansbridge *et al.* (2012, 1–2) write that

> no single forum, however ideally constituted, could possess deliberative capacity sufficient to legitimate most of the decisions and policies that democracies adopt. To understand the larger goal of deliberation, we suggest that it is necessary to go beyond the study of individual institutions and processes to examine their interaction in the system as a whole ... We thus advocate what may be called a systemic approach to deliberative democracy.

The move here is thus away from a focus on the design of ideally deliberative, perfectly legitimate (in the technical sense; see Parkinson 2006) mini-public formats, or on charting the problems with particular real world attempts at these, but rather to accept that individual fora will always be flawed. Deliberation, democratic engagement and legitimacy are distributed between different fora within a particular 'deliberative ecology'. No one process, space or forum – what Parkinson has called single 'deliberative moments' (2006) – is able to carry the weight of deliberative democracy's promise of open, multi-vocal discussion that leads to more robust outcomes; instead, both capacities and pathologies are spread around the system as a whole. The challenge becomes not designing a perfect process, but understanding the system around a particular issue, analysing its deficiencies, and suggesting ways to make it more robust. As Mansbridge *et al.* (2012, 4) note, a systemic approach can indicate 'where a system might be improved, and recommend institutions or other innovations that could supplement [it]'.

Within this systemic approach[5] to deliberation, citizenship, in the form of democratic engagement, is therefore not limited to traditional representative democracy plus engagement in formalized participatory processes. Instead one participates in issue deliberation within many different fora: the list that Mansbridge *et al.* (2012) provide – which they frame as 'nodes' within an issue system or ecology – includes everything from governments and NGOs to the mass media and informal 'kitchen table' discussion. They are careful, however, not to extend this indefinitely: deliberation, in their understanding of the term, is not any talk, anywhere. Deliberative discussions 'involve matters of common concern and have a practical orientation' (p9). The shared problem focus and decision-orientation that are characteristic of deliberative processes are therefore retained; the difference is that the site for such discussion need not be a government (or other elite)-sponsored, organized or structured process designed 'for' deliberation and participatory decision-making. In their model, then, deliberation is spread throughout societies: the challenge is to understand, and support, the ways in which such nodes relate to and inform one another.

It is important not to overstate the radical nature of these developments. What has been called informal 'political talk' has been valued as part of the workings of robust democracy in many political theory traditions (see Carpini *et al.* 2004; Searing *et al.* 2007); similarly, work on 'civic engagement' has attempted to theorize the value of participation in informal, non-policy-oriented public spaces (Campbell 2005). So it is certainly not new to understand citizenship as a practice that goes beyond government-sponsored activities (whether those that support representative or participatory democracy) to incorporate informal, even private, interactions. However, framing scientific citizenship as something that is not limited to engagement with formal participatory processes – or even uninvited participation – but is spread throughout a society, does have certain implications for how we think about the manifold ways in which publics engage with science. It alerts us to look for the ways in which science communication and other informal means of negotiating science may open up space for 'deliberative moments', on the one hand; on the other, it provides a bridge to my previous discussion of science communication as mundane affective practice.

First, then, Mansbridge *et al.*'s (2012) definition of deliberation – that is, talk that relates to matters of common concern and has a practical orientation – means that we can be generous in our understanding of where deliberative public participation occurs. Such talk will occur not only in spaces of invited (consensus conferences, upstream public engagement) and uninvited (protest, activism) participation, but also in sites where formal participation in science doesn't seem to be happening at all – including within different forms of science communication. To take again the example of nanotechnology, we might consider where such deliberative talk may occur: after a viewing of NISENet's 'Let's Talk about It', certainly; but also on the PBS TV series 'Small Talk'; in museums that have exhibits on nanotechnology; at informal dialogue events on the topic; in online spaces such as the comments on nano-related websites and news articles; at home, after reading Michael Crichton's *Prey*; and, of course, in the many participatory workshops and events organized by social scientists and dialogue practitioners (see Bowman and Hodge 2007). Scientific citizenship thus becomes something rather diffuse – something spread throughout society rather than localized in particular participatory spaces and practices. We can see this in practice if we return to hacking and making. In many ways, hackerspaces are archetypal instances of leisure activities: to use Roberts's (2011) term, they are largely 'inconsequential' in their large-scale societal effects. Hackerspaces do not (for the most part) federate, lobby or campaign. Indeed, the interviews I carried out showed that part of the appeal for many who use US hackerspaces is exactly their 'non-political' nature. At the same time, the imaginations of hacking that are expressed can be readily understood as relating to matters of common concern and having a practical orientation, and thus to Mansbridge *et al.*'s (2012) notion of deliberation. Hackers speak about their activities both as communal and, often, as having ramifications outside of the sphere of the private and personal. Karen, for instance, works at a DIYBio-oriented hackerspace. Coming from a non-scientific background, she was quickly absorbed by both the science and the 'altruistic' possibilities it presented:

> For me, it made me excited about science again in the way that I was excited when I was a kid … Nobody's ever done this [hackerspaces] before and it has the possibility for changing the world in a really positive way. Most of the people that are here honestly want to do something amazing … there really is some sort of altruistic, I'd really like to do something to change the world.

Note, again, that this is an affective relation to hackerspaces and the science they enable: Karen is 'excited' and sees the potential for 'amazing' things to happen. But it is also a relation that frames this excitement in the context of 'altruism' and 'changing the world'. In this context of hacking and making, personal enjoyment – of the kind that Winni described in the extract above – is not just something that is concerned only with individual pleasure and fulfilment; rather, it is construed as something that has wider implications, and which links to larger-scale effects. The science that is carried out, Karen says, has the potential for public good. Its practice is therefore something that relates intimately to questions of how laypeople should relate to science, what kinds of technoscientific development global societies require, and who should control that technology.

This is just one instance of the ways in which scientific citizenship, and deliberation, can be found in diverse locations: other kinds of deliberative moments, with other versions of citizenship, may be articulated in other instances of science communication and public engagement. But it provides a neat link to the second point I want to reflect on here, which is the way in which the identification of scientific citizenship with such material practices may lead us to push the definition of deliberation beyond its typical articulation as a form of talk.

Mansbridge *et al.* (2012) are resolutely talk-focused in their understanding of deliberation: to them, the deliberative society is fundamentally one in which citizens talk to each other, in various spaces and through different means. But it is useful here to return to that substantial literature, discussed above, that has argued for the need to go beyond reasoned argument and discourse, to allow for other means of persuasion (see, for instance, Sanders 1997; Sandercock 1998; Dryzek 2000). These discussions, in political theory and elsewhere, have resulted in an understanding of 'valid' participation in deliberative and participatory processes that includes a wide range of communicative activities: the use of rhetoric, storytelling, protest and disruption, humour and aesthetic, and creative expressions of position or perspective, for instance (Dryzek 2000). Such models of deliberation therefore start to incorporate 'alternative forms of communication', many of which are non-verbal (Bächtiger *et al.* 2010, 33–34). We have seen that science communication and related activities tend to foreground the affective, and that they are *practices* inevitably intertwined with particular material arrangements: might this, then, lead us to extend the notion of deliberation even further? Can we imagine deliberation on science as something that may in fact be embodied and affective, as well as discursive?

Such an understanding of deliberation – and, relatedly, the 'deliberative moments' within a particular ecosystem of participation – would divorce it completely from being a talk-based process. It would take seriously the notion of 'alternative forms of communication' to include all forms of embodied knowledge and meaning-making, and the ways in which these are transferred. Deliberation thus exactly becomes a further way of conceptualizing the material and affective practices around science communication I have discussed above, and scientific citizenship something that includes not just talk about scientific problems and futures, but also the practices that interrogate such issues through their doing and undoing. Again, we might illustrate this possibility with hacking and making, taking the (hypothetical) case of a participant in the same bio-oriented hackerspace in which Karen, quoted above, is involved. This individual stumbles across the space accidentally: they have no particular interest in science, but are intrigued by the excitement of the hackers, the lab aesthetic of the space, and the idea of being able to carry out genetic engineering in your garage. They learn to carry out simple DNA extractions, watching more experienced biohackers and mimicking their movements and tricks. Over time they become familiar with the practice of 'wet' lab science – its equipment, smells and concerns. Certainly they talk about the science they are engaged in, with other hackers and their friends and family, but they also interrogate it through the emotions it induces in them and through its particularized embodiment in and with them. The possibilities of a certain strand of technological development – synthetic biology, say – are thus *deliberated* through our hypothetical hacker's physical engagement with it. They examine the technology through their practice of it, and, further, become better equipped to engage in other moments of deliberation elsewhere – in public debate, for instance, or in informal discussion with those unfamiliar with the technology.

This is, of course, a thought experiment. It is as yet empirically unclear as to how such practices of informal engagement with science intersect with what we might call 'citizenly' capacities and behaviours (Selin *et al.* forthcoming). The broader point is to understand the potential of these kinds of science communication spaces to open up and question scientific knowledge and technological trajectories, and therefore to become instances of deliberation within a wider deliberative ecosystem. Such an ecosystem might then ultimately include anything from taking part in a consensus conference on nanotechnology to learning how to program your own open source software or reading Michael Frayn's play *Copenhagen*. Perhaps, we might speculate that all of these activities, and many others besides, equip us, in different ways, for the challenges our technological societies are likely to keep throwing at us.

Conclusion

I have used this chapter to reflect upon some ways in which the study of science communication – informal activities within which lay people engage with science and technology – may offer resources to think about public participation. In a necessarily abridged account, I have focused on two key lines of thought. I first argued that the affective orientation of much science communication – in which emotions such as pleasure, wonder or delight are foregrounded – offers a valuable corrective to the way in which many organized participatory activities have been stripped of emotional content. Science communication, I suggested, might offer resources to go beyond the discursive (and specifically beyond the emphasis, in traditional models of deliberation, on reasoned argument). In a further set of reflections I drew on recent work on systemic approaches to deliberation to argue that we can imagine deliberation on science, and therefore scientific citizenship, as something that is spread throughout society, and thus present in sites and encounters beyond the categories of invited and uninvited participation. All this has brought us a long way from Phoenix Maker Faire, fire-breathing robots and the Occupy movement. What I hope I have done, however, is to suggest how we might understand these two, seemingly very different, activities as complementary ways of performing citizenship. As I close, I want to reflect on one final, very practical implication. How might this model of scientific citizenship – as affective, embodied and diffuse – affect the way in which we, as STS scholars, work with and around practices of public participation?

Here we might, I think, build on Mansbridge *et al.*'s (2012) call to understand 'the democratic process as a whole, and therefore … the relationships of its parts to the whole' (p26). In the context of science this encourages us to look beyond the design or study of particular participatory moments or processes, and to explore how these may or may not connect to each other. An interest in participation in science might thus translate to an interest in deliberative moments on science, and to the task of tracing and connecting these so as to enhance societal deliberation as a whole. Specifically, it leads us to ask how, within particular issues and topic areas, we might map and make use of deliberative capacity within forgotten spaces of participation. How, in other words, can we connect Phoenix Maker Faire (for instance) to the uninvited participation of the Occupy movement, or the invited participation of a deliberative workshop? If we do indeed take the normative position that part of our role is to support citizen engagement with science, then a systemic approach to deliberation provides us with new resources to do this: we might, for instance, search out and map spaces where deliberative moments occur, assess – as Mansbridge *et al.* (2012) suggest – what pathologies exist within the system as a whole, and develop ways of countering these, for instance, through giving voice to under-represented moments of deliberation, or coupling different kinds of fora. Rather than primarily designing, running or analysing new kinds of participatory devices, we might see our role as being rather more exploratory, seeking out deliberation on particular issues and devising ways to raise the profile of, and connect, these moments and spaces.

In the end, then, we are left with a set of empirical questions. What publics are created, and how is scientific citizenship performed, in different instances of public engagement with science? Where are there 'deliberative moments', whether verbal or embodied, on particular technoscientific issues? Is it possible, and useful, to connect these? I have argued that science communication, from science theatre to bio-art or museum exhibitions, presents potential opportunities for the creation and negotiation of new forms of scientific citizenship, and that these practices suggest ways that we might re-make the imagination and practice of participation. But how does this happen? How do our imaginings of participation mesh or contrast with those of the publics who attend, use, consume or reject such activities? Only further study, and in-depth engagement with the pleasures and problems of science communication, can help us answer these questions.

Notes

1 This study involved visits to 14 different hackerspaces across the US, and a total of 30 interviews with users of the spaces. The interviews included questions about participants' use of hackerspaces, the projects they worked on, and the way in which the hackerspace functioned. More details on this work can be found in Davies (2016).
2 See www.nisenet.org.
3 See www.nisenet.org/catalog/programs/same_sides_lets_talk_about_it.
4 It is perhaps a little late in the day to mention that theories of participation and deliberation are not the same (though I have tended to elide these differences in my discussion thus far). Deliberation refers to a specific mode of interaction: one in which, traditionally, multiple perspectives are represented, where the assumed aim is the best outcome for as many actors as possible, and where rational argument is used to aim for consensus (Chambers 2003). Participation is a democratic form where publics or interest groups participate directly in democratic processes (see Fiorino 1990). Thus a deliberative process need not be participatory (e.g., parliamentary discussion within

representative democracies) and participation need not be deliberative (e.g., referenda). Within STS, however, differences between deliberation and participation have tended to be downplayed: the *participatory* formats that have been promoted and used, such as consensus conferences or citizen juries, have taken for granted that fair, thorough *deliberation* is integral to the process.

5 It is important to note that this is not a *systems* approach: the aim is not to comprehensively map, or indeed plan, a particular system (even if such a thing were possible), or to conceptualize it as something that is fixed, rigid or mechanistic. The ecology metaphor is perhaps more fruitful, as implying a degree of unpredictability and the sense of constant change.

References

Ahmed, S. 2008. Sociable happiness, *Emotion, Space and Society* 1(1): 10–13.
Babbage, F. 2004. *Augusto Boal*. London: Routledge.
Bächtiger, A., Niemeyer, S., Neblo, M., Steenbergen, M. R. and Steiner, J. 2010. Disentangling diversity in deliberative democracy: Competing theories, their blind spots and complementarities, *Journal of Political Philosophy* 18(1): 32–63.
Bennett, J. 2001. *The Enchantment of Modern Life: Attachments, Crossings and Ethics*. Princeton, NJ: Princeton University Press.
Bowman, D. M. and Hodge, G. A. 2007. Nanotechnology and public interest dialogue: Some international observations, *Bulletin of Science, Technology and Society* 27(2): 118–132.
Campbell, K. B. 2005. Theorizing the authentic, *Administration & Society* 36(6): 688–705.
Carpini, M. X. D., Cook, F. L. and Jacobs, L. R. 2004. Public deliberation, discursive participation and citizen engagement: A review of the empirical literature, *Annual Review of Political Science* 7(1): 315–344.
Chambers, S. 2003. Deliberative democratic theory, *Annual Review of Political Science* 6(1): 307–326.
Cohen, J. 1989. Deliberation and democratic legitimacy. In: A. Hamlin and P. Petit (eds) *The Good Polity*. London: Blackwell, pp17–34.
Davies, C., Wetherall, M. and Barnett, E. 2006. *Citizens at the Centre: Deliberative Participation in Healthcare Decisions*. Bristol: Policy Press.
Davies, S. 2016. *All Hackers Now: Hackerspaces and the Rise of the Maker Movement*. San Francisco, CA: Polity Press.
Davies, S., McCallie, E., Simonsson, E., Lehr, J. L. and Duensing, S. 2009. Discussing dialogue: Perspectives on the value of science dialogue events that do not inform policy, *Public Understanding of Science* 18(3): 338–353.
Dryzek, J. S. 2000. *Deliberative Democracy and Beyond*. Oxford: Oxford University Press.
Elam, M. and Bertilsson, M. 2003. Consuming, engaging and confronting science, *European Journal of Social Theory* 6(2): 233–251.
Fiorino, D. J. 1990. Citizen participation and environmental risk: A survey of institutional mechanisms, *Science, Technology & Human Values* 15(2): 226–243.
Harvey, M. 2009. Drama, talk, and emotion: Omitted aspects of public participation, *Science, Technology & Human Values* 34: 139–161.
Horlick-Jones, T., Rowe, G. and Walls, J. 2007. Citizen engagement processes as information systems: The role of knowledge and the concept of translation quality, *Public Understanding of Science* 16: 259–278.
Gregg, M. and Seigworth, G. J. 2010. *The Affect Theory Reader*. Durham, NC: Duke University Press.

Irwin, A. and Wynne, B. 1996. *Misunderstanding Science? The Public Reconstruction of Science and Technology*. Cambridge: Cambridge University Press.

Kearnes, M. and Wynne, B. 2007. On nanotechnology and ambivalence: The politics of enthusiasm, *NanoEthics* 1(2): 131–142.

Law, J. 2004. *After Method: Mess in Social Science Research*. Abingdon: Routledge.

Leys, R. 2011. The turn to affect: A critique, *Critical Inquiry* 37(3): 434–472.

Lezaun, J. and Soneryd, L. 2007. Consulting citizens: Technologies of elicitation and the mobility of publics, *Public Understanding of Science* 16(3): 279–297.

Mansbridge, J., Bohman, J., Chambers, S., Christiano, T., Fung, A., Parkinson, J., Thompson, D. F. and Warren, M. E. 2012. A systemic approach to deliberative democracy. In: J. Parkinson and J. Mansbridge (eds) *Deliberative Systems: Deliberative Democracy at the Large Scale*. Cambridge: Cambridge University Press, pp1–26.

Marres, N. and Lezaun, J. 2011. Materials and devices of the public: An introduction, *Economy and Society* 40(4): 489–509.

Massumi, B. 1995. The autonomy of affect, *Cultural Critique* 31(II): 83–109.

Parkinson, J. 2006. *Deliberating in the Real World: Problems of Legitimacy in Deliberative Democracy*. Oxford and New York: Oxford University Press.

Parkinson, J. and Mansbridge, J., eds. 2012. *Deliberative Systems: Deliberative Democracy at the Large Scale*. Cambridge: Cambridge University Press.

Powell, M. C. and Colin, M. 2009. Participatory paradoxes: Facilitating citizen engagement in science and technology from the top-down?, *Bulletin of Science, Technology & Society* 29(4): 325–342.

Roberts, K. 2011. Leisure: The importance of being inconsequential, *Leisure Studies* 30(1): 5–20.

Rowe, G. and Frewer, L. J. 2005. A typology of public engagement mechanisms, *Science, Technology & Human Values* 30: 251–290.

Sandercock, L. 1998. *Towards Cosmopolis: Planning for Multicultural Cities*. Chichester and New York: J. Wiley.

Sanders, L. M. 1997. Against deliberation, *Political Theory* 25(3): 347–376.

Searing, D. D., Solt, F., Conover, P. J. and Crewe, I. 2007. Public discussion in the deliberative system: Does it make better citizens?, *British Journal of Political Science* 37(4): 587–618.

Selin, C., Campbell Rawling, K., de Ridder-Vignone, K., Sadowski, J., Allende, C. A., Gano, G., Davies, S. R. and Guston, D. forthcoming. *Experiments in Engagement: Designing PEST for Capacity-Building*.

Stewart, K. 2007. *Ordinary Affects*. Durham, NC: Duke University Press.

Stilgoe, J., Lock, S. J. and Wilsdon, J. 2014. Why should we promote public engagement with science? *Public Understanding of Science* 23(1): 4–15.

Taylor, A. 2011. *Occupy! Scenes from Occupied America*. London: Verso Books.

Thorpe, C. and Gregory, J. 2010. Producing the post-Fordist public: The political economy of public engagement with science, *Science as Culture* 19(3): 273–301.

Turney, J. 1998. *To Know Science Is to Love It? Observations from Public Understanding of Science Research*. London: COPUS/Royal Society.

Warren, M. E. and Pearse, H. 2008. *Designing Deliberative Democracy: The British Columbia Citizens' Assembly*. Cambridge: Cambridge University Press.

Webster, S. 2005. Art and science collaborations in the United Kingdom, *Nature Reviews Immunology* 5(12): 965–969.

Wehling, P. 2012. From invited to uninvited participation and back? Rethinking civil society engagement in technology assessment and development, *Poiesis & Praxis* 9(1–2): 43–60.

Wetherell, M. 2012. *Affect and Emotion: A New Social Science Understanding*. London: SAGE.
White, P. 2009. Introduction: The emotional economy of science, *Isis* 100(4): 792–797.
Wilkinson, C., Bultitude, K. and Dawson, E. 2011. 'Oh yes, robots! People like robots; the robot people should do something': Perspectives and prospects in public engagement with robotics, *Science Communication* 33(3): 367–397.
Wilkinson, C., Dawson, E. and Bultitude, K. 2012. 'Younger people have like more of an imagination, no offence': Participant perspectives on public engagement, *International Journal of Science Education, Part B: Communication and Public Engagement* 2(1): 43–61.
Wynne, B. 2007a. Public engagement as a means of restoring public trust in science: Hitting the notes, but missing the music? *Community Genetics* 9(3): 211–220.
Wynne, B. 2007b. Public participation in science and technology: Performing and obscuring a political conceptual category mistake, *East Asian Science, Technology and Society: An International Journal* 1(1): 99–110.
Young, I. M. 2001. Activist challenges to deliberative democracy, *Political Theory* 29(5): 670–690.

9

THE TEMPORAL CHOREOGRAPHIES OF PARTICIPATION

Thinking innovation and society from a time-sensitive perspective

Ulrike Felt

Time is an essential feature of social life that not only enables us to structure and order our worlds but also to create and sustain the feeling of stability and belonging (e.g., Edensor 2006). Memory, anticipation, rituals, rhythms and tempo are but a few of the many ways in which time materializes (Adam 1998, 202). However, even though time is deeply entangled with questions of control and power, it tends to be all too easily naturalized and turned into the 'deep structure of taken-for-granted, unquestioned assumptions' (Adam 2003, 60). This definitely holds true with respect to performing and analysing public engagement with technoscientific issues. The goal of this chapter is thus to bring time to the forefront of debates on participation. Following the general line of questions spelled out in Chapter 1 of this book, the analysis will explore the multiple invisible temporal textures as well as the temporal choreographies[1] (i.e., the entanglements of different temporalities) of participatory practices, pointing at how time structures, moulds and guides any engagement with science and technology. In doing so, temporality is addressed on two interconnected levels: investigating, first, temporal structures of participatory exercises as such, thereby reflecting on the role that time plays in how people (can) gather around a public issue related to technoscientific developments; and, second, the temporalities embedded in contexts, objects or matters of concern (Latour 2005) that are identified and addressed in such participatory exercises.

At the cross-roads of innovation and participation discourses

In grappling with the intricate ways in which time and participation are interwoven, we find ourselves amidst two simultaneous developments. On the one hand, we observe the rising importance attributed to innovation in the development of contemporary societies, manifested in discourses on speed, pressure and promising directions to follow. On the other, we witness the emergence of policy discourses stressing the need to be more inclusive towards society when it comes to making technoscientific choices. The headline 'Europe in a changing world – Inclusive, innovative and reflective societies'[2] that can be found on the *Horizon 2020* webpage is but one of many examples marking the proliferating discourses on the entanglement between innovation and public engagement in the European Union.

Indeed, in European policy discourse, innovation is promoted more vigorously than ever 'as a way out of crisis and as a foundation for future prosperity' (Felt *et al.* 2013, 3). Europe's future is perceived as depending on its power to innovate combined with its capacity to create an innovation-friendly climate (EC 2013). Citizens are constantly reminded of the competitive pressure faced by Europe in the global race to innovate, and the speed of delivering innovations has become a major concern for policy-makers. A growth-focused mind-set has gained significant ground, with policies fostering specific, strategically selected innovation trajectories. 'Act now, before it's too late' has become a key slogan when imagining and performing European futures. The core of the European innovation narrative thus gravitates towards issues of tempo and timing, roadmaps, milestones and trajectories, windows of opportunity and futures to be enabled. A powerful European sociotechnical future is envisioned that will depend on both an ever-increasing flow of technoscientific innovations and a 'European public' supporting them (Felt 2010). This vision fits well with the broader diagnoses that we live in a time characterized by a 'breathless futurology' (Harrington *et al.* 2006) embedded in an 'economy of technoscientific promises' (Felt *et al.* 2007); a time when 'standing still means falling behind', when 'acceleration becomes an economic imperative' (Adam and Groves 2007), and when multiple 'anticipatory regimes' (Adams *et al.* 2009) are put in place to ensure the realization of the not-yet. In these debates on remaking Europe through innovation, we perceive a multitude of temporal orders at work that appear to be largely taken for granted.

Simultaneously, bringing societal actors on board to support this innovation-driven European future becomes a core concern (Irwin 2006). The reference to the creation of *inclusive and reflective societies* in the above-mentioned headline is meant to capture this preoccupation. A number of highly visible public controversies around technoscientific issues have sparked debates on the limits of classical forms of democracy and have triggered efforts to open new spaces to better accommodate the values and visions of broader sets of societal actors. Public participation has thus come to be perceived as an essential remedy against an allegedly missing public trust in science and technology (Wynne 2006) and against an insufficiently sustained innovation-friendly climate (Felt *et al.* 2007). A flurry of participatory activities varying in format, intensity and goal have been designed, tested and assessed, ranging from public consultation exercises and debates over citizen panels and consensus conferences to longer-term engagements between researchers and specific publics, as is the case for some patients' associations or civil society organizations. The notion of the 'participatory or deliberative turn' attempts to capture this mood of addressing new and potentially contested technoscientific developments in a more inclusive manner, thereby reinvigorating public debate and creating space for a more active citizenship.

However, this enthusiasm to 'democratize democracy' has been tempered by analysts pointing out the limitations and pitfalls in the execution of participatory ideals. Critics, among others, underline that participatory exercises have often (re)performed the classical deficit model of science and technology communication under a new guise (Wilsdon et al. 2005). They highlight the frequently quite narrow problem framings that characterise the ways that broader societal issues are addressed in the deliberation process (Irwin 2006; Stirling 2008). Analysts reflect how well-delimited sets of publics are created through these (often experimental) exercises, while others are marginalized or silenced (Wynne 2007; Braun and Schultz 2010; Felt and Fochler 2010). When comparing national contexts or different formats of participation within any single context, authors underline the need to more carefully consider the situatedness of such exercises and the difficulty of their standardization in the form of best practices. The latter would carry the risk of reducing participation to an exercise done 'by the book' and of developing a quasi-ritual character (Felt et al. 2013), thus pre-empting its creative potential. Others again point at the problem that the consensus-oriented nature of participation can limit the space for dissenting opinions (Horst and Irwin 2009) and thus silence potentially valuable minority positions.

Beyond repair work: it's time for a time-sensitive perspective

While these critical analyses emphasise the range of specific weaknesses of participatory exercises and trigger 'repair work' through redesigning participatory formats and developing new ones, less attention has been paid to more pervasive, often tacit structural features such as the prevailing temporal orders. In addressing these latter aspects, this chapter will ask how participatory practices are shaped by the ways in which time is scripted in innovation as well as how temporalities matter in the formation of publics, in the framing of issues, in the ways responsibility gets addressed and in multiple other aspects.

Time has previously been a concern in this context. Debates centring around upstream/downstream participation (Wilsdon et al. 2005) and thus around the best moment in an imagined and imaginary innovation trajectory when engagement should take place are one way of addressing time. Societal voices should be heard at a moment when the 'future direction of technological development is not yet established; the social and ethical impacts of [innovations] are uncertain; and public attitudes ... are not yet fixed' (Doubleday 2007, 60). Situating participation too far downstream was seen as leading to framing the issues at stake mainly in terms of the potential risks of applications while failing to address broader issues of societal choices. Other approaches have investigated the 'temporal coding within developmental discourses' that attend to emerging technological domains, showing how these 'are caught and constrained by ideas about expectations, good timing and opportune times' (Selin 2006, 122). There is also an extensive body of literature on projection work, futuring and expectations with respect to new technologies (e.g., Brown *et al.* 2000; Brown and Michael 2003). Furthermore, we can point at research converging around the notion of 'anticipatory governance' (Barben *et al.* 2008) being concerned with the 'rise of assessment regimes' (Kaiser *et al.* 2010) more broadly speaking or discussing the nature of 'foresight knowledge' (von Schomberg *et al.* 2005). What connects all these latter approaches is the attempt to develop anticipatory and more adaptive forms of governance as well as to connect public participation explicitly with the realm of making technoscientific futures. However, while these approaches triggered a reconsideration of participation, they neither led to a cross-cutting investigation of how these different temporalities frame participatory processes involving technoscientific issues nor to a systematic reflection regarding the combined effects of different temporal dimensions on technoscientific developments, democracy, citizenship and participation.

In performing such an in-depth time-sensitive analysis of participatory exercises, it will be essential to attend to the multiple (often tacit) temporal practices and imaginaries of organizers, participants and societal actors alike. Such an approach enables us to carefully reconsider how 'time horizons and time structures are constitutive for action orientation and self-relations' as well as how 'temporal structures form the central site for the coordination and integration of individual life plans and "systematic" requirements' (Rosa 2013, 5). Studying participation through the lens of temporalities will thus enable us to move beyond the macro/micro divide as well as beyond a choice in focus between structure and individual practices and to direct our attention to interaction between different scales.

The following analysis is inspired by Barbara Adam's (1998) concept of timescapes, which highlights the intertwined character of physical time, cultural time and more personal perceptions of time. Investigating the dynamic nature of timescapes, and not time as such, means taking a broader perspective, addressing complexities, giving space not only to physical notions of time but also being sensitive to the multiple, deeply culturally rooted time-related practices of recalling, projecting, anticipating, experiencing and imagining technoscientific and societal developments. Uncovering the co-presence and interplay of heterogeneous forms of time will allow us to examine limitations and frictions that occur and identify how they shape the participatory potential of any setting.

The observations and arguments presented in this chapter are based on extensive fieldwork conducted over the past decade. The material covers transcripts and field notes from a broad range of public engagement exercises with technoscientific issues primarily in the Austrian context, the majority of which have been conducted by the author and her colleagues.[3] These events range from a large number of focus groups, to specifically designed, card-based discussion workshops, to long-term round-table discussions bringing citizens and scientists together, to a citizen conference and to open debate methods called 'discourse days'. The topics covered were related to different aspects of nanotechnology, the life sciences, biomedicine and health. To contextualize these participatory events and to grasp their temporal framing, policy documents were studied. These two sets of data form the basis for an in-depth reflection on the temporalities present in different formats of engagement, how different times are entangled with each other and how they shape participation. Time will be addressed through four perspectives: clock-time, trajectorism, emplacement of time, as well as multiplicities and inconsistencies of time. The conclusion will then highlight the time-related ontological politics at work, the need for more care-oriented approaches to open up possibilities for and within participatory exercises, the ways in which temporal choreographies frame responsibility and, finally, the interrelatedness of time and citizenship (and thus of democracy).

Clock-time and the illusion of control and order

The first temporal perspective is centred on the theme of clock-time as a major ordering force in contemporary societies. Physical time appears to provide 'the external framework within which actions are planned and executed. It is a time that operates independent of human actions, an objective parameter that allows us to locate actions in a temporal grid and consider questions of timing and speed' (Adam 1998, 32). When approached from this angle, clock-time has, as Adam (2003, 63) convincingly argues, become 'decontextualized ... invariant, quantifiable and external'. It is a powerful filter 'through which reality is sieved and [a] lens through which all social relations and structures are refracted' (p64). In addition, it 'obscure[s] other forms of time, banishing them as unproductive and irrational' (Hassan 2009, 41).

Social theorists and science and technology studies (STS) scholars alike (e.g., Elias 1988; Latour 1993) call for an awareness that clocks should not be perceived straightforwardly as instruments that measure time independently of humans and our actions. Instead, clock-time is itself an instrument intended to provide orientation and regulation in our lives and to order actions. Thus, we are confronted with 'the solid facticity of time' while knowing about its social nature (Rosa 2013, 5): clock-time is man-made as much as any other temporal structure under which we live. At the same time, as Manuel Castells (2010) has clearly noted, the power of clock-time as a physical entity is omnipresent: it invites us to believe that we can squeeze an ever-increasing amount of activity into the same time unit. Or, to say it in Jeremy Rifkin's (1987, 3–4) terms, 'the idea of saving and compressing time has been stamped into the psyche of Western civilization and now much of the world'. This perspective supports the predominant mind-set of efficiency, which is highly valued in contemporary industrialized societies as a clear marker of successful, competitive behaviour. Consequently, specific modes of ordering society as well as regimes of monitoring and control are introduced to support and stabilize this ideal. With respect to conceptualizing technoscientific innovations, embracing a physical understanding of time also enables thinking in terms of an acceleration of flows to produce more innovations in ever-shorter amounts of time. The powerful figure of the global race thus perfectly fits with such a framing of time.

In participatory exercises, clock-time becomes apparent in multiple ways. To begin, what is regarded as an adequate duration and temporal structure of participatory events obviously impacts the possible ways in which matters of concern take form and are debated. This perception shapes what types of scenarios are elaborated and tested and whether and how the right to take time for deliberation can be exercised. This facet became clear when comparing discussions from a round-table event on genome research that lasted seven full days over a period of nine months with one-time, two-hour focus groups on similar topics. People participating in the latter generally tended to take less ownership over how an issue was exactly framed, were more pragmatic in their positioning work and showed a higher readiness to be guided by a prescribed format. Participants in the long-term engagement exercise particularly stressed the importance of disposing over enough time to speak not only about research itself but also to explore how one could make sense of that research more broadly speaking.

In this context, time is framed as an essential resource. In interviews made with participants in a long-term discussion reflecting on their experiences, they pondered the issue of 'time scarcity' hindering (their) engagement and the fact that any capacity to participate depends on the power and control one has over one's own time (Nowotny 1994). They further stressed that it was 'the time to really speak to scientists at length' that was essential for the quality of the engagement. In this vein, participants sometimes also speak of 'donating' their leisure time to engagements that serve the greater social good, framing this act as their contribution to society's necessary concern for how technoscientific and societal development relate. Yet, 'the proper amount of time' that was necessary to adequately address complex issues always remained undecided (Flaherty 2010); thus, which temporal structure would make any participatory event sufficiently robust to withstand public scrutiny also remained open.

While taking time to carefully explore the technoscientific issues at stake mostly holds a positive connotation for participants, policy-makers and also scientists, in part, framed this as (irresponsible) temporal luxury in situations of pressing public choices: taking time could potentially hinder scientific developments and endanger the country's/Europe's place in the innovation race. A broad opening-up of the issues at stake or deviations from what is regarded as the core topic thus run the danger of being classified, even by some of the participants, as 'losing time' (Callon et al. 2009). In a context of tight schedules and the feeling of external pressure to act immediately, the idea of participation can thus also be constructed as a 'waste of time' because policy decisions will have to be made long before any decent assessment can be rendered.

Temporality of participation is also essential with respect to asking whether and how participants constitute themselves as a collective or whether they prefer the role of affected individuals, citizens or consumers. The shorter the timeframe of any engagement exercise, the less people can conceive of their potential capacity to form a thought collective, to develop a group identity or to experiment with different modes of valuing issues at stake. This in turn affects how issues can be constructed and addressed. Thus, time to make an issue and time to become a specific collective (or to form a public) must be perceived as closely intertwined. This observation is in line with Noortje Marres's (2005) argument on the entanglement of issues and publics, underlining that publics are not simply 'sparked into being' but instead are always coproduced with issues – and both are related to the temporalities at work.

Finally, clock-time is also omnipresent when people reflect on the pace of technoscientific developments and how this impacts their participatory capacity. Technoscientific developments are generally described as extremely rapid, with a tendency 'to overwhelm all of us', as one participant would express it. Accordingly, researchers and industrial players were often pictured as 'quick and versatile' and ready to 'jump on any opportunity' offered by technoscientific advances, whereas regulators, policy-makers and publics were perceived as inherently slower, as lagging behind. These differences in pace made it difficult for participants to clearly identify where, when and how they could intervene in technoscientific developments, which in turn triggered reflections on how one could adequately address issues of responsibility. In addition, these reflections nourished the perception of policy-makers not being able to adequately respond to these rapid developments and promissory pressures of progress (e.g., Beynon-Jones and Brown 2011).

In this context, and embracing the broader policy narrative of competition and speed, participants would also explicitly ponder the fact that, whereas Austria could refrain from embracing certain technoscientific developments, 'the Germans or the Swiss, or I don't know who ... would do it'. In the end, one participant would explain, 'If we are faster, we do it [laughter], it's our chance', and this would at least be profitable. This attitude clearly shows the tacit assumption of the unavoidability of knowledge-related developments and points to the limitations of any participatory exercise in a globalized knowledge society.

Participation and the ambivalent love for trajactorism

An omnipresent trajectorial narrative of sociotechnical developments is the second temporal perspective to be investigated. Indeed, both within and around participatory exercises, we encounter multiple narratives on historical cases of successful innovation trajectories and on how societies gradually overcome natural limitations and impediments through technoscientific innovations. This manner of perceiving time resonates with Appadurai's (2012, 26) 'trajectorism', which he describes as

> a deeper epistemological and ontological habit, which always assumes that there is a cumulative journey from here to there, more exactly from now to then, in human affairs ... Trajectorism is the idea that time's arrow inevitably has a telos, and in that telos are to be found all the significant patterns of change, process and history.

Thus, time becomes aligned in a specific manner when constituting a phenomenon or an artefact, and specific causal connections are enacted along with this mode of ordering. Technoscientific developments and their intertwinement with societal developments are thus understood as coherently developing and are conceptualized as at least somewhat predictable phenomena that can be analysed and eventually managed accordingly.

How is trajectorism mobilized and how does it gather force in the practice of participation?

In much of the talk in participatory exercises and policy making, we encounter the persistent assumption that innovations follow a trajectorial development, 'starting from basic research, moving to applied research and then to product development' (Felt et al. 2013). This perception is attractive because it suggests clear causal relationships between input and output. It allows for the illusion that innovation flows can be quantified, which points to 'the significance of new regimes of measurement' (Espeland and Stevens 2008) at work. Problems with innovation flows can then be attributed either to a problem of knowledge transfer from basic research to more application-oriented environments or to the absence of an innovation-friendly societal climate.

Trajectorism is, however, also palpable inside the broader discourse of upstream engagement, as the very 'term "upstream" already displays the deterministic connotation of a necessary direction of flow' (Stirling 2008, 264). It tacitly reinforces the very idea of a linear innovation trajectory and supports the illusion that no further accompanying reflection is needed once a direction is chosen and the question of potential risks has been clarified (Editorial 2007). Expressing their concerns, participants highlighted the potential danger of having to advance along a single predefined trajectory with little possibility of escape. A participant would voice his concerns, mimicking an industrial actor who just realized that the development is going in the wrong direction. 'However, instead of stopping, reflecting, maybe even reversing,' the participant continues to argue, a new technology is developed, so that one 'can continue this dangerous pathway'. He thus describes a blind forward movement that closes down other potential directions: 'I don't know where to, but I advance.'

The idea of the innovation trajectory also matters when it comes to issues of responsibility. Participants strongly adhering to a trajectory model often conceptualize knowledge production as clearly separable and separated from the diffusion or application of knowledge. In such a model, basic research follows a solely internal rationale and is determined by what nature allows scientists to see, whereas values only appear when applications are envisaged and produced. A citizen participating in round-table discussions about the genomics of fat metabolism outlined this vision in a rather straightforward manner, stressing that 'findings are actually already there in some way, aren't they? I mean, they somehow all float around, and [the scientists] just discover them.' As a consequence, scientists 'probably could not blame themselves for doing something ... particularly negative; because it's there, anyway, and they just discover it'. Thinking that knowledge is always already there and only needs to be discovered thus indirectly exempts basic research from engaging with societal values. The ethical, social or legal aspects of innovation would only need to be addressed later, once knowledge is transformed into applications. This in turn allows the conceptualizing of knowledge as inherently apolitical and as only becoming political in a specific application context.

When being asked to deliberate on technoscientific choices, participants first aim at clarifying at which point in time on an innovation trajectory they perceive the development. To do so, they either work with analogies of past techno-trajectories or they construct fictitious products as an outcome of knowledge production that can then be assessed. In the Austrian nano-debates, participants used analogies to genetically modified organisms (GMOs) or nuclear energy to imagine a development in this emerging field, even though they were aware that these innovations would substantially differ (Felt 2015). In the case of engagement with research on fat metabolism, we found participants constructing the fiction of 'a fat pill' as an endpoint of the research trajectory, which in turn enabled them to build and assess scenarios of how research might impact society and how society could potentially deliberate on future research directions (Felt and Fochler 2011). Along with this reasoning, however, participants using the application scenario also transformed their frame of assessment from one focused on innovation governance to one that was more concerned with risk governance.

Even though some people explicitly wanted to avoid buying into the trajectorial mode of argumentation and referred more often to network-like innovation models, when they wanted to make a strong statement and wished to identify a moment where responsibility considerations should begin, they tended to switch to the linear model (Felt and Fochler 2011). Only in this modus did they feel capable of arguing for responsible innovation and were able to point at causal orders, which gave them the power to make claims. Thus, despite some feeling of ambivalence towards this rather simplified vision of innovation, it was the clear temporal order of the linear model that often made it attractive as an argumentative resource.

Finally, when examining participatory exercises and the policy-making around them, trajectorialism is also present in how participation events were understood and advocated. In many ways, Austria has constructed itself as lagging behind international developments with respect to public engagement. In that sense, the development of 'technologies of participation' is performed as a 'social innovation trajectory'. Some countries – those who are more advanced on the trajectory (e.g., the UK or, in the case of the consensus conference, Denmark) – become the leaders, whereas others are either set up or set themselves up as the followers/as in need of catching up (Felt et al. 2013). However, this attitude is not without consequences. In the context of the Austrian citizen conference, to take one example, we observed the organizers stressing the extent to which this type of engagement is new for Austria but had already been successfully performed in many other national contexts (Felt and Fochler 2010). The unintended consequences of such a framing was that participants in the exercise were more concerned about complying with this progress narrative of participation and with fulfilling the implicit challenge addressed to them: they should be ready to be good citizens (Michael 2009), ready to perform the expected engagement and to comply with the predefined format and its questions.

Putting time into place: relating situated pasts and futures

Combining the observations on clock-time with those on trajectorism clearly points to the strong impact of specific temporal structures on both how we see the world and how we imagine its development. It is this perception of time that enables us to imagine that we can – also through performing participatory exercises – 'colonize the future' (Giddens 1999) and to conceptualize it as open 'to exploration and exploitation, calculation and control' (Adam and Groves 2007, 2). This temporal ordering invites us to prioritize further thinking in terms of speed, acceleration, frequency and efficiency; it fosters highlighting time scarcity, wastes of time and the need to act immediately 'before it is too late'. This ordering highlights the political role played by time in debates and justifications of technoscientific and societal choices, in the formation and proclamation of urgent problems, and in requests for citizens' compliance with certain decisions.

Building on this analysis, we will now move on to reflect how senses of time, whether shared, contradictory or even incompatible, are both transgressive and increasingly global while simultaneously being deeply entangled with specific locations and their history. Although a substantial amount of the future-making observable at the science policy level appears to be preoccupied with making translocal claims and inscribes itself in the flow of European and international policy discourse, empirical observations of participants' debates show clear traces of how projections of technoscientific futures always carry traces of specific places, and thus of specific pasts and perceptions of tradition, that are used to construct futures. Laura Watts's (2008, 196–197) argument that 'different places – their temporality, topography, sociality and sensory experience – may lead to very different everyday practices, and therefore the creation of very different futures', should thus guide our reflections. Accordingly, the future as well as future-making practices are to be understood as situated (Suchman *et al.* 2008).

Tensions thus become palpable between a strongly emplaced vision of participation and the idea that spaces of governance and citizenship have expanded well beyond the nation state (e.g., expressed through notions such as multi-sited governance) (Barry 2001; Ellison 2013; Felt *et al.* 2013). Tensions are also apparent between the radical novelty of innovations and more culturally entrenched visions of technoscientific development. In what follows, I propose to reflect on how the situatedness and a certain continuity of our senses of time matters when engaging with technoscience. In doing so, I will specifically investigate the role of place in how past, present and future become related to each other and in how 'creatures of the future tense' (Selin 2008) can materialize and be dealt with in the framework of deliberations. This approach means moving our attention from the kinds of futures that are imagined, told and traded in participatory exercises to the work of relating temporal developments to a specific place, to collective memory practices and the capacities of imagination, to how future-oriented agency and relevant actors are mapped out, and to varying understandings of responsibility and its (non-) distributedness (Adam and Groves 2011).

Although the notion of 'future' proliferates not only in science policy discourses but also in various public arenas, the concrete conceptualizations of 'future' by participants in engagement exercises often remain vague and multiple. 'The future' is sometimes conceptualized as a specific event that is supposed (or not supposed) to happen in a specific span of time; at other moments, for example when people ponder generational justice, the future refers to a present situation that should be stabilized; then again, it appears as an attribute that people or a society can claim – they are 'future-able' ('*zukunftsfähig*'). Sometimes, it is described as a repetition of past futures under a new guise, for example when people make analogous references to past developments. Or when participants engage with 'retrospecting prospects' (Brown and Michael 2003) – that is, use their recall of locally rooted past futures to either demonstrate the limited anticipatory capacity of certain actors or to argue for their own legitimacy in an attempt to challenge specific dominant assumptions. In particular, in policy documents, the future is often described as commodifiable and empty, waiting to be realized by producing the right blend of sociotechnical innovations and thus dependent on society's belief in the capacity of foresight and prediction (Adam and Groves 2007). Finally, sometimes the future is conceptualized by policy-makers and scientists alike as simply 'waiting for us' out there, with every society's task being to arrive faster than others and to take advantage of that lead. Participation can and does take different forms depending on which conceptualizations of the future are considered to be important.

Connecting time and place further enables us to devote more attention not only to the always-new ways in which global and local technoscientific developments relate to each other but also to how identities, whether national or regional (e.g., European), are reconfigured through the temporal. Here, observations in diverse forms of participatory settings support Tim Edensor's (2006, 526) assertion that nations do not disappear through the predominance of global flows but that these 'global processes are accommodated and domesticated in the mundane spaces and rhythms of everyday national temporalities' – participatory exercises being one such space.

How place – in this case the Austrian context – matters when imagining futures can be clearly traced in participatory exercises. When comparing focus group discussions on biomedical technologies in different European countries, Felt and co-authors (2010) showed how broader national techno-political cultures frame participation. Using this notion implies nation-specific differences in the ways in which technologies are inscribed into and give shape to society, with time being an important element in such a process. Participation thus builds on the shared imagination of a national developmental trajectory, on the perceived place in the global innovation race, on recognized socio-political structures and practices and their inherent temporalities, as well as on orders of worth at work that seem specific to a particular place (Boltanski and Thevenot 2006). Thus, the narrative of being a nation that lags behind both with regard to the innovation flow but also with respect to engagement exercises shapes how participants imagine their agency. A lack of shared imagination on concrete past innovation futures that have become successful presents might also frame the potential futures to be constructed and assessed. Or, the idea that some technological innovations might disrupt specific arrangements, such as a nation's privileged relationship to nature, might also have a strong impact upon how engagement can evolve.

To capture this local framing, the concept of sociotechnical imaginaries as developed by Sheila Jasanoff (2015, 4) is helpful. She defines sociotechnical imaginaries as 'collectively held, institutionally stabilized, and publicly performed visions of desirable futures, animated by shared understandings of forms of social life and social order attainable through, and supportive of, advances in science and technology'. Such imaginaries encode not only what can or cannot be attained through science and technology but also what a good life is and how it ought to be lived. They are, as has been shown in the case of participatory exercises on nanotechnology-related issues in the Austrian context (Felt 2015, 104), an essential resource for how people connect pasts and futures in locally adapted manners. Creating a feeling of solidarity and cohesion, these imaginaries offer a shared frame of reference to the past; they are part of an 'invention of tradition' (Hobsbawm and Ranger 1983, 2): a tradition that enables, first, judging which innovations are worth addressing in some detail as they are understood to impact societal values; and, second, using past experiences with innovations to project or challenge specific futures. 'Tradition' is in this context not a stable, unchanged set of practices and values but something that is 'dynamic, contested ... and ... continually reinvented in the present' (Edensor 2006, 526).

Participants in anticipatory activities thus try to link up with and follow a set of time-related practices that perform values and norms assumed as shared..Being able to refer to such traditional – and thus legitimate – ways of connecting pasts and futures then offers a feeling of a stable point from which to make assessments; this is attractive to participants in a sociotechnical world that they otherwise describe as rapidly changing. In the Austrian case, the long history of rejecting both nuclear energy and GMOs enables participants to construct a sociotechnical imaginary of the absent – that is, to believe in the nation's capacity of 'keeping a set of technologies out of the national territory and becoming distinctive as a nation precisely by refusing to embrace them' (Felt 2015). They thus feel empowered to choose a local sociotechnical direction different from those of more powerful neighbours.

From these observations, the emplacement of time becomes visible, and it seems essential to acknowledge the situatedness of temporal imaginaries and rituals. Futures, with all of the norms and values embedded in them, may always differ across fields and nation states.

Multiplicities of times and 'temporal inconsistencies'

This last perspective on time and participation focuses on the coexistence of multiple forms of times and some of the resulting tensions or 'temporal inconsistencies' (Giesen 2004) that abound in participants' reflections when trying to anticipate sociotechnical developments. In doing so, inconsistencies are not regarded as the exception but much rather as the normal state of things. Concretely, this analysis is interested in four specific forms of inconsistencies because they seemed to matter most in participatory practice: non-contemporaneity, simultaneity, asynchronicity and divided memories (Giesen 2004).

The first inconsistency refers to the fact that both citizens and policy-makers alike classify certain ways of reasoning as no longer fitting within contemporary cultural frames and as inadequate for a time in which technoscientific progress is perceived as the pacemaker. In this vein, rejecting or questioning certain innovations is classified by some as backward-orientedness, as technophobia (a notion frequently used by Austrian policy-makers) or as holding on to a 'Stone-Age mentality', as one participant expressed it. Thus, the argument of being or not being 'in tune with time' or 'keeping up with time' becomes a powerful argumentative element advantaging some viewpoints over others or even some participants over others, thus opening up or closing down opportunities for deliberation.

The second inconsistency refers to people's narratives about the challenging feeling of 'simultaneity' (Nowotny 1994): too many technoscientific changes are perceived as happening in different places 'at the same time', yet evolving at different speeds and following different rhythms. This perception in turn adds to participants' uncertainty about how their deliberation could potentially fit into this overall temporal choreography of sociotechnical development. Certain technoscientific fields in particular (such as genomics or nanotechnology) are pictured as evolving at a breath-taking speed, and citizens thus report their feeling of confusion. They are uncertain regarding both what is actually happening *right now* in the field as well as about the constellation of relevant actors shaping the technoscientific development at any point in time. In short, they bemoan the spatiotemporal fluidity of the situation that tends to escape scrutiny whenever they attempt to pin it down. These circumstances seem to render participatory governance almost impossible. One participant captured his feelings with the metaphor of a 'machine in motion', one 'in which there are an incredible number of gears in motion. To stop that again is difficult or impossible.'

The third inconsistency is visible in participants' expressions of a feeling of synchronicity or asynchronicity. The idea that people around the table share the idea that certain temporal routines, including the pace and rhythms of technoscientific developments and institutional responses, are adequate and at least acceptable actually contributes to creating a feeling of belonging to an imagined community (e.g., Edensor 2006). At the beginning of a participatory event, citizens tended to assume that there was a shared tacit understanding that there is a good timing or correct speed of development that fits with a given society. Upon diving into the engagement exercise, however, new sets of complex interferences between different temporalities became palpable, and the feeling of a deep asynchronicity emerged. Frequently, that feeling was referred to as a lack of coordination, with some actors pushing innovation too fast when other sectors of society are not ready to follow. We encountered complaints about relatively slow, delayed or retarded responses of some parts of society compared with rapidly changing innovation systems or markets. And, we heard arguments that what is regarded as a 'good life' would not necessarily fit with the speed of innovation. Therefore, much more than simply accepting the temporalities of innovation as their starting point, participants reflected on their personal experiences, position and values and how these related to those observed for other generations, professions and technological fields. Temporalities of diverse developments in technosciences and societies in different places interfere in specific space-time points at which they either annihilate or reinforce each other, thus creating situated experiences difficult to anticipate. As a consequence, visions of how to govern these temporal interferences and to create a state of synchronicity differed quite substantially. While some would argue that society needs to catch up with the speed of innovation to fall into a pattern of synchronicity, others would argue against that suggestion, deliberately advocating a slower path and thus forcing innovation to adapt to society.

Finally, time also matters when it comes to participants' identity in the participation process. While identities might shift within the framework of such exercises (Callon *et al.* 2009), it is also essential to grasp that participants as members of different social groups are also carriers of different temporal horizons and collective memories. Giesen (2004) calls this phenomenon 'divided memories'. As a consequence, the moment in people's lives and their social attachments play an important role in the process of assessing sociotechnical developments and in the positioning work that people can perform. We therefore encounter some moments where different collective memories might collide, while at other times convergences are palpable. One such case was the above-mentioned sociotechnical imaginary of Austrian citizens being able to reject certain technologies if they perceived them as threatening an important part of their national identity (Felt 2015). Major disasters or deeply polarizing events in sociotechnical history then may leave an imprint on how debates can develop and deliberation can occur. It is thus essential to consider the stage of life at which citizens join in participatory exercises. This might considerably affect how they construct their positions, how they can draw on lived memories and how they can claim more plausibility and authenticity for their judgments regarding long-term developments.

Discussion and conclusion

The purpose of this chapter was to render time structures, which are so consequential for the development of contemporary societies, visible and to unpack them, and thus to transform them from 'habits of mind to moral and conceptual tools' (Adam 2004). The discussion aimed at showing how profoundly time is not a given and static entity but always a 'result of the connection among entities' (Latour 1993, 76) and how our engagement with time defines us. Employing the notion of 'choreography of temporalities' reminds us to be attentive to how different forms of time are connected, how they overlap and intertwine, and how they collectively shape participation.

The following conclusions will be grouped around four concerns essential for rethinking public participation. The first points to the *ontological politics of temporalities* at work in participatory practices, which shape how innovation is conceptualized, how problems get assembled, how publics are made and how potential action and responsibility is imagined. This ontology matters in how participation is framed by policy-makers, practitioners or researchers organizing such events as well as by the participants themselves. Linear innovation models as well as the need for speedy and steady innovation flows from a robust but seldom-questioned basis from which projection work is performed. This basis also infuses versions of potential futures with plausibility, creates the illusion of control and delimits the realm of possibilities.

What are the consequences of such an observation? When performing and analysing participatory exercises, it is essential to more explicitly address these temporal ontologies and to question them. This questioning means that the implicit understandings of the temporal developments of both technoscience and society should be challenged as much as the concrete issues at stake. This approach connects to a more general critique of the classical conceptualization of upstream engagement (Joly and Kaufmann 2008), which highlights the need for more fine-grained models of innovation, rendering visible the complex networked character of any innovation. Yet, a time-sensitive analysis pushes this critique further, underlining the need to render visible – and thus debatable – the multiple temporal regimes governing any innovation and their relationship to society. Innovation policy and participatory elements within it thus need to be assessed and understood in light of these often-tacit assumptions.

The second concern is best captured by analogy to Annemarie Mol's (2008) distinction of a logic of choice versus a *logic of care*, a distinction inhabited by a strong temporal order. Policy-makers appear to be quite attached to the idea that there is an ideal moment in the developmental trajectory when sociotechnical issues can be assessed once and for all; after that 'moment of engagement', research should be left on its own again. This temporal understanding of participation is intimately tied to the ideal of efficiency and planning and is in line with the idea that this mind-set allows innovators to grasp the windows of opportunity. In observing participatory moments, however, we note the complexity of developmental understandings at work. Paraphrasing Tim Ingold (2000) in stating that people know as they go and not before they go, my observations not only highlight (in line with other authors) the centrality that we need 'iterative, continuous and flexible processes of learning' (Owen et al. 2012, 755) but also draw specific attention to the importance of the temporal choreography of technoscientific developments and participation. If participation is not limited to taking up issues already preformed, then time is necessary to collectively carve out what is at stake, and it has to be admitted that any single actor involved might shift perceptions in the course of this process. Continuing this line of thought and assuming that issues and publics are always co-produced (Marres 2007), we have to consider that specific temporalities of participation perform specific publics while also determining which publics will never enter the realm of the possible (Felt and Fochler 2010).

If we do not conceptualize participation as part of a wider logic of choice, then caring for technoscientific developments in contemporary societies through participation would mean that detours or controversies in the context of participatory events should not be conceptualized 'as a waste of time that could be dispensed with' (Callon *et al.* 2009, 27). Rather, we should perceive them as valuable moments during which different perspectives are opened up, during which tacit orders are identified and can be explored, and during which the complexities of technoscientific issues can be more fully addressed. This means that we must move beyond a purely physical understanding of time, with the speed of decision-making not necessarily being a sign of efficiency or success in an engagement event. Understanding participation as part of a wider process of caring for how innovation shapes collective societal futures and considering the contextuality and complexity of this relationship thus also requires a different imaginary of the timescapes in which such undertakings are embedded.

The third concern addresses the relation between temporalities and ideals of responsibility. If participation is understood as a core element in realizing what is labelled 'responsible research and innovation' (Owen *et al.* 2012), it becomes essential to address what I call *responsibility conditions* – in this case, the temporal boundary conditions under which responsibility can be envisioned and exercised (Felt 2014).

Let us turn our attention to three exemplary ways in which time has been shown to matter with respect to thinking about the relationship between innovation, responsibility and participation. First, this chapter has argued how the short-term nature of many participatory exercises excludes the formation of specific collectivities, that time as a resource that an individual can dispose of is unequally distributed in society and thus not everybody can participate in an equal manner and, finally, that the complex temporalities of innovation dynamics are difficult to examine in a short one-time-only event. Second, we need to consider what Adam and Groves (2011) have called the 'timeprint' of both innovation imaginaries at work but also of the participatory exercises themselves. This viewpoint draws our attention to how much the present and its prevailing knowledge practices lead to a specific way of consuming 'the future potential' of any innovation and alerts us 'to the problematic relation whereby current future making extends far beyond any capacity to match our concern and responsibility to the temporal reach of our actions' (Adam and Groves 2011, 26). Thus, the way in which future-making takes place in innovation discourses but also in the accompanying participatory measures needs to be reflected not only as a way of performing foresight but also as an act of doing and of doing responsibility. Third, tight time schedules for participation or, more broadly, the lack of time resources for reflecting on issues of responsibility in research, as well as the omnipresence of discourses on speed, acceleration and competition, lead to a transformation from being 'response-able' to being 'account-able'. The former notion – 'response-able' – captures the ability to produce responses to the fluidity of the sociotechnical issues at stake in participation and thus to explore the future potential of innovations as well as the potential futures that come along with them. The latter notion of being 'account-able' reflects the transformation of these complex and open-ended ways of thinking into a much more standardized and time-efficient reaction to the call for responsibility. Being 'account-able' is then tied to standardized and 'form-ularized' (e.g., ethics forms are an excellent example for the widespread use of forms in accounting practice) relationships (Becker 2007) when addressing issues of sociotechnical change. In turn, this relationship generally also means accepting predefined power and authority structures and valuation regimes.

These exemplary ways of looking into responsibility conditions clearly point to the need to more closely consider temporal choreographies and the formative power they exercise to better grasp the dimensions of participatory exercises as spaces of knowledge-making and collective experimentation as well as part of a new regime of responsibility.

Finally, participatory exercises should not be solely directed at imagining specific technology-related futures and thus to practising a narrow perspective of anticipation; they should instead more carefully open up the issues at stake and invite the creation of wider connections between various elements available in our 'knowledge realms of perception, memory and anticipation' (Adam 2010, 362). Participatory exercises and the ways in which their temporalities relate to a specific place do not build solely on the specific imaginaries of *citizenship* (and thus of *democracy*), roles and obligations; instead, these exercises actively contribute to performing them. This means that attention needs to be paid to the stories told, the temporalities embedded in them and their role in making identities, whether group-specific, national or personal. This analysis showed the centrality of people's memory work in participatory settings as well as the broader sociotechnical imaginaries to which they can refer when trying to assess new technologies. Thus, taking participation seriously and making it an important part of societal learning would mean paying close attention in the realm of policy-making in particular but also to the argumentative repertoires across time, space and technologies. Thus a comparative gaze is essential for understanding the different temporal regimes at work and for fully grasping how such spaces of engagement are sites where traditions are made and enacted, where specific pasts and presents get connected, as well as where global and local temporalities need to find arrangements. In this sense, participatory experiments should not be regarded as the end point of a process or as an exercise in making choices but instead should become a locus where techno-cultural identities are made and unmade, where understandings of the relationship between the local and the global are negotiated, where sociotechnical memories are deployed and actualized and, thus, where the very meaning of citizenship, democracy and the idea of the nation state are continuously (re)shaped (Edensor 2006).

Embracing a time-sensitive perspective to participation thus not only allows participants to identify the different temporalities deeply inscribed into events, processes, places and things related to technosciences but also to widen the scope of what we can learn from and what can be learned within such exercises.

Notes

1   For the use of the concept choreography in participation see Moreira (2012); for a broader work on ontological choreographies see Thompson (2005).
2   See http://ec.europa.eu/programmes/horizon2020/en/h2020-section/europe-changing-world-inclusive-innovative-and-reflective-societies.
3   This chapter builds on observations of participatory exercises in Austria, but also on research conducted in the framework of the following projects: 'Evaluation of the Discourse Day on Genetic Diagnosis 2002', funded by the Austrian genome research programme GEN-AU; 'Challenges of Biomedicine: Socio-Cultural Contexts, European Governance and Bioethics', funded by the European Commission in the 6th framework programme, Contract No. SAS6-CT-2003-510238; 'Let's Talk about GOLD: Analysing the Interactions between Genome Research(ers) and the Public as a Learning Process', funded by the Austrian genome research programme GEN-AU as an ELSA project; and 'Making Futures Present: On the Co-Production of Nano and Society in the Austrian Context', funded by the FWF (Austrian Science Fund), grant number P20819. The project leader or coordinator for all projects was the author of this chapter. I would like to acknowledge the contribution of all colleagues involved in these projects, both as collaborators and advisers.

References

Adam, B. 1998. *Timescapes of Modernity: The Environment & Invisible Hazards*. London and New York: Routledge.
Adam, B. 2003. Reflexive modernization temporalized, *Theory, Culture & Society* 20: 59–78.
Adam, B. 2004. Minding futures: An exploration of responsibility for long term futures. ESRC PF Paper, www.cardiff.ac.uk/socsi/futures/mindingfutures.pdf (accessed 9 September 2015).
Adam, B. 2010. History of the future: Paradoxes and challenges, *Rethinking History* 14(3): 361–378.
Adam, B. and Groves, C. 2007. *Future Matters – Action, Knowledge, Ethics*. Leiden and Boston, MA: Brill.
Adam, B. and Groves, C. 2011. Futures tended: Care and future-oriented responsibility, *Bulletin of Science, Technology & Society* 31(1): 17–27.
Adams, V., Murphy, M. and Clarke, A. E. 2009. Anticipation: Technoscience, life, affect, temporality, *Subjectivity* 28: 246–265.
Appadurai, A. 2012. Thinking beyond trajectorism. In: M. Heinlein, C. Kropp, J. Neumer, A. Poferl and R. Römhild (eds) *Futures of Modernity: Challenges for Cosmopolitical Thought and Practice*. Bielefeld: Transcript Verlag, pp25–32.
Barben, D., Fisher, E., Selin, C. and Guston, D. H. 2008. Anticipatory governance and nanotechnology: Foresight, engagement, and integration. In: E. J. Hackett, O. Amsterdamska, M. Lynch and J. Wajcman (eds) *The Handbook of Science and Technology Studies*, 3rd edition. Cambridge, MA: MIT Press, pp979–1000.
Barry, A. 2001. *Political Machines: Governing a Technological Society*. London: Athlone Press.
Becker, P. 2007. Le charme discrete du formulaire. In: M. Werner (ed.) *Politiques et usages de la langue en Europe*. Paris: Editions de la maison des sciences de l'homme, pp217–241.

Beynon-Jones, S. M. and Brown, N. 2011. Time, timing and narrative at the interface between UK technoscience and policy, *Science and Public Policy* 38(8): 639–648.

Boltanski, L. and Thevenot, L. 2006. *On Justification: Economies of Worth*. Princeton, NJ and Oxford: Princeton University Press.

Braun, K. and Schultz, S. 2010. '… a certain amount of engineering involved': Constructing the public in participatory governance arrangements, *Public Understanding of Science* 19(4): 403–419.

Brown, N. and Michael, M. 2003. A sociology of expectations: Retrospecting prospects and prospecting retrospects, *Technology Analysis and Strategic Management* 15(1): 3–18.

Brown, N., Rappert, B. and Webster, A., eds. 2000. *Contested Futures: A Sociology of Prospective Techno-Science*. Aldershot: Ashgate.

Callon, M., Lascoumes, P. and Barthe, Y. 2009. *Acting in an Uncertain World: An Essay on Technical Democracy*. Cambridge, MA: MIT Press.

Castells, M. 2010. *The Rise of the Network Society*. Chichester, West Sussex; Malden, MA: Wiley-Blackwell.

Doubleday, R. 2007. The laboratory revisited: Academic science and the responsible development of nanotechnology, *NanoEthics* 1(2): 167–176.

EC 2013. *Innovation Union: A Pocket Guide on a Europe 2020 Initiative*. Brussels: European Commission.

Edensor, T. 2006. Reconsidering national temporalities: Institutional times, everyday routines, serial spaces and synchronicities, *European Journal of Social Theory* 9(4): 525–545.

Editorial 2007, 5 July. Enough talk already, *Nature* 448(7149): 1–2.

Elias, N. 1988. *Über die Zeit. Arbeiten zur Wissenssoziologie II*. Frankfurt am Main: Suhrkamp.

Ellison, N. 2013. Citizenship, space and time: Engagement, identity and belonging in a connected world, *Thesis Eleven* 118(1): 48–63.

Espeland, W. N. and Stevens, M. L. 2008. A sociology of quantification, *European Journal of Sociology* 49(3): 401–436.

Felt, U. 2010. Vers la construction dun public européen? Continuités et ruptures dans le discours politique sur les cultures scientifiques et techniques, *Questions de communication* 17: 33–58.

Felt, U. 2014. On the 'responsibility conditions' of contemporary academic research: Changing meanings of living and working in academia. Working paper.

Felt, U. 2015. Keeping technologies out: Sociotechnical imaginaries and the formation of Austria's technopolitical identity. In S. Jasanoff and S.-H. Kim (eds) *Dreamscapes of Modernity: Sociotechnical Imaginaries and the Fabrication of Power*. Chicago, IL: Chicago University Press, pp103–125.

Felt, U. and Fochler, M. 2010. Machineries for making publics: Inscribing and describing publics in public engagement, *Minerva* 48(3): 219–238.

Felt, U. and Fochler, M. 2011. Slim futures and the fat pill: Civic imaginations of innovation and governance in an engagement setting, *Science as Culture* 20(3): 307–328.

Felt, U., Fochler, M. and Winkler, P. 2010. Coming to terms with biomedical technologies in different technopolitical cultures: A comparative analysis of focus groups on organ transplantation and genetic testing in Austria, France, and the Netherlands, *Science, Technology & Human Values* 35(4): 525–553.

Felt, U., Barben, D., Irwin, A., Joly, P.-B., Rip, A., Stirling, A. and Stöckelová, T. 2013. *Science in Society: Caring for our Futures in Turbulent Times*. Policy Briefing 50. Strasbourg: ESF.

Felt, U., Wynne, B., Callon, M., Gonçalves, M. E., Jasanoff, S., Jepsen, M., Joly, P.-B., Konopasek, Z., May, S., Neubauer, C., Rip, A., Siune, K., Stirling, A. and Tallacchini, M. 2007. *Taking European Knowledge Society Seriously*. Luxembourg: Office for Official Publications of the European Communities.

Flaherty, M. G. 2010. *The Textures of Time: Agency and Temporal Experience.* Philadelphia, PA: Temple University Press.

Giddens, A. 1999. *Reith Lecture 2: Risk.* London: BBC.

Giesen, B. 2004. Noncontemporaneity, asynchronicity and divided memories, *Time & Society* 13(1): 27–40.

Harrington, A., Rose, N. and Singh, I. 2006. Editor's introduction, *BioSocieties* 1(1): 5.

Hassan, R. 2009. *Empires of Speed: Time and the Acceleration of Politics and Society.* Leiden and Boston, MA: Brill.

Hobsbawm, E. and Ranger, T., eds. 1983. *The Invention of Tradition.* Cambridge: Cambridge University Press.

Horst, M. and Irwin, A. 2009. Nations at ease with radical knowledge: On consensus, consensusing and false consensusness, *Social Studies of Science* 40(1): 105–126.

Ingold, T. 2000. *The Perception of the Environment: Essays on Livelihood, Dwelling and Skill.* London/New York: Routledge.

Irwin, A. 2006. The politics of talk: Coming to terms with the new scientific governance, *Social Studies of Science* 36(2): 299–320.

Jasanoff, S. 2015. Future imperfect: Science, technology, and the imaginations of modernity. In: S. Jasanoff and S.-H. Kim (eds) *Dreamscapes of Modernity: Sociotechnical Imaginaries and the Fabrication of Power.* Chicago, IL: Chicago University Press, pp1–47.

Joly, P.-B. and Kaufmann, A. 2008. Lost in translation? The need for upstream engagement with nanotechnology on trial, *Science as Culture* 17(3): 225–247.

Kaiser, M., Kurath, M., Maasen, S. and Rehmann-Stutter, C., eds. 2010. *Governing Future Technologies: Nanotechnology and the Rise of an Assessment Regime.* London and New York: Springer.

Latour, B. 1993. *We Have Never Been Modern.* Cambridge, MA: Harvard University Press.

Latour, B. 2005. From realpolitik to dingpolitik or how to make things public. In: B. Latour and P. Weibel (eds) *Making Things Public.* Cambridge, MA: MIT Press, pp14–41.

Marres, N. 2005. Issues spark a public into being: A key but often forgotten point of the Lippmann–Dewey debate. In: B. Latour and P. Weibel (eds) *Making Things Public.* Cambridge, MA: MIT Press, pp208–217.

Marres, N. 2007. The issues deserve more credit: Pragmatist contributions to the study of public involvement in controversy, *Social Studies of Science* 37(5): 759–780.

Michael, M. 2009. Publics performing publics: Of PiGs, PiPs and politics, *Public Understanding of Science* 18(5): 617–631.

Mol, A. 2008. *The Logic of Care: Health and the Problem of Patient Choice.* London and New York: Routledge.

Moreira, T. 2012. *The Transformation of Contemporary Health Care: The Market, the Laboratory, and the Forum.* New York/London: Routledge.

Nowotny, H. 1994. *Time: The Modern and Postmodern Experience.* Cambridge, MA: Polity Press.

Owen, R., Macnaghten, P. and Stilgoe, J. 2012. Responsible research and innovation: From science in society to science for society, with society, *Science and Public Policy* 39(6): 751–760.

Rifkin, J. 1987. *Time Wars: The Primary Conflict in Human History.* New York: Henry Holt & Co.

Rosa, H. 2013. *Social Acceleration: A New Theory of Modernity.* New York: Columbia University Press.

Selin, C. 2006. Time matters: Temporal harmony and dissonance in nanotechnology networks, *Time & Society* 15(1): 121–139.

Selin, C. 2008. The sociology of the future: Tracing stories of technology and time, *Sociology Compass* 2(6): 1878–1895.

Stirling, A. 2008. 'Opening up' and 'closing down': Power, participation, and pluralism in the social appraisal of technology, *Science, Technology & Human Values* 33(2): 262–294.

Suchman, L., Danyi, E. and Watts, L. 2008. *Relocating Innovation: Places and Material Practices of Future-Making Project Description*, www.sand14.com/relocatinginnovation/download/RelocatingInnovation_ResearchDescription.pdf (accessed 9 September 2015).

Thompson, C. 2005. *Making Parents: The Ontological Choregraphy of Reproductive Technologies*. Cambridge, MA: MIT Press.

von Schomberg, R., Pereira, A. G. and Funtowicz, S. 2005. *Deliberating Foresight Knowledge for Policy and Foresight Knowledge Assessment*. Brussels: European Commission.

Watts, L. 2008. The future is boring: Stories from the landscape of the mobile telecoms industry, *21st Century Society: Journal of the Academy of Social Sciences* 3(2): 187–198.

Wilsdon, J., Wynne, B. and Stilgoe, J. 2005. *The Public Value of Science*. London: Demos.

Wynne, B. 2006. Public engagement as means of restoring trust in science? Hitting the notes, but missing the music, *Community Genetics* 9(3): 211–220.

Wynne, B. 2007. Public participation in science and technology: Performing and obscuring a political–conceptual category mistake, *East Asian Science, Technology and Society: An International Journal* 1: 99–110.

PART III
Remaking participation

10

AN 'EXPERIMENT WITH INTENSITIES'

Village hall reconfigurings of the world within a new participatory collective

Claire Waterton and Judith Tsouvalis

Introduction

This chapter explores the experience of carrying out what the post-humanist scholar Rosa Braidotti calls an 'experiment with intensities' (Braidotti 2013, 190) – an experiment that drew on 'philosophical' ways of thinking more centrally than most participatory initiatives in order to open up and re-chart the framing of an assumed environmental problem in the English Lake District, Cumbria. The place in which this experiment took place is Loweswater, a hamlet of houses and farmsteads scattered around a small freshwater lake in the English Lake District, Cumbria. Here, for three years, a 'new collective' called the Loweswater Care Project (LCP) carried out its own epistemological and ontological experiments in response to the proliferation of harmful cyanobacteria (or blue-green algae) in the local lake.[1] The influence of research in science and technology studies (STS) (especially that of Latour 2004) within this forum was complemented by that of feminist STS and feminist post-human scholars (especially that of Barad 2007). Both of these theoretical approaches inspired an appreciation of the radical relationality of people and things, and a new approach to doing collective politics with non-human things. These approaches also brought the participatory experiment towards a kind of 'post-human ethics', a new kind of ethical sensibility that embraces an enlarged sense of community, opening out questions about the 'basic unit of common reference for our species, our polity and our relationship to the other inhabitants of this planet' (Braidotti 2013, 2).

In what follows we first offer a brief history of the LCP, highlighting how the particularities of the situation in Loweswater in the late 1990s and 2000s prepared the way for an experimental participatory forum to take place, from 2007 onwards. We then explore some aspects of this forum, particularly the STS and feminist philosophical insights that informed its routine practices and ways of working. Third, we show how the LCP worked to understand cyanobacteria and their complex relations. This involved the reframing of questions, the creation of alliances with many different things, and the acknowledgement of the LCP as a collection of various kinds of 'scientist' that create knowledge in a kind of ongoing relationship with elements of the non-human world, a relationship 'with intensities' where what empowers understanding is not entirely human (Stengers 2005; Braidotti 2013). Many different investigations combined within the LCP to make up an 'ethics of experiment with intensities'. We conclude with some reflections on this public experiment.

The Loweswater Care Project

In this chapter we narrate the history of the LCP in two different ways. The first centres on the human actors that preceded the LCP and set the tone for philosophical experimentation. The second centres on cyanobacteria themselves and describes them as drawing a public around themselves. These two stories interconnect, but it is significant that we can tell them differently.

As we have rehearsed elsewhere, the LCP is a participatory forum that evolved from a farmer-led group in 2001 to address the problem of the eutrophication[2] of Loweswater, a small lake in the English Lake District (see Tsouvalis and Waterton 2012; Waterton et al. 2015a and b). The Loweswater Improvement Project (LIP), as this group of farmers called themselves, was initiated by a farmer, the late Danny Leck, with support from the North West Regional Development Agency. His initiative was born, in part, out of a sense of frustration: in the early 2000s, potentially toxic blue-green algae in the lake at the centre of this quiet rural hamlet seemed to be flourishing. The statutory organization with responsibility for lake water quality, the Environment Agency, was of the opinion that the blame for this resided with Loweswater farmers; farmers were receiving warning letters and facing Environment Agency fines as well as stigmatization from the rest of the community. Other local residents had little or no sense of responsibility for what was occurring in the lake. What is more, Loweswater looked set to fail the standards required by a new European water quality directive, the EU Water Framework Directive. And, yet, no change seemed likely. Intermittent, but increasingly frequent, 'blooms' of potentially toxic blue-green algae continued to rise to the surface of the lake, surprising everyone with their slimy, lurid green appearance.

Progress in tackling this problem was slow, partly because arrangements for the 'management' of the land and water in Loweswater were complex and incorporated a wide range of actors. Furthermore, the practices and responsibilities of individual land managers and non-governmental and governmental organizations (Natural England, the Environment Agency and the Lake District National Park Authority) did not 'add up' to a comprehensive approach that would tackle the issue of 'diffuse pollution' and the eutrophication of fresh water bodies like Loweswater. Lastly, the institutions involved were acting in relatively insulated ways and at a distance from the Loweswater community. Scientists were monitoring the lake with funding from the Environment Agency; those responsible for the implementation of the new and legally binding European regulatory framework were gathering data – but knowledge was not shared. A narrow framing of the 'algae problem' predominated. There were limited and wary interactions between agencies and local people; and mistrust, blame and lack of communication were common between different parties in and around Loweswater (Waterton et al. 2015a).

From 2001 to 2004, the farmers' LIP group worked with ecologists from the Centre for Ecology and Hydrology (CEH) at Lancaster University to address this issue. Farmers, in particular, were resolved to act before the 'algae problem' attracted the attentions and fines of the Environment Agency. They were aware that human and animal waste products as well as grassland that is fertilized or 'improved' in Loweswater were thought to contribute substantial amounts of the soluble element phosphorus (the element that both feeds and 'limits' cyanobacteria[3]) to the waters of the lake. And so farmers, working collectively as part of the LIP, gained small grants in order to test soils for nutrient levels; create buffer zones of semi-natural vegetation to capture flows of phosphorus; re-route phosphorus-rich waste waters through farmyards; and install new septic tank and waste water systems on some properties adjacent to the lake. As the funding sources utilized by the LIP dried up, around 2004 a small grant from the Rural Economy and Land Use Programme (RELU) was obtained to explore the possibility of expanding the LIP to include relevant stakeholders and local residents (Waterton et al. 2006). In 2007, this expansion of the group formally took place through a further grant gained for the period 2007–2010 by a group of natural and social scientists based at the CEH, Lancaster, and Lancaster University.[4] This grant supported the initiation of a new participatory forum, named the Loweswater Knowledge Collective (LKC) by the Lancaster University and CEH researchers – soon to be renamed the Loweswater Care Project (LCP) by a vote of all participants in February 2008.

Recalling this short history above, what we are emphasizing here is that the LCP, as an experimental participatory forum, had a very strong foundation in human actions – that is, through the farmers' strong critique of the status quo, and through the LIP's energetic sense of agency. The farmers had paved the way for a relatively radical approach from 2001 onwards. They had already rejected having a predefined idea of what 'the algae' meant. They had worked proactively to avoid an overly narrow 'framing' of the issue of lake water pollution, insisting that this was not only about farm-yard 'runoff' or the excessive application of fertilizers, but was also about domestic sewage infrastructures, the pressures on farmers to produce, and so on. They had also shown that they were good at working collectively from a critical but grounded, 'bottom-up' perspective. As we have argued elsewhere, for the farmers, water quality in Loweswater was considered not just as a physical measure, involving an excess of nutrients that needed to be curbed, but as a complex problem involving serious thought about farm infrastructure/system change, and the possibility of sustaining viable farming scenarios and a healthy lake (Norton et al. 2011; Tsouvalis and Waterton 2012).

Collective politics with things

During the period 2007–2010, natural and social science researchers, farmers, institutional stakeholders and local residents built on the LIP's strong foundation of, first, a critical reframing of the issues at stake; and, second, a commitment to bottom-up collective work. In addition to this foundation, the newly named Loweswater Care Project was built around an explicit critique of participatory projects that were seen to be too tightly framed, and that assumed a hierarchy of expertise and neglected the ongoing redefinition of the problem by participants (examples of this kind of critique include Leach *et al.* 2005; Wynne 2007; Felt and Fochler 2008; Reed 2008; Chilvers 2010; Marris and Rose 2010; Lane *et al.* 2011; Welsh and Wynne 2013). The LCP was carefully nurtured by the researchers involved, together with all of the participants, as a forum that would critically question different forms and 'frames' of expertise, while simultaneously trying to construct and bring together appropriate forms of knowledge, data, understanding and experience for examination, scrutiny and possible use. As such, the LCP was an experiment in collective knowledge-making. As researchers and participants, the Lancaster University and CEH researchers called it a new participatory 'social mechanism', and considered it to be a critical, experimental space for the co-production of knowledge (Callon 1999) in ways that created deliberate connections between epistemological, ontological and political questions.

This harnessing together of questions of epistemology with those of ontology and politics is core to many aspects of science and technology studies and will be familiar to many readers. In this section we narrate how the researchers at Lancaster University and CEH used the ideas of STS and feminist STS/post-humanism to experimentally enact what it might be like to do politics, and participation through epistemological and ontological collective work. We focus on two works – Bruno Latour's *Politics of Nature* (2004) and Karen Barad's *Meeting the Universe Halfway* (2007) – since both of these were used most directly to forge new ways of questioning and creating knowledge within the LCP.

Perhaps one of the most controversial, challenging and inspiring moves that STS scholars have made has been to foreground the role played by non-human actors or actants in the ongoing politics and reconfiguring of the world. This theoretical demand for the acknowledgement of non-human agency derives from the work of actor-network theorists and, in particular, the research of Bruno Latour, Michel Callon, John Law and others. And of course this is a theory that overturns many conventional assumptions about the way that the world works. In what is by now a classic STS text, *We Have Never Been Modern* (1993), Latour both illustrates and contests the manner in which non-humans have historically been conceptualized, institutionalized and represented through the constitutional form and associated juridical system characteristic of Modernity, arguing that this is a system built on dichotomous modes of ordering that have kept nature and society, human and non-human, and subject and object, firmly apart (Latour 1993). A decade later, in his book *Politics of Nature: How to Bring the Sciences into Democracy* (2004), Latour describes how contemporary politics, built around these dichotomous modes, has become somewhat impoverished as a result. This impoverishment stems, he argues, from the exclusivity of the modern 'constitution' whereby participation is limited to carefully selected humans who meet specific and restrictive criteria.

Latour (2004) thus argues that the contributions of many non-humans are conventionally neglected in the modern polity. The problem that he identifies is that the only way for non-human actants to enter conventional arenas of politics or decision making is in the form of representations of nature presented to politics by their spokesperson, 'S'cience. Here, he is defining 'S'cience, with a capital 'S', as the *politicization* of the sciences. This politicization takes place, he argues, because 'S'cience tends to stand in for an 'incontestable nature', a nature that tells us what to do (this sometimes carries a capital 'N', as in 'N'ature). But 'S'cience, he also suggests, contrasts markedly from the 's'ciences (with a small 's') whose task is much less threatening, less powerful and much more interesting. In an interesting move, Latour supports the role of (little 's') 's'cience in creating collective propositions with which to constitute the world.[5] Rather than allowing 'S'cientists to become the supposedly non-political representatives of 'things' (cells, genes, viruses, bacteria and other life forms and biophysical processes), what he advocates is a democratic arrangement whereby *objects* might be given rights of expression and representation. This means that we need to find ways for them to figure in our politics as actants that extend, mediate and translate the actor-networks in which they are enrolled. In his musings on these issues, Latour (2005, 4) asks: 'What would an *object-orientated* democracy look like?' His answer is: 'A democratic republic of heterogeneous associations' (Latour 2005, 4).

As researchers involved in the emerging LCP, we were interested in these ideas and in Latour's suggestion that we should do away with the old 'constitution' and create new 'collectives' where politics *with* things can be done. These ideas were exciting and seemed to intersect with our commitment to build upon previous critiques of participatory initiatives. In Latour's (2005) new democratic fora, we saw that 'things' were understood not as something that brings an assembly of people together because they agree on something (and are looking for the facts to prove they were right), but, rather, as something that gathers an assembly around itself, 'triggering new occasions to passionately differ and dispute' (p5). Second, what seemed crucial about these new imagined collectives was their approach to fact-making. As Latour described it, things have the power to call forth assemblies or new collectives that co-mingle in 'complicated entanglements' (p31) where there is 'no unmediated access to agreement; no unmediated access to the facts of the matter' (p12). In such collectives it is held that:

- Nature is not self-evident.
- Knowledge and expertise have to be debated.
- Uncertainty is the main condition humans are in (rather than a condition of having knowledge).
- What is important is the creation of connections between people and things.
- Doubt and questioning must be extended to our own representations.

This commitment to interrogating 'facts' but also to the making of new connections became an important guiding principle of the working practices of the LCP.

But as we have suggested, these ideas were complemented by those coming from a slightly different perspective, that of feminist post-humanism.

Intra-active collective politics

The notion of 'intra-action' comes from a reworking of ontology and agency by the physicist and philosopher of science Karen Barad (2003, 2007), a reworking which she refers to as post-humanist agential realism. For her, while inter-action assures that there are separate individual agencies to people and things, 'intra-action' signifies the mutual constitution of entangled agencies. Intra-action 'recognises that distinct agencies don't precede, but rather emerge through, their intra-action' (Barad 2007, 33).

The reason why the notion of intra-action is important in our endeavours here is that it signals an ontology where matters and 'facts' are seen as 'discursive practices [that are] always already material (i.e., they are ongoing material (re)configuring of the world)' (Barad 2003, 822). Barad highlights how processes of continuous emergence and the becoming of multiple forms of reality through material intra-actions take on particularity at specific times and in specific places.

And so, just as Latour outlines a methodology for opening out the 'facts of the matter', the notion of 'intra-action' goes beyond representationalism (in the sense of representation as a mirror of nature), and beyond realist assumptions that the material world exists independently out there waiting to be discovered and reported/represented through human endeavours. In Barad's work, things come to be what they are through their intra-actions with others assembled as a collective. And things spawn new things. For her, intra-action and the dynamism of matter are 'generative not merely in the sense of bringing new things into the world but in the sense of bringing forth new worlds, or engaging in an ongoing reconfiguring of the world' (Barad 2003, 170).

Barad's analysis offered the LCP a way of understanding 'factual' representations through their emergence in relations with other entangled agencies, just as Latour's principles for understanding the 'facts of the matter' meant questioning how such facts came into being through mediations and connections. In particular, Barad's work supported a way of thinking about how things come into being through instruments and devices already in play. Scientific and other forms of 'apparatus', for example, rather than simply being considered as neutral instruments or inscription devices that are 'set in place before the action happens' (the action being to objectively observe, discover and record nature), are themselves 'inside the project' (Barad 2003, 810). In other words, apparatuses intra-act with the phenomena they are designed to 'discover' and thus in themselves constitute 'specific material practices through which local semantic and ontological determinacy are intra-actively enacted' (Barad 2003, 820). For the LCP this meant understanding how the kinds of images, graphs and statements brought into the forum could themselves be seen as materialities that participated in our politics.

Our aim was not to elaborate a new constitution. Our aim was, rather, to take STS and post-humanist ideas seriously and explore how in a specific location (Loweswater) and over a specific time (three years) fact-making, politics-making and policy-making could be done differently and in a way that, first, did not allow 'S'cience to take the place of politics; and, second, encouraged interpretation of the ongoing intra-actions underway in this place. We felt that the different perspectives and foci of Latour and Barad were complementary. Latour's project played directly into the idea of doing democratic politics differently, offering a set of principles by which we could explicitly act. Barad's perspective, in a less overtly political way, sensitized the researchers to the importance of understanding the ongoing politics as populated by apparatuses, materialities, graphic representations, machines and so on that were enabling certain intra-acting realities to participate in the forum. By the same token, this perspective allowed us to see what was becoming less visible, and which realities, therefore, were excluded (Barad 2007).

Rethinking the participatory collective

Taking Latour and Barad seriously, we were tasked with the problem of giving objects, quasi-objects and things credit for the role they were playing in shaping worlds or, as Barad might put it, in 'worlding'. Second, and in direct connection to critiques of participation, the task was to see if we could progress towards a democratic republic of heterogeneous associations in which such associations might thrive and proliferate, displacing a more conventional constitution where 'facts' and their spokespersons in 'S'cience usurp the role of politics. Instead of focusing on the agencies of farmers and environmental organizations ('humans first'), as we did earlier, we therefore need to retell this story in a way that highlights cyanobacteria's central role, but that also shows how these organisms are already part of existing 'intra-acting' assemblages of people and things.

From the early 1990s on, ever more frequent blooms of cyanobacteria had begun to draw people and things around them. From 2000 onwards, blooms even began to occur during the cold winter months. The blooms were often perceived negatively by those who witnessed them, and their vitality was interpreted as threatening to quash other elements of Loweswater's socio-ecology. Algae were seen to be not only 'killing' the lake, but the local economy as well. And yet, interpreted in another way, cyanobacteria was not only a symbol of destructive relations but, rather, a catalyst for new, productive ones. Cyanobacteria, analysed within a core of sediment taken from the bottom of the lake, for example, gave strong clues as to the complex processes and heterogeneous relations of their own past making. They revealed to the ecologists an ecological present which had been shaped by many other relations and things – including the effects of wars, the onset of chemical fertilizer manufacturing, changing cycles of farming prosperity, and an aging human waste disposal infrastructure (Winchester and Bennion 2010). In this way, cyanobacteria, we found out, were not only very good story tellers, but also inspired curiosity in other things: they were capable of convening different publics and objects around them, inspiring a lively politics in this seemingly unchanging valley (Latour 2005; Marres and Lezaun 2011; Marres 2012). The brilliance of cyanobacteria in this respect was that they continually achieved this – by floating to the surface intermittently, alerting residents to their presence, prompting the sending of letters to the owners of the lake-bed, prompting the erection of signs around the lake, and even stirring up an international policy public, since their presence was one of the factors that caused Loweswater to fail the (then new) European Water Framework Directive. They became pivotal, in fact, in bringing together the heterogeneous assemblage of the Loweswater Care Project where farmers, landowners, ecologists, policy-makers, infrastructures, policies, local residents, small business owners and social scientists, amongst a plethora of other human and non-human elements, worked collaboratively together.

Participatory practices

Within this assemblage, though, what did the philosophical commitments towards a more object-oriented democracy mean in practice? How did the LCP as a participatory initiative committed to finding places for previously excluded humans and non-humans work collectively to decentre the de-politicizing role of 'S'cientific facts? In practice, from 2007 onwards, the LCP took on the form of a heterogeneous group.[6] Members were not pre-selected, making this a very open forum. The LCP met 15 times for a full evening (5:30 pm to 9:00 pm) roughly every two months over a period of three years. It typically attracted between 25 to 35 participants, including 3 to 6 natural/social scientists from Lancaster University and the Centre for Ecology and Hydrology, 2 to 5 agency representatives from Natural England, the National Trust, the Lake District National Park Authority and the Environment Agency, and local residents, business owners and farmers, among others. As we shall see below, the agenda for each meeting was driven by human *and* non-human participants, and there was not a single strong 'leader' or entity within the group. Rather, the group worked collectively, generating ideas and future proposals from within. Meetings were usually chaired by Lancaster University/CEH researchers or by a Loweswater farmer employed one day per week on the research project.

From early meetings in the LCP we agreed to use the key principles drawn from Latour's *Politics of Nature* (referred to above) as a guide to help us structure the way in which our participatory forum approached the facts about the algae problem. The principles helped us to question: 'nature is not self-evident'; 'knowledge and expertise has to be debated'; and 'doubt and questioning needs to be extended to all the LCP's representations, including scientific representations'. But they also helped us to think about fact-making and knowledge in different, more provisional ways: 'uncertainty is the main condition humans are in (rather than a condition of having knowledge)'. They also, importantly, helped the LCP build connections and create its own knowledge: 'what is important is the creation of connections between people and things'.

To complement these principles we paid attention to how 'things', already part of assemblages of other people and things, came to matter (Barad 2003, 2007). We paid special attention to the way in which things organized and structured our conversations and deliberations within the LCP. Constant questioning of how things came to have power and to assemble other people and variables around them (examples are cyanobacteria themselves; the classification of the Water Framework Directive; and phosphorus) helped us to ask new questions and make new connections as a group. We embarked upon a kind of experiment that opened out questions of human and non-human agency, assumptions of control, and ideas about the future of socioecological relations in this place. Through this experiment, we forged a sense of an enlarged sense of community – one that included territorial, non-human, Earth and environmental connections (Braidotti 2013). To give an indication of what this meant in practice, we describe below how the LCP's questioning and connecting around cyanobacteria developed over time.

Cyanobacterial relations

The lake in Loweswater is about 500 metres north of the village hall in which LCP meetings take place. In early LCP meetings, from 2007 onwards, questions about the lake were common. Some of the questions asked in our first few meetings concerned what was *really happening* in the waters of the lake: 'How has the lake changed?' 'What is now in the lake that wasn't before?' 'Which organisms used to be in the lake and are no longer present?' Some questions were focused more directly on the blooming 'algae': 'What *are* green-blue algae or cyanobacteria?' 'How did they come to thrive so vigorously here?' 'Can we get rid of them?'

One participant of the LCP who was also a researcher on the Lancaster University/CEH team, a professor of lake ecology from the Centre for Ecology and Hydrology, began to field these early questions. At the second LCP meeting, Stephen Maberly reported some of what he knew about cyanobacteria at Loweswater. It would be very unlikely, he suggested, that there would be one single species of cyanobacteria in the lake. Cyanobacteria come in different forms, there are many species, and there may be different species thriving in the lake at different times of the year. He acknowledged that, to date, nobody had attempted to identify which species were present and when. He also described the way in which Loweswater's very specific form, as a lake, might provide a particularly good niche for various kinds of cyanobacteria. Loweswater, he described, was not a very big or deep lake (the lake is 2 kilometres long, almost 1 kilometre wide and 16 metres deep), but has a slow average 'retention time' of over 190 days. This slow retention time meant that, theoretically speaking, the time it takes from a molecule of water entering the lake at the inflow end, to the time that it takes for that molecule of water to leave at the outflow end, is a long time, relative to other similar-sized or even bigger lakes. In Loweswater this had been estimated at over 190 days. For comparison, the much larger nearby lake, Bassenthwaite, has a retention time of around 19 days. So, he suggested, Loweswater's particular form and topology and the slow rate at which water 'flushes' through it is likely to provide a good home for certain kinds of cyanobacteria. The kinds of factors that he was describing – the specific topological configuration of the lake itself and the rate of water throughput – were recognized as factors that one could not easily alter or change.

But Stephen Maberly was not delivering 'S'cience with a capital 'S' to the LCP in a way that would foreclose politics. Rather, he was speaking about the provisional knowledge that 's'cientists (with a small 's') had of the organisms that were partly responsible for convening this meeting. As a participant of the LCP, he was outlining to the remaining participants some of the things that we do not know about cyanobacteria (which species are present) and some of the things that we cannot control (the form and the throughput of the water in the lake). Learning a little about the 'blue-greens', as participants of the LCP we began to find out some interesting ambiguities and questions about cyanobacteria and their ability to thrive in Loweswater. Reading the work of ecologist Colin Reynolds, we became aware that these bacteria/algae are amazing organisms![7] We learned that they are known as 'autotrophic' – that is, they fix inorganic elements such as carbon, nitrogen and phosphorus in order to support themselves. Through their ability to do this, they also support all life on Earth! But in order to do this, they have certain requirements. First they must have a means of suspension in the upper illuminated layers of the water. For net growth to occur, their daily residence in this zone must permit photosynthetic carbon fixation to exceed their losses in breathing and excretion. But the blue-greens must also be able to move out of this zone by sinking. Through his observations and experiments, Reynolds had worked out how they have the facility to do this: what was understood was that all cyanobacteria have gas-filled structures called vesicles which enable them to float and sink. At lower light intensities the gas vesicles fill up, bringing the blue-green algae nearer to the surface. At higher light intensities this gas is released and they are able to sink in the water column. Some studies suggest that blue-green bacteria/algae themselves *control* this, an insight that seems to give the algae a lot of agency. But there are also other things which they cannot control, and one of these – the availability of phosphorus – became key to the inquiries of the LCP. What LCP participants came to learn, and to enquire more about, was how phosphorus moved from the land around the lake into the lake, thus nourishing and supporting cyanobacteria. A simple cause–effect model would suggest that if we could 'turn off' the supply of phosphorus to the lake, then cyanobacteria might cease to thrive. But as Stephen Maberly was able to explain, the relation of cyanobacteria are not as simple as that:

> Yeah, I mean it's such a complicated thing, so it [phosphorus] is coming in, it's being taken up by the algae, it's being taken up by the bacteria, it's being used, reused and recycled, which you can't always measure. It's being taken up by fish, and so you can't always see the change in concentrations [in the water]. Some of it gets washed out of the lake; some of it gets attached to sediment. Under certain circumstances some of that phosphorus [i.e., that attached to sediments in the lake] is released back into the water column.[8]

Over a series of meetings, LCP participants began to appreciate the real complexities of knowing what happens to phosphorus in soil and in water. The LCP began to make connections and investigate them through small research projects. In turn, it had many surprises that opened out more connections. The connections that the LCP investigated didn't lead to stable ground (Hinchliffe 2007) and they didn't give us the answers; rather, they opened out more questions and they suggested more work to be done together.

In time, as LCP participants began to work together, we began to ask less questions about what is really happening in the lake and to ask many more questions about what ecological and socioecological assemblages the algae are a part of, what their relations have been in the past, and what they are at present. These questions extended well beyond an understanding of cyanobacteria themselves, and opened out into questions about how they had come to be there, their historicity, their materiality and their relations. The assemblages that cyanobacteria were part of came to matter. From the minutes of LCP meetings,[9] one can see how LCP participants began to connect cyanobacteria to other matters, such as:

- a concern for the spawning grounds of brown trout (third LCP meeting);
- the use of household detergents amongst Loweswater residents (fourth LCP meeting);
- a new piece of European policy legislation – the EU Water Framework Directive (2000) (fourth LCP meeting);
- the maintenance of the banks of streams ('becks') that flow into Loweswater (fifth LCP meeting);
- the existence of well-functioning and less-well-functioning septic tanks within the Loweswater catchment (sixth LCP meeting);
- changing rainfall patterns (eighth LCP meeting);
- changing patterns of, and futures for, farming in Loweswater, Cumbria and Europe (ninth LCP meeting);
- the presence of vast numbers of species of midge larvae that thrive alongside cyanobacteria in de-oxygenated water in Loweswater (tenth LCP meeting).

Not only was the LCP able to raise these matters, it was also able to investigate aspects of them. During 2009 a sum of money included in the funding proposal was made available to the LCP to commission its own small-scale research studies. This budget of UK£35,000 enabled LCP participants to find out more about the issues that concerned them. Five studies were designed by LCP participants, picking up on the research linkage and themes noted above. For example, a survey of the functioning and use of all septic tanks in the catchment was carried out by local resident Leslie Webb (Webb 2010). A study of attitudes to tourism and the economic future of the valley was carried out by local residents David Davies and Emer Clarke (Davies and Clarke 2010). A study to juxtapose physical diatom data from lake sediment samples against historical data for land-use change in the catchment (thus relating the presence of cyanobacteria and other algae to land-use practices over the last 200 years) was carried out by Helen

Bennion and a historian at Lancaster University, Angus Winchester (Winchester and Bennion 2010). A study to collect more data to understand how agricultural phosphorus fertilizer applications relate to phosphorus losses from agricultural soils to watercourses was carried out by Stephen Maberly, with the help of local farmers. And a study of the large-scale hydrological movements in the entire catchment was carried out by a consultant hydro-geomorphologist in collaboration with the National Trust (Haycock 2010). As each study came to an end, those that had been involved wrote a report for the LCP. During two different evening meetings in 2010, these reports and the results they contained were presented to LCP participants. As always, the facts that were presented were questioned and scrutinized by LCP members: nature was not held to be self-evident; knowledge and expertise had to be debated; and doubt and questioning was extended to all the LCP's representations. Uncertainties were appraised. But what such studies also showed was that the LCP was building on the connections it saw as related to the algae in the lake, and was creating its own knowledge around these connections. The LCP, in other words, was extremely active in the creation of connections between people and things.

In spawning these discrete studies as well as a host of other smaller investigations, the LCP was coming to know cyanobacteria more closely through connections that were understood to matter. The algae in the lake became more real through these connections: they also became more multiple (Hinchliffe 2007, 175). This participatory forum was starting to know and understand cyanobacteria in a way that did not allow these organisms to be silently represented by 'S'cience. The participants in the forum, to put it another way, eschewed 'S'cience, but became a group of scientists! In doing so they learnt how to ask questions about the algae, and their relations, in ways that somehow brought these organisms and human participants together into a kind of creative encounter. We interpret this as a kind of experiment with intensities. We conclude with some thoughts on what it means to suggest this.

An ethic of experiment with intensities

In her writing about philosophy and science, Isabelle Stengers describes the way in which science can be seen as a local, selective process where links and knots are made between two parties (Stengers 2005). The two parties are non-humans that can be seen as experimentally reliable witnesses; and humans, their competent colleagues. She suggests that this is a kind of marriage, but that this marriage is not a conventional scientific fairy tale – a convergence of 'man' and 'nature'. Rather, she suggests, it is a divergence. Each scientific event produces a different kind of 'adventure' between the human and the non-human. The adventure itself produces very strong and specific obligations and loyalties.

At Loweswater, over three years, many of these kinds of adventures were entered into between non-human witnesses and their human colleagues. The original basis for doing so, and the rationale for the LCP as a whole, rested on a critique of many previous public participation initiatives (Tsouvalis and Waterton 2012). What the LCP tried to resist was 'framing' the issue of lake water pollution too narrowly, or allowing it to be defined and dealt with solely by the relevant environmental agencies such as the National Trust (who owns the lakebed) and the Environment Agency (who are the responsible authority for water quality in the lake (ibid.)). The LCP was committed to an ethos of critique and dissent, partly introduced through the LIP, as vital to ways of thinking, and doing, differently. It manifested a fulfilment of the hope, expressed by Stengers (2005, 160), that 'there may be a small, precarious possibility, part of our epoch, that a new kind of public is emerging'. This new public, as she saw it, was composed of 'emerging and stuttering objecting minorities' who were producing 'the power to object and to intervene in matters which they discover concern them'.

Some of the interventions of the 'minority publics' at Loweswater have been described above. The interventions of the participants of the LCP opened out many new lines of enquiry, which in turn were generative of new adventures, new loyalties and new contracts between 'nonhuman witnesses' and human colleagues. But, of course, what is notable about these adventures is that they do not add up to a consensus, or to an overarching conclusion about cyanobacteria, and what humans should do about them, at Loweswater. Instead of finding answers to the very first questions of the LCP ('How has the lake changed?'; 'What *are* green-blue algae or cyanobacteria?'; 'How did they come to thrive so vigorously here?'; 'Can we get rid of them?'), the LCP, and the matters that it made a home for, spawned new, multiple, questions and connections. In some ways the LCP became a ferociously active mechanism of enquiry, concerned about matters related to the algae.

We found, as we carried out the research, that our approach was attracting the attention of other groups and environmental agencies. The LCP was asked to explain its ways of working to other community-led groups, to local Rivers Trusts and, via a workshop for national representatives, to organizations with responsibility for managing catchments under the Water Framework Directive. The LCP provided inspiration for the UK's Department for Environment, Food and Rural Affairs (Defra) who subsequently launched the Catchment-Based Approach (CaBA) which now has coverage across England. Thus, in some ways, the LCP approach – the creation of a 'new republic of heterogeneous associations' – seemed to be an approach that people were willing to try elsewhere. This take-up of the LCP model seemed to chime with a mood towards accepting different versions of collective experimentation (e.g., as seen in Defra's CaBA).

An interesting point, in this context, was that the kinds of questions that excited LCP participants were not *only* analytical or managerial questions (e.g., how do 'modern' farming systems and cyanobacteria co-produce one another, and what allies do they have in this co-production? What part do the practices of ordinary householders play in Loweswater's water quality? What can we do about that?). Rather, the way in which facts and representations, and their coming into being through apparatuses, were questioned and opened out by the LCP allowed for a particular kind of intensity in the relationship between the non-human witnesses and their human colleagues. Thus, loyalties were created between non-humans and humans that did not require proclamations of certainty, direct cause–effect relationships or even epistemological grounding. The LCP came to acknowledge how much we humans do not know, perhaps cannot know, about cyanobacteria, their lives, their livelihoods and their relations with others. In some ways, the LCP even began to think positively about their future with cyanobacteria. At a recent LCP meeting that one of us attended in 2013, one participant quite happily, if a little fatalistically, remarked as we were leaving the village hall: 'Oh, *they'll* [cyanobacteria] always be here. And *we'll* always be here. We'll *always* be here ... thinking about the lake ... and the algae!'

This shift, perhaps towards what Haraway calls 'becoming with' (Haraway 2008), supports our tentative suggestion that the LCP began, in its three years of experimentation, to move towards a post-human sensibility. And this, perhaps, may be where the LCP differs from other examples of experimentation – perhaps other examples have not gone this far? This is not a post-human or 'post-everything' move towards 'one single life-affirming ecological totality' which some authors worry about (Colebrook 2012, 14). Rather, it is more like a letting go of many inherited ways of thinking 'in order to make sense of the complexities we find ourselves in' (Braidotti 2013, 11). And perhaps this was the most challenging and risky part of our collective work together. 'Letting go' of our inherited ways of thinking in order to acknowledge complexity is an extremely demanding thing to do individually, let alone as part of a collective of people and things that are themselves in flows of relations, obligation and responsibility. Even *within* an acknowledgement that the humanism and anthropocentrism that have dominated Western thought need to give way to a new relationship to the environment (Colebrook 2012), it is hard to make that leap. In the LCP it did feel as if we had begun to think of ourselves less as 'self-determining subjects whose relation to the world is one of representations (knowledge) or use', and a move towards supplanting that idea with relations of care, concern or respect (Colebrook 2012, 14). It felt like we were making the move, perhaps, to 'becoming with' others in some kind of companionate relations (Haraway 2008). But these have to be seen as relations that are unfolding, or becoming. As Haraway (2008) notes, they are not the *result* of scientific understanding, they are the '*condition* of understanding' (p308, note 19, emphasis in original). Thus, in a sense, in opening up this experiment, we could never fully know where we were heading, or how many other participatory initiatives might embrace such open-endedness.

What we are saying, then, is that we recognize that the 'post-humanist' label has to be approached cautiously – not so much as a jettisoning of previous ways of thinking, but as an opening out into what it may become. As Braidotti (2013) suggests: 'At this particular point in our collective history, we simply do not know what our enfleshed selves, minds and bodies as one, can actually do. We need to find out by embracing an ethics of experiment with intensities' (p190). Through this chapter, and through the example of the LCP, we can see that such an ethic has to be an adventure, a journey, an opening out into new relationships. Whether other participatory initiatives within the CaBA supported by Defra will take on the philosophical approach of the LCP, we do not know, but the LCP does currently continue as a Defra-sponsored, community-run participatory forum working in partnership with the West Cumbria Rivers Trust. Through our own continued participation in this forum, we sense that this experiment with intensities, this particular 'adventure' between the human and the non-human, may be ongoing for the foreseeable future.

Notes

1 We developed the concept of a 'new collective' from Latour's *Politics of Nature* (2004).
2 The term 'eutrophication' refers to a process where an increase of nitrate or phosphate in fresh water bodies encourages excessive biological productivity in the lake. A cycle ensues of growth and decay, which can lead to de-oxygenation of lake waters and sediments, and hence to loss of biodiversity and life within the water body.
3 Lake ecologists studying the type of problem observed at Loweswater often refer to phosphorus as the 'limiting factor' of cyanobacteria. If phosphorus is not present, in other words, cyanobacteria cannot flourish.
4 The research reported in this chapter was supported by the RELU programme through the project 'Understanding and Acting in Loweswater: A Community Approach to Catchment Management', RES-229-25-0008. We would like to thank the other researchers on the team for their contribution to the research reported in this chapter: Ken Bell, Dr Lisa Norton, Professor Stephen Maberly, Dr Nigel Watson, Dr Ian Winfield.
5 For further elaboration see glossary entry in Latour (2004, 249).
6 These practical details are also rehearsed in Waterton *et al.* (2015a).
7 See Reynolds (1984).
8 Stephen Maberly was able to explain the complex processes of the way that phosphorus can attach to sediment and later be released back into the water column several times within LCP meetings but this quotation comes from a later interview in 2014. Interview with Stephen Maberly, Centre for Ecology and Hydrology, Lancaster, 11 September 2014.
9 These can be seen on www.lancaster.ac.uk/fass/projects/loweswater/noticeboard.htm.

References

Barad, K. 2003. Posthumanist performativity: Toward an understanding of how matter comes to matter, *Signs* 28(3): 801–831.
Barad, K. 2007. *Meeting the Universe Halfway: Quantum Physics and the Entanglement of Matter and Meaning*. Durham, NC and London: Duke University Press.
Braidotti, R. 2013. *The Posthuman*. Chichester: John Wiley & Sons.
Callon, M. 1999. The role of lay people in the production and dissemination of scientific knowledge, *Science, Technology & Human Values* 4: 81–94.
Chilvers, J. 2010. *Sustainable Participation? Mapping Out and Reflecting on the Field of Public Dialogue on Science and Technology*. Report commissioned by ScienceWise, London.
Colebrook, C. 2012. Introduction: Framing the end of the species. In: C. Colebrook (ed.) *Extinction*. Open Humanities Press, www.livingbooksaboutlife.org/books/Extinction/Introduction (accessed 2 December 2014).

Davies, D. and Clarke, E. 2010. *Community and Culture: Tourism in a Quiet Valley*. Report to the Loweswater Care Project. Lancaster: Lancaster University.

Felt, U. and Fochler, M. 2008. The bottom-up meanings of the concept of public participation in science and technology, *Science and Public Policy* 35(7): 489–499.

Haraway, D. J. 2008. *When Species Meet*. Minneapolis: University of Minnesota Press.

Haycock, N. 2010. *Hydrogeomorphological Investigation of the Main Streams Feeding Into and Out of Loweswater*. Report to the Loweswater Care Project. Lancaster: Lancaster University.

Hinchliffe, S. 2007. *Geographies of Nature: Societies, Environments, Ecologies*. London: SAGE.

Lane, S. N., Odoni, C., Landström, S., Whatmore, N., Ward, S. and Bradley, S. 2011. Doing flood risk science differently: An experiment in radical scientific method, *Transactions of the Institute of British Geographers* 36: 15–36.

Latour, B. 1993. *We Have Never Been Modern*. Trans. Catherine Porter. Cambridge, MA: Harvard University Press.

Latour, B. 2004. *Politics of Nature: How to Bring the Sciences into Democracy*. Cambridge, MA: Harvard University Press.

Latour, B. 2005. From realpolitik to dingpolitik or how to make things public. In: B. Latour and P. Weibel (eds) *Making Things Public: Atmospheres of Democracy*. Karlsruhe and Cambridge, MA: Center for Art and Media Karlsruhe and MIT Press, pp4–31.

Leach, M., Scoones, I. and Wynne, B., eds. 2005. *Science and Citizens: Globalization and the Challenge of Engagement*. London: Zed Books.

Marres, N. 2012. *Material Participation: Technology, Environment and Everyday Publics*. Basingstoke: Palgrave Macmillan.

Marres, N. and Lezaun, J. 2011. Materials and devices of the public: An introduction, *Economy and Society* 40(4): 489–509.

Marris, C. and Rose, N. 2010. Open engagement: Exploring public participation in the biosciences, *PLoS Biology* 8(1)1: e1000549.

Norton, L., Elliott, J. A., Maberly, S. C. and May, L. 2011. Using models to bridge the gap between land use and algal blooms: An example from the Loweswater catchment, UK, *Environmental Modelling and Software* 36: 64–75.

Reed, M. S. 2008. Stakeholder participation for environmental management: A literature review, *Biological Conservation* 141: 2417–2431.

Reynolds, C. S. 1984. *The Ecology of Freshwater Phytoplankton*. Cambridge: Cambridge University Press.

Stengers, I. 2005. Deleuze and Guattari's last enigmatic message, *Angelaki* 10(2): 151–167.

Tsouvalis, J. and Waterton, C. 2012. Building participation upon critique: The Loweswater Care Project, Cumbria, UK, *Environmental Modelling & Software* 36: 111–121.

Waterton, C., Maberly, S.C., Norton, L., Tsouvalis, J., Watson, N. and Winfield, I. 2015a. Opening up catchment science: An experiment in Loweswater, Cumbria, England. In: L. Smith, K. Porter, K. Hiscock, M. J. Porter and D. Benson (eds) *Catchment and River Basin Management: Integrating Science and Governance*. London: Earthscan/Routledge, pp183–206.

Waterton, C., Maberly, S.C., Tsouvalis, J., Watson, N., Winfield, I.J., Norton, L.R. 2015b. Committing to place: The potential of open collaborations for trusted environmental governance, *PLoS Biology* 13(3), e1002081–e1002081.

Waterton, C., Norton, L. and Morris, J. 2006. Understanding Loweswater: Interdisciplinary research in practice, *Journal of Agricultural Economics* 57(2): 277–293.

Webb, L. 2010. *Survey of Local Washing Practices and Septic Tank Operation in Relation to Domestic Phosphorus Inputs to Loweswater*. Report to the Loweswater Care Project. Lancaster: Lancaster University.

Welsh, I. and Wynne, B. 2013. Science, scientism and imaginaries of publics in the UK: Passive objects, incipient threats, *Science as Culture* 22(4): 540–566.
Winchester, A. and Bennion, H. 2010. *Linking Historical Land-Use Changes with Paleolimnological Records of Nutrient Changes in Loweswater Lake*. Report to the Loweswater Care Project. Lancaster: Lancaster University.
Wynne, B. 2007. Public participation in science and technology: Performing and obscuring a political-conceptual category mistake, *East Asian Science, Technology and Society: An International Journal* 1(1): 99–110.

11
AGAINST BLANK SLATE FUTURING

Noticing obduracy in the city through experiential methods of public engagement

Cynthia Selin and Jathan Sadowski

Making futures visible: raising obduracy in technology assessment

Most people interact with cities on a daily basis – the arteries that move traffic, the grid that energizes communication, the buildings that prevent and direct action – yet, this familiarity, and the careful tucking away of infrastructure (Star 1999), renders the technologies that underpin the city invisible. Such obdurate structures are pushed to the background but nevertheless are resistant to change and 'only seem to attract attention when they fail' (Hommels 2005, 325). It is easy to take for granted something that seems to have always been there, but the consequence of this is that we place the intricate socio-technical systems that constitute cities into a black box (Guy *et al.* 1997). Cities are complex, dynamic patterns, yet also immobile and stable, opposing change in numerous ways.

This black-boxing curtails opportunities to reflexively 'see' and redesign urban socio-technical infrastructures. At a time when it is imperative that cities are rejuvenated (or freshly built), with attention to sustainability, obdurate systems should come into the equation, and be brought from the background to the foreground. As such, we argue that obduracy is an important, but overlooked, conceptual lens that broadens and deepens the possibilities for grounded critique of emerging technologies and thus should be incorporated into efforts designed to publically reimagine technological systems. As we shall explore in this chapter, directing the gaze toward obduracy in public engagement practices offers leverage for citizens to consider what changes are not only desirable, but also which futures are plausible (and not just possible).

This work is rooted in the understanding that technology assessment, whether grounded in public participation or expert elicitation, is *prospective* and thus inevitably tied to notions of time and change (see also Chapter 9 in this volume). Reimagining change is a central component of anticipatory governance, which, alongside other approaches to the governance of emerging technologies, seeks to nurture 'the ability of a variety of lay and expert stakeholders, both individually and through an array of feedback mechanisms, to collectively imagine, critique, and thereby shape the issues presented by emerging technologies before they become reified in particular ways' (Barben *et al.* 2008, 992). Note that the rationale for anticipatory governance also rests on this notion of obduracy, highlighting that the reification of technological systems broaches the Collingridge dilemma (1980): governance is caught between lacking enough information in the present and waiting for perfect knowledge, thus inviting the risk of systems becoming too entrenched for effective change to occur.

When publics are asked to critically appraise emerging technologies, they are implicitly asked to tend to both things that might change and those that are more stagnant. However, it is more common in practice to see open-ended changes relayed to citizens, asking them to express their future desires without a more nuanced understanding of the temporalities of socio-technical change. While the public engagement with science has been a mainstay of STS work at large (Stirling 2008; Delgado *et al.* 2011), and technology assessment methods more particularly (Barben *et al.* 2008; von Schomberg and Davies 2010), few studies have worked to isolate the role of temporality and the future. To be sure, there has been some progress on this front (Felt and Fochler 2009; Davies 2011), but it is nonetheless the norm to see the future treated as a blank slate. There is a glaring need for a finer, more rigorous incorporation of time and futures into public engagement theory and praxis. Beyond a simple accounting of new methodologies, these enquiries, following the charge of this book, should look to the way in which attention to temporality is mediated through the framing, procedures, technologies and institutional infrastructures of public engagement.

We propose that Anique Hommels's (2005) three conceptual dimensions of obduracy offer clues for how to approach and study different aspects of obduracy. Hommels sees that obduracy – as frames, as embedded and as traditions – provides a way to more richly and reflexively account for technology and advance the grounded appraisal of it. While Hommels has been criticized for lapsing too far into technological determinism (see Kirkman, 2009, for a discussion), we contend the wholesale relinquishing of obduracy shields critical analysis from having to robustly look at 'the future' and how it maps onto what already exists. In the context of public engagement, there is much worth preserving and reviving in Hommels's work.

Take, as a case, technology assessment of nanotechnologies. Nanotechnology is most often associated with high-tech lab equipment or new materials; however, a range of different applications have the potential to either become obdurate, moulding future trajectories for better or worse, or else be blocked from realization due to stubborn material, economic, social or technological systems that resist integration with new nanotechnologies. Nanotechnology is still arguably in its early phases of research and development (R&D), but firmly in place is a mythic momentum around what it can and will accomplish. Associated with the term nanotechnology are, according to Christine Peterson, director of Foresight

Institute, 'huge expectations, as a long-term, exotic, extreme technology' (as quoted in Selin 2007, 212), and an unwavering commitment to the idea that new technologies equal progress (Marx 1987), as evidenced in policies such as the National Nanotechnology Initiative, which continue to pump billions of dollars into R&D. The persistent rhetoric fuelling investments in nanotechnology has already begun to construct new institutional, political and economic obduracies, all ripe for interrogation through public engagement.

As R&D continues, nanotechnologies will, in some form, become embedded in the urban environment. Nanotechnologies are presented as both an 'enabling technology (on top of other technologies) or a platform (below other technologies) to deliver complementary technologies' (Wiek *et al.* 2012, 16), which suggests an easy integration into prevailing systems. Wiek *et al.* (2013) and Shapira and Youtie (2012) describe a number of such nanotechnology applications: coatings for buildings that make them 'self-cleaning'; advanced photovoltaic materials that power structures; paints that eliminate glare and do not absorb heat; water filtration systems for individual households; and tailpipe membranes that reduce pollution from vehicles. What most of these applications have in common is how they embed into present sociotechnical networks: they are expected to be layered, painted or otherwise coated onto the already existing world. However, the *ease* at which nanotechnology will be seamlessly layered into the city is questionable, and tending to obduracy helps tell a fuller, more complicated, story about the promised ubiquity of nanotechnologies.

It is with this concern for appreciating obduracy and the constraints, obstacles and surprises attending nanotechnologies that researchers at the Center for Nanotechnology in Society at Arizona State University (CNS-ASU) designed a public engagement exercise to interrogate the future of nanotechnology and the city. As it stands, nanotechnologies have a potential for impending and pervasive ubiquity, yet this also means that they are hard to 'see' and draw into the foreground, and thus provide a good case study for how to draw out obduracy. In autumn 2012, CNS-ASU deployed the pilot for the 'Futurescape City Tours' (FCT), a public engagement research project centred on nanotechnology that featured an urban walking tour of Phoenix, AZ.[1] The three-tiered intervention – an orientation meeting, a walking tour of the city and a deliberative session – mixed ideas of obduracy in concert with imaginative speculation and community-based visioning about the future of nanotechnology in the urban and built environment.

The remainder of this chapter hones in on this public engagement research project and describes the novel set of methods that approach temporality and obduracy in a studied fashion. After illustrating obduracy in an urban context, leading to a disentanglement of Hommels's three dimensions of obduracy, we then describe the Futurescape City Tours, plucking out how obduracy was treated and brought to the fore. The FCT involves an interactive urban walking tour, using photography, reflective writing and group dialogue to support critical debate about nanotechnology in the city. In this chapter we look particularly at the way in which the temporal gaze of the participants was structured in such a way to draw attention to obduracy. We will show how the FCT invites citizens to engage with the future in a tempered fashion, informed by current constraints and material circumstances, thus setting the stage for fuller engagement with temporality.

We argue that tending to obduracy is a neglected area of STS generally and relevant to technology assessment and public engagement. In dissecting temporal dispositions in practice, we also tie to perennial questions about the import and consequences of assembling participation and the variety of inputs at stake in engagement. More broadly, this chapter opens up a public engagement practice, by interrogating the underpinning ideas that inform design and looking to the practicalities of structuring future-oriented inquiries. Practices of public engagement are always imbued with thorny questions about framing and boundaries, about what's up for discussion and what is hidden from view, and this piece contributes to broader discussions in STS about the co-production of participation. Tending to obduracy and how it is rendered in a participatory process is part of the larger impulse to develop a critical engagement with science and technology, which necessitates a reflexive stance about the politics, power and mediation of such processes.

Understanding obduracy in an urban context

Humans design values into city structures and through them forge social orders. As the sociologist Thomas Gieryn (2002) puts it: 'Buildings stabilise social life. They give structure to social institutions, durability to social networks, and persistence to behaviour patterns. What we build solidifies society against time and its incessant forces for change' (p35). In understanding obduracy it is useful to keep in mind the ever-present ways that technologies of all kinds mediate social life (Latour 1992, 1994, 2002). Buildings, for instance, are technological artefacts, and the work of architects, designers and engineers has a direct impact upon human behaviours and social relationships – an impact that imbues buildings with moral characteristics (Verbeek 2006, 2011).

It is well understood that material reality shapes experience. Foucault's (1995) exploration of Bentham's panoptical prison relays how architecture is fraught with power relationships that discipline subjects. Winner's (1986) description of the 200 low-hanging overpasses on Long Island that were designed by Robert Moses, New York's 'master builder', provides another well-worn example in STS. These bridges of hard edges and concrete were built by Moses 'according to specifications that would discourage the presence of buses on his parkways' which, as Winner explains, limited 'access of racial minorities and low-income groups to Jones Beach' (p23). Regardless of the parable qualities of Winner's tale (Joerges 1999), this case study and others like it show how obdurate structures are capable of imposing long-lasting limitations on the autonomy and decisions available to entire classes of people (Kirkman 2009).

What these examples illustrate is that lasting social orders are readily instilled and enforced through 'architectures of control' (Lockton 2005) – perhaps narrated by just one powerful person, like Robert Moses, but smeared out in a cultural ethos of a particular time and place. As values and norms change over time, obdurate structures resist transformation and continue to urge and direct behaviour in specific ways. Even in those cases where citizens might welcome decay or destruction because it allows a fresh start, urban structures – whether freeways, canals or zoning laws – remain built to last.

Obduracy is not a characteristic restricted to urban infrastructures. Many other everyday technologies resist change and adaptation and exert social influence. There are many well-trod examples of artefacts persisting against the odds and how people get locked in to a technology due to the systems that sprout up around it, from the *VHS versus Beta* saga (Arthur 1990) to the battle between gas and electric refrigerators (Cowan 1985). Often these are talked about in economic terms such as 'path dependency' and 'lock-in' (Liebowitz and Margolis 1995; Tiberius 2011). Path dependency refers to the way in which starting conditions or choices can lead to one outcome over another. In the words of the economist Paul David (1985, 332): 'One damn thing follows another.'

While path dependency and lock-in are commonly discussed in the economics literature, couched in terms of markets and efficiency (i.e., Technology X hit the market first, and even though it is less efficient than Technology Y, it gained a larger consumer base), studying obduracy also requires looking at a wider array of cultural contexts, cognitive perspectives, material relations, sociotechnical systems and enduring traditions.

Untangling obduracy: three levers

Urban-focused public engagement projects, inherently concerned with change and transition, must consider obduracy or else risk forging an incomplete picture of the technology in question. As the philosopher Robert Kirkman (2004, 213) explains:

> It is relatively easy to articulate a grand vision of the city of the future, one that is radically different from what we have today; it is not nearly so easy to get there from here. Whatever high principles we may invoke, however artfully we may strike the theoretical balance between individual well-being and the demands of justice, the real world is likely to resist, to complicate, and ultimately perhaps to frustrate our attempts to put theory into practice.

We propose that visioning in public engagement exercises without consideration of obdurate sociotechnical systems is dangerously incomplete and probably wasteful. While there are many competing visions of nanotechnology (or other emerging technologies from synthetic biology to geoengineering), there is little consideration to, first, what existing products, systems and values are in place that might thwart the vision; or, second, how that new technology might create a new set of obduracies that stymie adaptability in the longer term. Instead, visioning processes often yield ungrounded laundry lists of concerns. Technology assessment in most forms is about creating 'participatory, deliberative processes [that] will stimulate efforts to enhance desirable impacts and mitigate undesirable ones'

(Guston and Sarewitz 2002, 106), and thus should go beyond regurgitating 'grand visions' or list-making, and critically engage with normative values with the ballast of obduracy for grounding.

Anique Hommels (2005, 2008) has laid out much of the foundational work to study the creation and effects of obduracy. Her framework offers a starting point to consider how obduracy can be knitted into technological assessment and public engagement. Hommels (2005) describes three different conceptual dimensions that together represent a 'productive fusion between technology studies and urban studies' (p1). The following subsections describe the concepts, approaches and traditions that constitute each of the dimensions.

Frames

The primary focus for frames is on the entrenched worldviews, frameworks and 'mental models' (e.g., Gorman and Carlson 1990) that different actors who design and implement technologies bring to the table. Often actors who represent different frames, such as architects and engineers, clash with each other, which can impede or halt technological changes that require joint decision-making. Hommels (2005) argues that 'as an interactionist conception of obduracy, this category highlights the struggle for dominance among groups of actors with diverging views and opinions' (p331). So in this sense the dimension centres on conflicts that people and groups face when dealing with each other.

This dimension of obduracy is derived from the work done by Wiebe Bijker on 'technological frames' as well as related concepts associated with the social construction of technology (e.g., Pinch and Bijker 1987; Bijker 1995). In relation to the study of obduracy, Bijker (1995) argues that social groups can invest so much into an artefact that it becomes part of their 'technological frame' – that is, their methods, values, goals, theories, perspectives and key problems as they relate to technologies in the world. Once this happens, the artefact's meaning becomes fixed – 'it cannot be changed easily, and it forms part of a hardened network of practices, theories and social institutions' (p282). In addition, it is important to note for whom the artefact is obdurate. An actor with 'high inclusion' in the frame is unable to see any other kind of design, while an actor with 'low inclusion' must choose to either accept the artefact or abandon it.

Understanding obduracy through technological frames draws attention to how worldviews, values and convictions can cause the meanings associated with artefacts to become ingrained and inflexible. A defining feature of this dimension is that obduracy is not just a material characteristic; it also arises from the social interactions of groups who have different frames that constrain their thinking.

Embeddedness

Technologies do not exist in a vacuum and it is important to take into account the relational aspects of how they are interconnected in ways that can cause change in one system to translate into change in another. Understanding this dimension of obduracy involves a focus on 'technology's embeddedness in sociotechnical systems, actor-networks, or sociotechnical ensembles' (Hommels 2005, 334). Like the roots of a tree that tangle further into the earth, the deeper an artefact is embedded in a network or system of some kind, the more resistant it is in the face of any attempts to change it.

Hommels builds up this embeddedness dimension from Actor-Network Theory (e.g., Callon 1986; Latour 1987, 2005). As more actors – both human and nonhuman – become linked together into complex sociotechnical networks, they can stabilize into obduracy. New dependencies form, and diverse actors and actants become reliant on the relations. Hommels (2005) draws particular attention to Latour's study (1988) of the subway network that snakes beneath Paris. In Latour's account, over time the subway grew, but the original design remained static. Seventy years later the railroad companies and the subway companies decided it would benefit them to link their networks together, but by this time the material structure of the subway had become complex, large and obdurate. It was not enough for the executives, engineers and politicians involved to agree to this new plan of changing the subway so that it linked with the railway. Hommels quotes Latour's apt description:

> What could have been reversed by election seventy years ago, had to be reversed at higher cost. Each association made by the socialist municipality with earth, concrete and stones had to be unmade, stone after stone, shovel of earth after shovel of earth.
> (Latour 1988, 37, as quoted in Hommels 2005, 335)

Latour's study helps to illustrate how increasing degrees of obduracy can arise as an emergent characteristic of different actors becoming interlinked and deeply embedded in strong sociotechnical networks.

Persistent traditions

The last feature of obduracy that Hommels (2005) summons focuses on how 'long-term shared values and traditions keep influencing the development of a technology throughout a longer period of time' (p338). This dimension takes its cue from scholars in the history of technology and explores how wider structural contexts, enduring traditions and large-scale sociotechnical systems contribute to the enduring quality of obduracy. The conception of technological momentum described by T. P. Hughes (1987, 1994) illustrates how technical systems acquire obduracy as they become further entrenched in society and culture. When a system is young it is easier for societal and cultural pressures to shape and change how the system's characteristics develop. But as the system matures it picks up momentum – or as Hughes called it, 'dynamic inertia' – and grows resistant to attempts at changing it. That is, the larger and more 'socio' a technical system becomes, the harder it is to affect the system.

Adding further detail to Hughes's framework of technological momentum, Hommels (2005) draws on Anders Gullberg and Arne Kaijser (1998) who develop the concept of 'City Building Regime'. This idea emphasizes the complex interactions between an array of different systems within the city, such as infrastructure, cultural attitudes, political actors and economic circumstances. Moreover, consistent themes or 'archetypal designs' can direct how features of the urban environment are planned and built (Kitt Chappell 1989). Such established archetypes have a lasting influence on architectural design, which creates obduracy by, in part, causing the original values embodied in the design to persist over time.

With this, the question becomes: how can Hommels's dimensions of obduracy help us to design and understand public engagement with science and technology that approaches anticipation in a more nuanced fashion?

Novel methods for noticing obduracy

As we have seen, breaking down obduracy to make way for the new or different (in terms of things such as an urban garden, or community values like sustainability) is a feat that often requires inordinate quantities of energy and intention as sociotechnical systems settle into a semblance of permanence. We argue that the first step to overcoming obduracy is broader awareness of what are otherwise invisible features of the city and that public engagement practices should find ways to explicitly inject such hidden features into the conversation. Any effort to assess, shape or direct technological development has to take the multiple facets of obduracy into account, yet current models of public engagement with such goals regularly fail on this front. Since we recognize that obduracy can be abstract and hard to pin down, we propose that efforts to engage the public in these matters can be fruitfully situated in experiential visual and digital methods.

The CNS-ASU, through the 'anticipatory governance' framework (Barben et al. 2008; Guston 2008, 2014b), designs, implements and assesses novel forms of public participation. Like many of CNS's other public deliberation efforts, the Futurescape City Tours focuses on a critical appraisal of nanotechnology (Cobb 2011; Guston 2014a). However, the FCT differs in its direct connection to urban landscapes and focuses on beginning with citizen concerns and voices rather than starting the enquiry with attention to technological promises and risks. The FCT is also marked by moving away from developing policy recommendations, to instead suggest that capacity-building is a worthwhile outcome in and of itself (Selin et al. forthcoming). While there is much to be said about these features, and how the tours handled recruiting citizens, integrating experts and managing expectations, this description touches only lightly on those important concerns to focus on how obduracy was treated in the FCT as a means to more robustly account for alternative future trajectories. It is worth noting that the FCT is meant to prototype a new method of public engagement, and thus the work at the CNS provides a test and demonstration of the methodology. Even within this context of experimentation, however, we worked to assemble a group of citizens roughly representative of the city and managed to gather a group who roughly matched the local demographic in light of age, gender, ethnicity, income and education level, etc. We were, however, excluding those with limited mobility (in our reliance on walking) and with limited sight (in our reliance on photography). Thus there are also some hidden obduracies built into a public engagement method that relies upon walking and seeing.

During the walking tour, the FCT participant group visited particular sites in the city, derived from their collective 'concerns and curiosities' unearthed during an initial three-hour-long orientation session. Critically important for us as researchers (and worth a longer discussion) was not to over-determine the citizens' gaze by suggesting particular future applications of nanotechnology worth critical appraisal; instead, we began eliciting participant's own 'concerns and curiosities' about the future of the city. At each stop along the tour, citizens met with community leaders, scientists and engineers, politicians or other stakeholders, and engaged in mutual sharing, learning and questioning in a variety of different settings – canal, metro light rail, rooftop solar installation, science museum, public square, graffiti alley – and in different formats – panel discussion, plenary discussions, small group dialogues, one-on-ones, DIY demonstration and so on. In addition to discrete 'stops', citizens were encouraged to pause along the way and reflect on what they were noticing and take photographs of how they saw the role of technology in transforming the urban environment.

Unlike traditional public engagement exercises[2] that are far removed from the obduracies and possibilities of technology in the city, this exercise was designed for lay publics to experience an intimate and experiential connection with the city. We asked participants to touch, hear, smell and move with technology in the city. We designed interactions to enable the citizen participants to embrace the fact that the places we inhabit orient our ways of – to put it in phenomenological terms – being in the world (Casey 1996). Noticing obduracy was primed in a number of ways: the experiential design of the tour; the framing of technology as social, embedded and systemic; the use of photography; and the prompts during the deliberative session. We will now pluck out a few exemplary instances from the 2012 pilot of the tour itself where attention to obduracy was evident.

At the start of the tour, each participant was given a notebook for use during the daylong tour that contained maps, writing space, portraits of nanotechnology in the city, and information about our stops. The notebook also contained old pictures of locations they would visit on the tour route, each with prompts to promote reflection about temporality and change. For instance, one photo from 1911 showed a street corner in Phoenix; that same street corner is now replete with a sports stadium in sight, the light rail, Urban Outfitters and banks. We asked participants to consider: 'What's here today? How has this street changed? What will this area look like 100 years from now?' Another historic photo featured the canals that ripple through Phoenix. The turn-of-the-century photo featured a bucolic scene and asked: 'Canals have remained, but they are no longer lined with trees. How has the lack of trees changed how we use the space around the canals?' These open-ended questions did not address obduracy head on, but sought to draw attention to what changes and what stays the same and with what sorts of consequences.

One of the first stops of the urban walking tour involved taking a closer look at solar energy technologies. For the most part, technologies such as photovoltaic (PV) panels can be integrated into the urban environment with relative ease. There's a wide spectrum of options and possible applications: parking lot arrays; rooftop installations of many different sizes; and even as part of public art projects. Although urban technologies such as PV are all around, they are often unseen or unnoticed as we go about our daily lives; they are either hidden from view or they blend into the cityscape. A goal of the experiential tour was to point out such technologies and discuss their embedded qualities. As PV advances and becomes nano-enhanced, engineers anticipate that they will become cheaper and more flexible (Regan *et al.* 2012; Park *et al.* 2013), which might then make them easier to install or integrate directly into structures. However, the following vignette from FCT helps to illustrate how in some cases the built environment resists change from technologies, like PV, that are designed to seamlessly integrate into current structures.

To observe issues surrounding energy and the city we first went to the Bioscience High School (HS), a STEM-oriented magnet school in downtown Phoenix. Recently, a large energy firm, NRG Solar, donated a PV array to the school. As the citizen group examined the school's new installation and spoke to an engineer and the school's principal about solar energy and how it fits into the cityscape, we found out that the Bioscience HS was not the first intended location for the PV array. A representative from NRG Solar told us that they initially wanted to donate and install the panels on a food bank or a women's shelter. But without extensive renovations the rooftops of these places were not strong enough to bear the PV's weight for the 20 (or more) years that were required. Hence, the locations that could have perhaps made the most use of PV were unable to take advantage of a charitable donation and incorporate the technology into their structure. This proved to be a tangible example of Hommels's embeddedness dimension of obduracy; in this case, the material relations presented a barrier to desirable technological change.

After finishing at the Bioscience HS the citizen group walked down a nearby alley that was covered in elaborate, and often politically charged, graffiti. We used this opportunity to talk to the citizen group about the existence of nano-technology coatings – designed to be applied directly onto buildings – that prevent things like spray paint or dirt from sticking to surfaces. The participants began to question how these coatings might change the city, and discussed how certain cultural values are preserved or rejected, aided by a new technology. They asked: is graffiti undesired vandalism or an artistic community expression? A popular photograph of the graffiti taken by the participants, one that garnered much discussion in the deliberative session, was a negative portrayal of the local (infamous) Sheriff Joe Arpaio and Arizona's governor Jan Brewer, both of whom are very conservative public figures. By walking through an interstitial alley populated with graffiti, which many people did not know existed, questions bloomed about cultural aesthetics, political resistance and the role of technology in modulating expression. Citizens questioned how the cultural values obdurately persist, or do not, through the quiet use of technology.

At another stop along the walking tour, the group paused to observe a parking lot that had been treated with a 'cool slurry' nano-coating for concrete and asphalt that reflects, rather than absorbs, solar rays. The coating, developed by a company called Emerald Cities Cool Pavement, had been applied to a couple of parking lots in Phoenix, Arizona, including a 90,000 square feet temporary lot, with plans to roll it out in 100 other cities. The idea here is that the reflective properties will help mitigate the ambient temperature and cut down on urban heat effects (Nusca 2011). The nano-coating seems simple enough, but questions still remain about what happens when the coating degrades from wear and tear and disaster. On top of potential ecological issues, who will pay to maintain and reapply the coating, especially when it is easier to just not pay any attention to the care of a large concrete lot? According to anecdotal evidence, the high tech nano-coating on the lot we observed had been relatively quickly recoated with ordinary layers due to maintenance procedures the city already had in place. In this way, the citizens got a first-hand look at obduracy in action: there are social practices which persisted (maintenance schedules and procedures) and prevented the intended value of the new-fangled material from being reaped.

This story illustrates the ways in which a positive social value (sustainability) pursues a known environmental problem (urban heat island) with a technological solution (nano-coated pavement). What are ignored, in this case, are the broader structures that embed the parking lot. This is a story of obduracy the citizens could actually see, helping them to realize that a vision for a technology must be matched by revamping an entire system. Solving the problem is not only a matter of installation, but also of maintenance, which comes with its own human and material dimensions, mired in habit, values and administrative procedures. Without taking into consideration the full 'life' of the parking lot, any technological fix is doomed to not 'stick'.

The final stop on the tour – a walking and biking trail near the intersection of a major road in downtown Phoenix with the Grand Canal waterway – offers another illustration of the obduracy in the city. Here participants learned from engineers about new bacteria-based and nano-enabled technologies for cleaning dirty water and from a social scientist about local water politics. Phoenix is a desert city and it really only exists because of the canals, which were first built almost 150 years ago. They are intricately embedded into innumerable sociotechnical networks. As such, the canals have an important influence on how the city does and does not develop. In addition, the decision-making that surrounds water in Phoenix, and Arizona in general, is fraught with political actors who inhabit different frames, which impedes much of the change – both social and technological – to the water system. Along the canal, as with other stops during the tour, the citizens were confronted with stakeholders who maintained different epistemological or professional frames, causing the participants to think about how the historical weight of these different frames persist in modern-day politics of water in the valley.

Designing attention to obduracy

In order for citizens to experience how the urban environment is constrained by legacy systems, persistent materiality and sticky ideologies, the FCT implemented some unusual methods. Realizing the limits of workshop settings and conventional discourse, while developing FCT, we built up the concept of 'material deliberation'. This laid a theoretical foundation for the methods we deployed, which enabled the participants to get close to the city and notice obduracy first hand. Material deliberation rejects public participation based solely on straight rational discourse, and instead also seeks to include attention to affect, sensorial experiences, multiple forms of expression, embodied relationships and place-based sense-making. Material deliberation, Davies et al. (2012, 353) argue, 'show[s] a sensitivity to the situated nature of all encounters, deliberative or not, as embedded in particular spaces, material configurations, and temporalities'. Towards this, our research team devised multiple mediums through which the participants could record, reflect and represent their impressions and proposals about the future of the city.

For instance, the participants in FCT were instructed to take photographs during the entire walking tour with a specific eye towards capturing the past, present and future of Phoenix. While Phoenix is a newer city, the cityscape provided many instances of the juxtaposition between old historic preserved buildings and new ones. The very act of walking through the city with a visual focus on temporality raised awareness of the tensions surrounding obduracy. The photography was intended to be a tool for enhancing obduracy-awareness. As the French writer Emile Zola's famous aphorism relates: 'In my view, you cannot claim to have really seen something until you have photographed it' (quoted in Sontag 2001, 87). In this sense photography enhanced the citizens' ability to see the city in a new and more focused way. This mindfulness aided participants in noticing and giving attention to ways that the past doggedly persists into the present as well as how new features of the city will push forward.

In order for the citizens to further iterate on their impressions from the tour, we invited participants back for a three-hour deliberative session a few days after the walking tour. This gave them the opportunity to have a richer conversation with explicit reference to the experiences and photographs from the tour. The discussions during this session made it clear that the participants were much more adept at recognizing many kinds of persistence in the urban environment. During the deliberative session, participants were asked to select photographs (from the lot that they took during the tour) that signalled the past and were invited to come to a consensus about the three most important ways in which the past persists into the present. They selected a broad range of images from those of a postman delivering mail, old buildings and PV panels, and they discussed the limits and possibilities of nanotechnology within this more grounding structure.

Through the tours, the citizens could 'see', capture and discuss how, as nanotechnologies become part of society, there will inevitably be tensions amongst different technological frames and worldviews about how best to implement them so they align with desired values. Debates around sustainability, especially pertaining to urban environments, already exemplify these clashes where some groups – often scientists and engineers – see nanotechnologies as being a quick fix, while others – often philosophers and social scientists – argue that these problems are far more complex and require social and political solutions (see Seager et al. 2012; Wiek et al. 2013). Interacting with the different experts and, importantly, with each other, the citizens recognized that whichever frame prevails will greatly influence future developments and methods for addressing such problems.

The place-based, material and experiential method of engagement that FCT represented helped participants to combine different frames, see how the past traditions affect the present, and actually pay attention to the sociotechnical systems that constitute the urban environment. As one participant mentioned, the tour helped them 'to be able to see the interconnectivity of all the systems – to connect them visually and physically'. This, in turn, helped them to gain a better awareness and understanding of their city and how obduracy is relevant.

At the end of the walking tour one participant remarked on how striking it was to become a tourist in his own city. He was born and raised in Phoenix and knew the city very well; however, he took the city for granted. By taking on a different perspective he was able to confront the city in a new way. This general sentiment was expressed by many of the participants. One person stated: 'I was surprised about what I didn't know about the things I see every day.'

Politics of obduracy in STS and public engagement

If we take public engagement as a critical exercise – as a process with normative and/or substantive goals of 'improv[ing] social outcomes in a deeper sense' (Wilsdon and Willis 2004, 39) – then the politics of these projects are important. For the most part, STS, especially in its participatory vein, emphasizes a political commitment to widespread deliberative democracy (Durant 2011). In fact, there is much in common with what political philosopher Joshua Cohen (2009) describes as 'radical-democratic' ideas, which seek to expand participation and deliberation, as well as resolve the tensions thereof. These commitments are enacted through a variety of public engagement methods – consensus conferences, citizen juries, participatory design and technology assessment, among others – but are founded on shared democratic principles and aims (Hamlett 2003). There are open questions about how effective deliberative democracy can be (Ryfe 2005) and how to account for different deliberative cultures (Sass and Dryzek 2014). Still, the 'age of engagement' in STS remains vital (Delgado et al. 2011).

While much work has been done in theorizing the politics of participation (cf. Campbell 2005; Fung 2006; Evans and Plows 2007; Stirling 2008), we argue that there is room for further reflection on the temporal elements. When public engagement exercises attempt to understand the present and envision the future, they are necessarily dealing with politics. However, we argue, such exercises are unbalanced and weakened if they do not also incorporate the myriad of ways the past persists.

During the previous few decades, social constructivist frameworks, which arose to combat technological deterministic principles, have become mainstream in STS – and rightfully so. Yet the dominant trends in STS scholarship have, in a sense, radicalized STS theory and practice: analyses and conclusions that emphasize the staying power and influence of sociotechnical systems, rather than wide-open possibilities for the future, are easily cast as deterministic. In effect, as Hommels (2005, 330) explains, 'Little space has been left for exploring the social and technical limits to what is indeed possible.' And, as Kirkman (2009, 242) notes, 'contemporary technology studies are still haunted by the spectre of determinism, the anxiety that even one admission that physical artefacts might have some inherent power of their own will set analysts on a slippery slope to Heideggerian despair'. So rather than embrace and probe sociotechnical obduracies, our political commitments to contingency and democratic intervention – worthwhile goals, to be sure – construct blinders to anything that, as Bijker (1995, 281) puts it, 'suggests that all interventions are futile'.

Indeed, STS scholars have, under the rubric of a 'politics of hope' (Coutard and Guy 2007), decried theories of urban techno-politics (Graham and Marvin 2001) that seek to show how powerful actors, institutions and interests enact sociotechnical systems that shut down options and disempower groups of people. For the hopeful, these 'pessimistic' and 'alarmist' theories amount to '"soft" forms of economic, institutional and technological determinism' (Coutard and Guy 2007, 720) when, arguably, those theories are merely describing the unfortunate techno-political economic facts of reality (cf. Graham 2002).

> The promise of durability has attracted kaisers, kings, mayors, and other megalomaniacs to the built environment. The physical-technical ensemble of the city – buildings, sewers, roads, monuments, transport networks – conveys a sense of fixity and obduracy that appeals to the political desire to make strong, lasting statements.
>
> (Weber 2002, 519)

A sure-fire way to cripple participatory democratic intervention and limit possible futures is to overlook the power dynamics and political regimes that become embedded within and ossified by sociotechnical systems, large and small.

The future is not a wide-open frontier. As we have argued thus far, the obdurate past – the concrete entanglements, dynamic inertias, sticky ideologies and clashing frames – make it so that futures are already in the making. The 'struggle over what the past could mean in the present is at the same time a struggle for the future', writes political theorist Wendy Brown (2005, 13–14). If we take a step back and look around, we can see where the past manifests all around us and how it structures society (Gieryn 2002). The obduracies of ubiquitous urban infrastructures, for example, are often invisible because they exist in the background as a platform required for other types of systems and actions (Star 1999). When they are working fine, infrastructures can be functionally invisible because people no longer notice their relationships and interactions with them. Or they can be physically invisible because they are intangible or hidden. Regardless, the infrastructures are fraught with politics as they 'mediate exchange over distance, bringing different people, objects, and spaces into interaction and forming the base on which to operate modern economic and social systems' (Larkin 2013, 330).

Deeper considerations of the persistent politics of the sociotechnical systems we live in, with and through can reveal hidden obstacles, which, in the case of public engagement, can impede participatory democratic alternatives and outcomes. It should not be surprising that politics play a role in shaping possible futures. And it is common knowledge in STS that sociotechnical systems are thoroughly political by virtue of the ways they are, among other reasons, embedded with values and visions (Winner 1980). They can mediate our lives in numerous ways (Verbeek 2011); allow, afford, deter and block behaviours (Latour 2004); be used as tools of social control, to nudge or corral people (Sadowski and Selinger 2014); and encourage technocratic or democratic ends (Feenberg 1994, 2011). However, once entrenched – materially, systemically or cognitively, to use Hommels's models (2005) – the politics and their effects are often not so easily uprooted and overwritten. While concerns about obduracy might be more apparent in the urban context – perhaps a reason why our own project, the FCT, was primed to study them – any participatory project with substantive goals should account for obduracies.

When participatory practices stemming from STS focus on (radical) possibility and contingency – sometimes exclusively, to their detriment – they neglect the obdurate world around them. The resistant and persistent characteristics of technologies, systems and worldviews are pushed out of sight and in their place stands a blank slate waiting for scholars and participants to write their own preferred futures on it. It's easy when designing and producing public engagement projects to make promises and allusions to democratic change. But theory and practice can disconnect, and efficacy is not so forthcoming. The obdurate world bogs things down, and feelings of frustration and inadequacy are the result (cf. Delgado *et al.* 2011). The point, of course, isn't a return to determinism, technological or otherwise; nor is it to throw up our hands in despair. Instead, the point is to temper our visions, manage expectations and, at least in part, redirect engagement efforts. Hope and promise are wonderful things; they fuel political will and action. But efficacy requires seriously accounting for obduracy – that is, knowing what we are pushing up against – and what's pushing back. Stubborn structures can be torn down and powerful institutions toppled, but not if we ignore or merely accept the totality of their existence.

Conclusion: beyond blank slate futuring

To be sure, we are not under the illusion that FCT will be the lone wrecking ball that knocks down obduracy and clears the path for deliberate reshaping. There are limits to our proposal – hidden durabilities in methods of public engagement more broadly – such as the one-off nature of the event, and the initial necessity of the backing of the National Science Foundation which enables the expenditure of much social and intellectual capital. Rather, we see these types of efforts as crucial first steps – hammers and chisels that chip away at obduracy and enable a more fruitful approach to futuring in public engagement exercises. Hommels's frames, embeddedness and traditions were important anchoring points for the experiential design of the tour, and drew ideas of path dependency and obduracy away from only material artefacts and the built environment to also consider the ways in which cultural values, institutional procedures (like the maintenance of roads and parking lots) and competing professional agendas persist. Appreciation of these obduracies complemented the FCT participants' more creative imaginings about the possibilities afforded by emerging technologies.

This may be a modest proposal for public engagement projects that look at the future of urban spaces and other built environments: obduracy is clearly relevant for thinking through the ways in which public spaces are transformed. However, following Hommels, obduracy is key for looking at most emerging technologies, whether health monitoring jewellery, climate adaptive building codes or big data-enabled surveillance systems. Indeed, even with FCT, citizens explored topics as diverse as graffiti art, DIY biofuels, asphalt and bicycle lanes in ways that took up obduracy.

The principal conceit that obduracy offers is that the future is not open ended for new technologies to freely populate, but is always already conditioned by contemporary social, material and economic circumstances, some of which are quite resistant to change. Regardless of social values or political will – or, for that matter, citizens wishing it to be so – there are structures of technological systems that do not yield. Encouraging engagement with obduracy – like we did through the multi-modal methods deployed in the Futurescape City Tours – tempers blank slate futuring. In doing so, tempered futuring leads us away from a linear notion of progress that implies continual improvement sparked from technological fixes, and instead invites a more critical stance. This chapter thus offers a corrective to the idea of an empty future by proposing a helpful starting point to enrich the future-oriented gaze of public engagement practices by stressing that new technologies do not emerge in a vacuum, but must fit in or tear down existing structures, routines and value systems.

Technology assessment is bent towards the future – whether under the rubric of 'anticipatory governance', 'constructive TA', 'upstream engagement' or 'pTA' – but often in un-reflexive ways. By describing how the constitution and interrogation of obduracies were explicitly and implicitly taken up in this novel public engagement exercise, which used an experiential, place-based urban walking tour focused on emerging technologies, this chapter opens up attention to the politics of the future that are usually covertly inscribed into public engagement methods.

Acknowledgements

Without the other members of the Futurescape City Tours research team at the Center for Nanotechnology in Society and the School of Sustainability at Arizona State University, this research would not have been possible. The authors wish to thank researchers involved in the 2012 Pilot of the FCT in Phoenix: Kelly Campbell Rawlings, Kathryn de Ridder Vignone, Mindy Kimball, Gretchen Gano, Carlo Altamirano Allende and David Guston.

This research was funded by the US National Science Foundation under cooperative agreement #0937591. The findings and observations in this chapter are those of the authors and do not necessarily reflect the views of the National Science Foundation.

Notes

1 This engagement exercise was the pilot for a larger project that involved conducting similar urban walking tours in five different cities across the United States during autumn 2013.
2 While lacking a studied encounter with temporality, there are other emerging practices of PEST that involve a focus on materiality and intimate encounters with the urban environment; see, for instance, citizensense.net and Marres (2012).

References

Arthur, W. B. 1990. Positive feedback loops in the economy, *Scientific American* 262: 92–99.

Barben, D., Fisher, E., Selin, C. and Guston, D. 2008. Anticipatory governance of nanotechnology: Foresight, engagement and integration. In: E. Hackett, O. Amsterdamska, M. Lynch and J. Wajcman (eds) *The Handbook of Science and Technology Studies*, 3rd edition. Cambridge, MA: MIT Press, pp979–1000.

Bijker, W. E. 1995. *Of Bicycles, Bakelites and Bulbs: Toward a Theory of Sociotechnical Change*. Cambridge, MA: MIT Press.

Brown, W. 2005. *Edgework: Critical Essays on Knowledge and Politics*. Princeton, NJ: Princeton University Press.

Callon, M. 1986. Some elements of a sociology of translation: Domestication of the scallops and the fishermen of St Brieuc Bay. In J. Law (ed.) *Power, Action and Belief: A New Sociology of Knowledge?* London: Routledge, pp196–233.

Campbell, K. B. 2005. Theorizing the authentic: Identity, engagement and public space, *Administration & Society* 36(6): 688–705.

Casey, E. 1996. How to get from space to place in a fairly short space of time. In: S. Feld and K. Basso (eds) *Senses of Place*. Santa Fe, NM: School of American Research Press, pp13–52.

Cobb, M. D. 2011. Creating informed public opinion: Citizen deliberation about nanotechnologies for human enhancements, *Journal of Nanoparticle Research* 13: 1533–1548.

Cohen, J. 2009. Reflections on deliberative democracy. In: T. Christiano and J. Christman (eds) *Contemporary Debates in Political Philosophy*. Malden, MA: Wiley-Blackwell, pp247–263.

Collingridge, D. 1980. *The Social Control of Technology*. New York: St. Martin's Press.

Coutard, O. and Guy, S. 2007. STS and the city: Politics and practices of hope, *Science, Technology & Human Values* 32(6): 713–734.

Cowan, R. S. 1985. How the refrigerator got its hum. In: D. MacKenzie and J. Wajcman (eds) *The Social Shaping of Technology*. Philadelphia, PA: Open University Press, pp202–218.

David, P. A. 1985. Clio and the economics of QWERTY, *The American Economic Review* 75(2): 332–337.

Davies, S. R. 2011. How we talk when we talk about nano: The future in laypeople's talk, *Futures* 43(3): 317–326.

Davies, S. R., Selin, C., Gano, G. and Pereira, A. G. 2012. Citizen engagement and urban change: Three case studies of material deliberation, *Cities* 29(6): 351–357.

Delgado, A., Kjølberg, K. L. and Wickson, F. 2011. Public engagement coming of age: From theory to practice in STS encounters with nanotechnology, *Public Understanding of Science* 20(6): 826–845.

Durant, D. 2011. Models of democracy in social studies of science. *Social Studies of Science* 41: 691–714.

Evans, R. and Plows, A. 2007. Listening without prejudice? Re-discovering the value of the disinterested citizen, *Social Studies of Science* 37(6): 827–853.

Feenberg, A. 1994. The technocracy thesis revisited: On the critique of power, *Inquiry* 37: 85–102.

Feenberg, A. 2011. Agency and citizenship in a technological society. Lecture presented to the Course on Digital Citizenship, IT University of Copenhagen, pp1–13.

Felt, U. and Fochler, M. 2009. Between the fat-pill and the atomic bomb: Civic imaginations of regimes of innovation governance, *Science as Culture* 20(3): 307–328.

Foucault, M. 1995. *Discipline & Punish: Birth of the Prison*, 2nd edition. New York: Vintage Books.

Fung, A. 2006. Varieties of participation in complex governance, *Public Administration Review* 66: 66–75.

Gieryn, T. 2002. What buildings do, *Theory and Society* 31: 35–74.

Gorman, M. E. and Carlson, W. B. 1990. Interpreting invention as a cognitive process: The case of Alexander Graham Bell, Thomas Edison and the telephone, *Science, Technology & Human Values* 15(2): 131–164.

Graham, S. 2002. On technology, infrastructure and the contemporary urban condition: A response to Coutard, *International Journal of Urban and Regional Research* 26(1): 175–182.

Graham, S. and Marvin, S. 2001. *Splintering Urbanism: Networked Infrastructures, Technological Mobilities and the Urban Condition*. London: Routledge.

Gullberg, A. and Kaijser, A. 1998. City building regimes in post-war Stockholm, *TRITA-HST Working Paper* 98/3. Stockholm: Royal Institute of Technology, Department of History of Science and Technology.

Guston, D. H. 2008. Innovation policy: Not just a jumbo shrimp, *Nature* 454: 940–941.

Guston, D. H. 2014a. Building the capacity for public engagement with science in the United States, *Public Understanding of Science* 23(1): 53–59.

Guston, D. H. 2014b. Understanding 'anticipatory governance', *Social Studies of Science* 44(2): 218–242.

Guston, D. H. and Sarewitz, D. 2002. Real-time technology assessment, *Technology in Society* 24: 93–109.

Guy, S., Graham, S. and Marvin, S. 1997. Splintering networks: Cities and technical networks in 1990s Britain, *Urban Studies* 34(2): 191–216.

Hamlett, P. W. 2003. Technology theory and deliberative democracy, *Science, Technology & Human Values* 28(1): 112–140.

Hommels, A. 2005. Studying obduracy in the city: Toward a productive fusion between technology studies and urban studies, *Science, Technology & Human Values* 30(3): 323–351.

Hommels, A. 2008. *Unbuilding Cities: Obduracy in Urban Sociotechnical Change*. Cambridge, MA: MIT Press.

Hughes, T. P. 1987. The evolution of large technological systems. In: W. E. Bijker, T. P. Hughes and T. J. Pinch (eds) *The Social Construction of Technological Systems*. Cambridge, MA: MIT Press, pp51–82.

Hughes, T. P. 1994. Technological momentum. In: M. R. Smith and L. Marx (eds) *Does Technology Drive History? The Dilemma of Technological Determinism*. Cambridge, MA: MIT Press, pp101–113.

Joerges, B. 1999. Do politics have artefacts?, *Social Studies of Science* 29(3): 411–431.

Kirkman, R. 2004. The ethics of metropolitan growth: A framework, *Philosophy & Geography* 7(2): 201–218.

Kirkman, R. 2009. At home in the seamless web: Agency, obduracy and the ethics of metropolitan growth, *Science, Technology & Human Values* 34(2): 234–258.

Kitt Chappell, S. A. 1989. Urban ideals and the design of railroad stations, *Technology and Culture* 30(2): 354–375.

Larkin, B. 2013. The politics and poetics of infrastructure, *Annual Review of Anthropology* 42: 327–343.

Latour, B. 1987. *Science in Action: How to Follow Scientists and Engineers through Society*. Cambridge, MA: Harvard University Press.

Latour, B. 1988. The prince for machines as well as for machinations. In: B. Elliott (ed.) *Technology and Social Process*. Edinburgh: Edinburgh University Press, pp20–43.

Latour, B. 1992. Where are the missing masses? The sociology of a few mundane artifacts. In: W. E. Bijker and J. Law (eds) *Shaping Technology/Building Society: Studies in Sociotechnical Change*. Cambridge, MA: MIT Press, pp225–258.

Latour, B. 1994. On technical mediation – Philosophy, sociology, genealogy, *Common Knowledge* 3(2): 29–64.

Latour, B. 2002. Morality and technology: The end of the means, *Theory, Culture & Society* 19(5/6): 247–260.

Latour, B. 2004. Nonhumans. In: S. Harrison, S. Pile and N. Thrift (eds) *Patterned Ground: Entanglements of Nature and Culture*. London: Reaktion Books, pp224–227.

Latour, B. 2005. *Reassembling the Social: An Introduction to Actor-Network-Theory*. New York and Oxford: Oxford University Press.

Liebowitz, S. J. and Margolis, S. E. 1995. Path dependence, lock-in and history, *Journal of Law, Economics & Organization* 11(1): 205–226.

Lockton, D. 2005. *Architectures of Control in Consumer Product Design*. Unpublished paper, http://ssrn.com/abstract=908493 (accessed 11 December 2012).

Marres, N. 2012. *Material Participation: Technology, the Environment and Everyday Publics*. London: Palgrave Macmillan.

Marx, L. 1987. Does improved technology mean progress?, *Technology Review* 90(1): 33–41.

Nusca, A. 2011. In downtown Phoenix, cooler pavement for parking lots, *Smartplanet*, 31 May, www.zdnet.com/article/in-downtown-phoenix-cooler-pavement-for-parking-lots (accessed 19 February 2013).

Park, H., Chang, S., Jean, J., Cheng, J. J., Araujo, P. T., Wang, M., Bawendi, M. G., Dresselhaus, M. S., Bulović, V., Kong, J. and Gradeča, S. 2013. Graphene cathode-based ZnO nanowire hybrid solar cells, *Nano Letters* 13(1): 233–239.

Pinch, T. J. and Bijker, W. E. 1987. The social construction of facts and artifacts. In: W. E. Bijker, T. P. Hughes and T. J. Pinch (eds) *New Directions in the Sociology and History of Technology*. Cambridge, MA: MIT Press, pp11–44.

Regan, W., Byrnes, S., Gannett, W., Ergen, O., Vazquez-Mena, O., Wang, F. and Zettl, A. 2012. Screening-engineered field-effect solar cells, *Nano Letters* 12(8): 4300–4304.

Ryfe, D. M. 2005. Does deliberative democracy work? *Annual Review of Political Science* 8: 49–71.

Sadowski, J. and Selinger, E. 2014. Creating a taxonomic tool for technocracy and applying it to Silicon Valley, *Technology in Society* 38: 161–168.

Sass, J. and Dryzek, J. S. 2014. Deliberative cultures, *Political Theory* 42(1): 3–25.

Seager, T. P., Selinger, E. and Wiek, A. 2012. Sustainable engineering science for resolving wicked problems, *Journal of Agricultural and Environmental Ethics* 25(4): 467–484.

Selin, C. 2007. Expectations and the emergence of nanotechnology, *Science, Technology & Human Values* 32(2): 196–220.

Selin, C., Rawlings, K., de-Ridder Vignone, Sadowski, J., Altamirano, C., Gano, G., Davies, S. and Guston, D. forthcoming. Experiments in engagement: Designing PEST for capacity building, *Public Understanding of Science*.

Shapira, P. and Youtie, J. 2012. The economic contributions of nanotechnology to green and sustainable growth. Paper presented at *International Symposium on Assessing the Economic Impact of Nanotechnology*, Organisation for Economic Co-operation and Development and the US National Nanotechnology Initiative, Washington DC, 27–28 March.

Sontag, S. 2001. *On Photography*, 1st edition. New York: Picador.

Star, S. L. 1999. The ethnography of infrastructure, *American Behavioral Scientist* 43(3): 377–391.

Stirling, A. 2008. 'Opening up' and 'closing down': Power, participation, and pluralism in the social appraisal of technology, *Science, Technology & Human Values* 33(2): 262–294.

Tiberius, V. 2011. Path dependence, path breaking, and path creation: A theoretical scaffolding for futures studies?, *Journal of Futures Studies* 15(4): 1–8.

Verbeek, P.-P. 2006. Materializing morality: Design ethics and technological mediation, *Science, Technology & Human Values* 31(3): 361–380.

Verbeek, P.-P. 2011. *Moralizing Technology: Understanding and Designing the Morality of Things*. Chicago, IL: University of Chicago Press.

von Schomberg, R. and Davies, S. 2010. *Understanding Public Debate on Nanotechnologies: Options for Framing Public Policy*. Report from the European Commission Services. Brussels: European Commission.

Weber, R. 2002. Extracting value from the city: Neoliberalism and urban redevelopment, *Antipode* 34(3): 519–540.

Wiek, A., Foley, R. W. and Guston, D. H. 2012. Nanotechnology for sustainability: What does nanotechnology offer to address complex sustainability problems?, *Journal of Nanoparticle Research* 14(9): 1093–2013.

Wiek, A., Guston, D., van der Leeuw, S., Selin, C. and Shapira, P. 2013. Nanotechnology in the city, sustainability challenges and anticipatory governance, *Journal of Urban Technology* 20(2): 45–62.

Wilsdon, J. and Willis, R. 2004. *See-Through Science: Why Public Engagement Needs to Move Upstream*. London: Demos.

Winner, L. 1980. Do artifacts have politics?, *Daedalus* 109(1): 121–136.

Winner, L. 1986. *The Whale and the Reactor: A Search for Limits in an Age of High Technology*. Chicago, IL: University of Chicago Press.

12

REFLEXIVELY ENGAGING WITH TECHNOLOGIES OF PARTICIPATION

Constructive assessment for public participation methods

Jan-Peter Voß

Introduction

In this chapter, I do two things. First, I briefly reconstruct how political participation becomes technologized and argue that there is a modal shift in how constitutions of democracy are built: from politics to technoscience. I discuss how this modal shift is accompanied by reflexive engagement practices that counter technoscientific closure and seek to open up and re-politicize methods of public participation. Second, I give a more detailed account of a recent interactive assessment exercise on the future development of citizen panels. It was an attempt to apply methodological considerations of constructive technology assessment (CTA) to the 'social technology' of participation methods. The chapter discusses how the exercise engages with the innovation of citizen panels, but also how, as an expertly devised method, it may itself be conceived of as a further instance of the technoscientization of governance. In conclusion, I return to the overall innovation dynamics of public participation methods. I argue that technoscientization and reflexive engagement make a precarious balance in coping with ambiguities of innovation, and I briefly discuss what this means for wider areas of 'social innovation' and their links with issues of 'responsible research and innovation'.

Technoscientization of politics

What do I mean by the awkward term 'technoscientization' and by saying that public participation becomes 'technoscientized'? Very briefly, I refer to a process that makes the reality of public participation an object of scientific analysis and technological control. That implies the application of a technoscientific mode of innovation for the remaking of political order (Voß and Freeman forthcoming). This particular mode of innovation works through the configuration of phenomena in 'secluded research', by a collective of trained experts and in protection from the uncontrolled interference of the wider world (Callon *et al.* 2009, 46–70). The results of secluded research, however, are presented to the public, not as a proposal whose reception and expansion may be considered politically, but as insights into the nature of things and a neutral mirroring of 'the' reality. The technoscientific mode of innovation thus draws on epistemic authority. Scientifically demonstrated functions become realized not as a deliberate remaking of collective order, but as the application of knowledge about nature and more or less passive adaptation to the conditions that it holds ready.[1]

For the case of public participation, technoscientific innovation means that model realities of participation are expertly constructed and then presented as insights into the nature of politics and the conditions for effectively articulating concerns and views of 'the' public. To replicate the experimentally demonstrated and theoretically explained effects, however, requires the model reality to be rebuilt in different places and on a broader scale. Locally configured political reality has to be technically fixed and made transportable. This is what methods of public participation do. In order to realize particularly theorized functions, they prescribe certain ways of enacting political reality.

Technoscientific innovation can be highly productive. It allows contentious questions of collective ordering to be sorted and negotiated in small and disciplinary-aligned groups of experts. Their construction of proto-orders in strategically purified environments, if presented as a discovery of functional patterns in reality *as it is*, does not appear as an intervention to change and reorder the world, but rather as a clever way of coping with naturally given conditions. The replication and up-scaling of laboratory realities can thus rely on epistemic authority; it does not require the construction of political authority in order to orchestrate collective action.

In application to questions of democracy the technoscientific mode offers a way to avoid cumbersome and contingent political processes of reconfiguring political order. Composing political phenomena and functional effects (such as the legitimacy of representative procedures) in the mode of secluded research provides protection from the wider world of politics 'in the wild', with diverse and irreconcilable worldviews, values and interests that may become mobilized in non-transparent and uncontrollable ways. By shifting questions of political order into the laboratory, by turning them from matters of concern into matters of fact, the design of representational procedures becomes a technical problem whose solution is to be justified on the basis of objective functionalities, not on grounds of collective autonomy and will (Arendt 1979 [1969]; Latour 2003; Rosanvallon 2006; Disch 2012). Politics, as a process of performatively representing collective subjects and their common will, can thus be bypassed in defining the functions and methods of political representation.

The technoscientific mode can be productive for the remaking of political reality; just think of the engagement of political science, law, economics and sociology in matters of state, democracy and governance (cf. Ezrahi 1990; Desrosières 1998; Osborne and Rose 1999). As much as technoscience substitutes for political modes of ordering political reality, however, it does so at a cost. That is linked to a shift from political to epistemic authority. The remaking of political order is no longer justified by collective will, but by reference to factual conditions and functional necessity. This changes the way in which people can relate to and engage with the process of collective ordering. While participation in *political* ordering requires a voice to articulate a subjective opinion or to refuse incorporation into a proposed representation of the collective, participation in the negotiation of the *factual* conditions of politics requires expert status and the wielding of an experimental apparatus to assert alternative political realities. The problematization of a technoscientific displacement of politics in science and technology studies (STS) was mostly with a view to natural sciences and engineering, not so much with regard to the social sciences and how they are involved in producing collective order (see Irwin 2008; Camic *et al.* 2011). Shifting over to the making of political reality, however, we mainly get to grips with the work of the social sciences and the establishment of 'social technology'.[2]

We may more specifically speak of public participation methods as technologies of politics, democracy or community (Rose 1999, 188; Irwin 2001; Barry 2002; Laurent 2011b). They promise to provide a true view of 'the public' for reference by those who seek to legitimize actions as 'collective' or 'public' rather than private, partial or partisan (Grönlund *et al.* 2014). They prescribe procedures for articulating a collective will of the public.

As such they determine the purpose and product of political interactions, the eligible issues, duration and location of meetings, the composition of panels and required qualifications and forms of conduct by participants and moderators, tools of facilitation and the input of information into the process. This makes them 'machineries for making publics' (Felt and Fochler 2010). If successfully established they may come to be seen as the natural way for publics and their collective will to exist, just as national elections, parliaments and opinion polls are now.

The problematization of public participation methods as 'technologies' corresponds with a gradual shift in patterns by which they are articulated, advocated and spread. Their development comprised a shift from practices that were embedded in local political cultures and issue areas and which were normatively justified and regulated in situated negotiations over the purpose and design of particular participation projects, to the design of standard methods in transnational expert networks, and epistemic-technical justifications on the basis of evidence about 'universal' functionalities from laboratory experiments, along with their global marketing as new tools of democracy (for emphatic accounts see Sulkin and Simon 2001; Carpini *et al.* 2004; Karpowitz and Mendelberg 2011; for more sceptical accounts see Laurent 2009, 2011b; Bogner 2012). In the course of this process political participation becomes increasingly objectified as a tool of governance. Even leading political scientists who discuss the prospects of democratizing global governance defer the design of public participation procedures to experts who are trusted to care that 'there is technology for that'.[3]

In order to reliably reproduce an expertly modelled function of participation, political subjects have to be disciplined to perform the model – for example, by practising communicative rationality in order to produce public reason that transcends particular subjective positions and can thus legitimize public action (Habermas 1981, 1993). Democracy is reinvented along particular expertly devised models. This implies a technoscientization of political culture. Collective 'imaginaries of democracy' (Ezrahi 2012) are constructed with the help of the laboratory. In this way new methods of participation establish alternatives to incumbent technologies of liberal-representative democracy like elections and parliaments. The latter were established in extended political struggle and are territorially and culturally anchored (Heurtin 2005). The new technologies are mobile – they can flexibly be deployed as ready-made political devices for specific issues and decision problems (cf. 'instant democracy', Sloterdijk and Mueller von der Haegen 2005). Their circulation and installation silently punctuates the cultural infrastructures of representative democracy.

The interrogation of 'technologies of participation' is linked to a concern for the 'collateral realities' that they produce (Law 2012). This includes, for example, the negation of creative political agency of participants in defining their roles or of situated judgments on appropriate and effective procedures. It is cautioned that if experts assume the power to define political subjects and adequate forms of conduct (Rose 1999; Braun and Schultz 2010), this may, paradoxically, undermine democratic empowerment rather than enhance it (e.g., Levidow 1998; Irwin 2001; Lezaun and Soneryd 2007; Wynne 2007; Felt and Fochler 2010; Bogner 2012; Chilvers 2013). The irony is that the establishment of expertise about public participation originally started as an anti-technocratic project: the attempt to work against expert rule with regard to issues of policy decision. But it gave rise to a new technocracy of political agency and procedure that shapes the innovation respective methods (Voß and Amelung forthcoming).

The innovation of citizen panels

I turn to the interlinked innovation of 'citizen panels' as a set of resemblant public participation methods that, throughout the last decades, expanded globally across jurisdictions and issue areas (for a more extensive account see Amelung 2012; Voß and Amelung forthcoming).[4] The umbrella term comprises the methods of citizens' jury, planning cell and consensus conference. They all prescribe the convocation of groups of 10 to 25 randomly selected citizens to produce a public judgement on a given issue of concern. Participants are provided with factual information and expert statements on the issue, and a moderator facilitates their deliberation. The procedure usually takes a few days and the resulting consensus is reflected in a report with policy recommendations. Originally, citizens' juries, planning cells and consensus conferences emerged independently in different contexts in the 1970s and 1980s. At the beginning of the 1990s their innovation journeys became entangled, and since about 2000 they have been discussed and further developed under the umbrella term 'citizen panels' (Hörning 1999; Brown 2006).

The overall process of their development can be interpreted as a truncated version of an 'aggregation' pattern which describes a type of technological innovation processes in which situated, practical knowledge gradually becomes explicated, objectified, codified and thereby decontextualized, meaning that local technological practices become part of cosmopolitan regimes of knowledge (Disco *et al.* 1992; Deuten 2003; Geels and Deuten 2006). From a relating of different local knowledges 'bottom up' follows the building of generic categories and frameworks for comparison, which take on a life of their own and develop into abstract global models that come to define local practices 'top down'. Design practices that emerged locally in particular historical settings are successively drawn into the laboratory in order to define a global technical standard. In the case of citizen panels we find an incomplete version of this pattern (see Figure 12.1): incipient technoscientific dynamics did not lead directly into a global regime and a dominant model. Instead, they gave rise to forms of engagement that counter the looming shift from open, situationally embedded innovation to a configuration of universal methods in transnational expert networks.

Local practices

Citizen panels began as dispersed and unconnected local practices of organizing procedures to involve citizens in the deliberation of public policy issues. Planning cells developed in the context of municipal planning in the German state of North-Rhine Westphalia (Dienel 1971, 1978), the citizens' jury emerged in a civic education context in the state of Minnesota in the US (Crosby 1974, 1975), and the consensus conference developed in a context of technology assessment in Denmark (Joss and Durant 1995; Andersen and Jäger 1999). The different approaches agreed that citizens should provide constructive input to authoritative decision-making on contested issues. They sought to make public engagement productive by offering organizational support for citizens to articulate a coherent view of the public. Methods were developed through practical tinkering, guided by general philosophical considerations. They took shape in the course of pragmatically coping with circumstances and opportunities, such as alliances with local politicians and activist groups. There was no explicit functional theory on how participatory procedures work but the general idea of facilitating the articulation of a consensus among ordinary citizens. The know-how of doing citizen panels was embedded in local communities of practice which were led by entrepreneurial figures. Learning was a matter of socialization and practical experience on the job (Gastil and Keith 2005; Hendriks 2005; Vergne 2010).

Proliferation

Towards the end of the 1980s, all three methods were dislodged from their niches and they proliferated into new areas. Procedures were documented in books and articles and started to circulate across cultural and political contexts and issue areas (Stewart

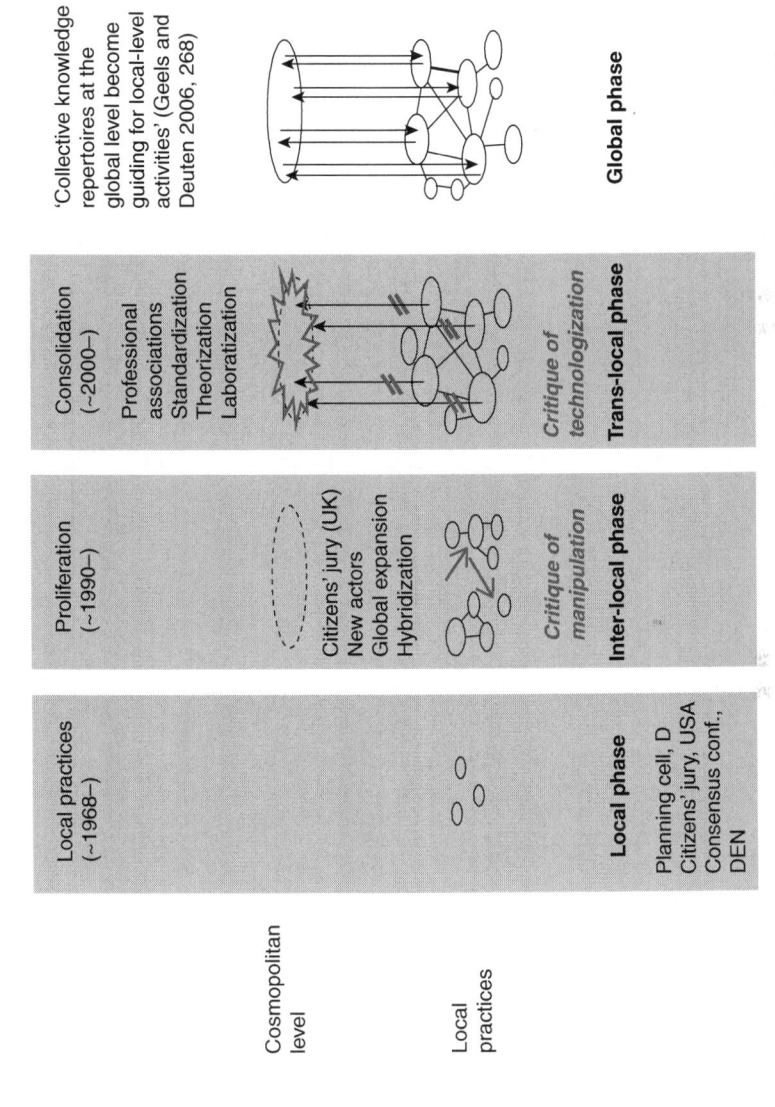

FIGURE 12.1 The innovation journey of citizen panels as a truncated process of 'aggregating' technical knowledge and practices of participation

Source: adapted from Geels and Deuten (2006, 269).

et al. 1994; Renn *et al.* 1995; Coote and Lenaghan 1997). Around the year 2000, several reports and overviews listed citizen panels as elements of a universal toolkit for public participation (e.g., OECD 2001; Elliott *et al.* 2005). With surging political demand for public engagement services in the second half of the 1990s, new actors from commercial consultancy and marketing entered the field. A hot spot was the UK (Chilvers 2008). Citizen panels were hybridized with polling, focus groups and public relations methods. The cultural embedding and coherence of citizen panel practices eroded and they lost trust in the media discourse of the wider public (Wakeford *et al.* 2007). Citizen panels were criticized for 'whitewashing' governmental strategies by manipulating citizens to produce views that were aligned with predefined policy decisions (Levidow 1998; Parkinson 2004, 2006; Hendriks and Carson 2008).

Technoscientific consolidation

In the first half of the 2000s, partly in response to problems with wider public acceptance, efforts increased to systematize design knowledge and regulate the wild spread and modification of public participation methods. Trans-local frameworks were developed to compare a variety of practices that had emerged from inter-local exchanges and hybrid developments. Their working was to be made transparent, objective and reliable, ultimately with a view to professionalize and discipline the practices of doing participation as a move to regain trust (cf. Porter 1996). In this context, the term citizen panels became established as an umbrella term to align the methodological development of planning cells, citizens' juries and consensus conferences (Hörning 1999; Brown 2006).[5] Academics provided systematic comparisons and evaluations of public participation exercises (Rowe and Frewer 2000, 2005; Fung 2006). Internet platforms, academic journals and professional associations were established to develop a shared discourse and establish global knowledge of quality criteria and standards of good practice.[6] Governmental institutions supported the development of a shared body of knowledge and design standards which could provide devices that could be reliably used to compensate the alleged deficits of liberal-representative democratic legitimation.[7] Transnational entrepreneurs of participation methods set up experiments to demonstrate the applicability of citizen panels also for issues of global governance.[8]

The establishment of standards, quality criteria and design specifications required more explicit theoretical explanations of the functioning of citizen panels (Brown 2009; Renn and Schweizer 2009; Lövbrand *et al.* 2011). This made a productive link with research on theories of deliberative democracy which were perceived to lack empirical grounding (Smith and Wales 2002). Citizen panels thus became incorporated as practical exemplars of deliberative democracy and, in exchange, could draw on a theoretical apparatus, which was developed mainly in interaction with Habermas' theory of communicative action and ethics of discourse, for explicating their functionality. The tinkering with designs in practice was complemented by the testing of design features in laboratory experiments to put 'Habermas in the lab' (Sulkin and Simon 2001; Carpini *et al.* 2004). The move to the laboratory made it possible to establish unified and coherent definitions of the function of citizen panels as well as quality criteria and design specifications. Procedures were increasingly designed in controlled experimental settings, far off from particular sites of application. Authoritative expertise on procedural matters of political participation could be built up on the basis of empirical evidence. Citizen panels came to be organized with a view to experimentally demonstrate a general theory of participation (Laurent 2009, 2011b; Bogner 2012; Ureta 2015).

Reflexive engagement

The gradual technoscientization of citizen panels became problematized early on. One form of reflexively engaging with the development was in an emerging critical discourse. Academic work exposed the social dynamics, contextuality, contingency and politics of participatory methods. It deconstructed objective functionality by empirically studying participation methods in the making and at work (Irwin 2001; Gomart and Hajer 2003; Lezaun 2007; Lezaun and Soneryd 2007; Chilvers and Burgess 2008; Braun and Schultz 2010; Felt and Fochler 2010; Horst and Irwin 2010;). This showed the situational embedding and inherent bias in any particular procedure for constructing a public view (Gomart and Hajer 2003) and it included the warning that the instrumentalization of citizen participation can undermine rather than promote the legitimacy of collective decisions (Wynne 2006; Felt and Fochler 2010). The natural link of participation with theories of deliberative democracy was challenged by reference to alternative political ontologies which emphasized situated sense-making (Dewey 2012 [1954]), different identities, irreconcilably diverse rationalities and hegemonic discourses (e.g., Freire 2000 [1970]; Laclau and Mouffe 2001; Dewey 2012 [1954]). The critical discourse worked to deconstruct the notion of participation methods as politically neutral, functional tools. By highlighting underlying ontological assumptions, non-exigent design decisions, situated agency and latent impacts, it disrupted the expertocratic immunization of their design and created openings for political engagement.

Another form of reflexively engaging with the establishment of a dominant professional discourse was the development of alternative designs which were explicitly geared to grant citizens agency in defining the issues and designing the procedures of their engagement. A prominent example was the 'Do-It-Yourself' citizens' jury that was proposed by PEALS at the University of Newcastle to support citizens in organizing themselves and articulating marginalized viewpoints to counteract dominant discourse (Wakeford 2003; Wakeford and Singh 2008).

Yet another form comprised protest actions, in similar ways as known from other areas of contested technology development. A silent form of protesting, for example, was to stay away and refuse being instrumentalized in organized exercises of public participation (Maier 2009). More active contestation could be found in legal challenges of public consultation exercises, such as Greenpeace accusing the UK government and the judicial judgement of participation in nuclear energy of being 'seriously flawed' and 'procedurally unfair' (Greenpeace 2007; Chilvers and Burgess 2008).

More overt resistance and sabotage against technologies of participation can be observed in the strategic disruption of organized 'public debate' (on nanotechnology and synthetic biology) in France. The protest movement Pièces et Main d'Oeuvre (PMO) pursued a radical critique of technologization by seeking to break down positioning preconfigured dialogue and participation exercises as a 'social technology' that works to co-opt publics for the technological discourse, and by disrupting organized participation with loud shouting (Laurent 2011a; see also Ehrenstein and Laurent, Chapter 6 this volume). Taken together, these different forms of engagement contributed to the ongoing innovation process. They brought dimensions into view that were beyond narrowly conceived functions and they stimulated wider and more controversial public debates about methods of participation. As such they worked as informal technology assessment for emerging technologies of participation (cf. Rip 1987).

In addition to that, more explicit forms of assessment for increasingly technoscientized participation methods started to be discussed: 'Now that forms of public dialogue are a site of innovation and professionalisation as part of a global public engagement industry ... there is a need for anticipatory assessment of these social (science) "technologies of participation" themselves' (Chilvers 2013, 306). Accordingly, 'reflexive learning about public dialogue' would require

> actors ... to actively acknowledge, reflect on, and openly express to others their underlying assumptions, motives and commitments relating to the forms of public dialogue they orchestrate or are exposed to, rather than treating dialogue and engagement (and learning for that matter) as a homogeneous, reified, and acontextual technical procedure.
>
> (Chilvers 2013, 301)

The workshop series from which this book has grown is an example of this kind of engagement (as introduced in the Preface and noted in Chapter 13 of this volume). In the following section we turn to another example: an experiment with adopting methods of constructive technology assessment (CTA) to stimulate reflexive interactions on the future development of citizen panels among practitioners.

For a brief summary of the innovation journey of citizen panels we may hold that, over the last four decades, it went on as a gradual process of technoscientization. As such it resembles an 'aggregation' pattern observed in other areas of technology development, where local practices become connected, then overarched and finally controlled by abstract functional representations that are composed in laboratorized practices that are configured by global centres of expertise. So far, the innovation journey of citizen panels depicts a truncated version of this process: the technoscientization of public participation is met by critical academic discourse, alternative designs, direct protest actions and dedicated assessment exercises. Together these different forms of reflexive engagement counteract the reification of political participation by a global theory of the function of public deliberation and a dominant procedural design.

'Challenging Futures of Citizen Panels'

Let's now have a closer look at the mentioned constructive assessment exercise for citizen panels. It is here presented as another instance of reflexive engagement with technologies of participation. I have been practically involved with this case as initiator and organizer of the exercise. I reflect on this engagement with regard to how it is related with the ongoing innovation process. The exercise went under the title 'Challenging Futures of Citizen Panels'. The approach was inspired by concepts such as constructive technology assessment (Rip et al. 1995; Schot and Rip 1997; Rip and Schot 2002), hybrid fora (Callon et al. 2009), real-time technology assessment (Guston and Sarewitz 2002) and anticipatory governance (Barben et al. 2008; Guston 2014). An overall orientation was to stimulate interactions across ongoing strands of activities which shape the design and development of citizen panels (Mann and Voß forthcoming). Various concerns and requirements were to be articulated and confronted with each other in order to identify critical issues for a robust approach to innovation. The exercise sought to open the debate on particular modes and designs for developing citizen panels in the future. To this end it brought actors beyond the usual in-groups of experts together and demonstrated the challenges of negotiating the diverging and irreconcilable constructions of reality, the attributions of purpose and functionalities, as well as various situated practices of doing participation. It contributed to the reflexive pluralization of conceptions and appraisals of citizen panels and raised awareness of fundamentally political issues connected to apparently technical design problems (Stirling 2008).

At the centre was a workshop with 25 actors who were in different ways practically involved in the development of citizen panels, both in more affirmative and critical perspectives. Participating actors were identified on the basis of our preceding research into the historical development of citizen panels. Some actors were engaged in the academic design and theorization of methods, others as professional operators and commissioners, or as activists and critical commentators. The selection did not aim to be comprehensive or representative, but it was oriented by the attempt to create a marginal opening in the debate by providing for interesting encounters, contrasting realities and the articulation of controversial issues. Table 12.1 lists the workshop attendees.

In the run-up to the workshop, participants were presented with a set of three scenarios which exemplified the dynamics and possible tensions that may shape the future innovation process (Mann et al. 2013). They were constructed on the basis of historical innovation dynamics. Each scenario presented the rampant expansion of a particular rationality that was at play in shaping the innovation in the past. Table 12.2 gives an overview of the scenarios.

At the workshop, participants were prompted to articulate challenges that may come up in the future development of citizen panels, against both the background of their own experience and the presented scenarios, and to discuss them with a view to identify 'critical issues' for the innovation process. The moderation aimed to have issues articulated from different angles and to have propositions made on how they should be dealt with.[9] The discussion reflected ambiguities and controversial aspirations in the design and strategic development of citizen panels (e.g., purpose and function, selection of participants, quality control). An overview of the agenda of the workshop is provided in Table 12.3.

TABLE 12.1 Attendees of the Challenging Futures of Citizen Panels workshop, 26 April 2013, Berlin

Number	Name	Organization
1	Abels, Gabriele	Eberhard Karls Universität Tübingen, Department of Political Science
2	Banthien, Hennig	IFOK Institut für Organisationskommunikation (Bensheim)
3	Brown, Mark	California State University, Department of Government
4	Chilvers, Jason	University of East Anglia, School of Environmental Sciences
5	Crosby, Ned	The Jefferson Center for New Democratic Processes (St Paul, Minnesota)
6	Dienel, Liudger	NEXUS, Institute for Cooperation Management and Interdisciplinary Research (Berlin)
7	Font, Joan	Instituto de Estudios Sociale Avanzados, Córdoba
8	Galiay, Philippe	European Commission, Directorate-General for Research and Innovation
9	Gastil, John	Pennsylvania State University, Department of Communication
10	Hennen, Leonhard	Office for Technology Assessment at the German Bundestag
11	Huitema, Dave	VU University Amsterdam, Institute for Environmental Studies
12	Joss, Simon	University of Westminster, Department of Politics and International Relations
13	Lietzmann, Hans J.	Bergische Universität Wuppertal, Research Centre in Public Participation
14	Lopata, Rachel	Community Research and Consultancy Ltd (Leicester)
15	Masser, Kai	German Research Institute for Public Administration (Speyer)
16	Prikken, Ingrid	Involve (London)
17	Rauws, Gerrit	King Baudouin Foundation (Brussels)
18	Schweizer, Pia-Johanna	Universität Stuttgart, Department of Social Science
19	Shinoto, Akinori	Junior Chamber International Japan
20	Soneryd, Linda	University of Gothenburg, Department of Sociology
21	Sturm, Hilmar	Society for Citizens' Reports (München)
22	Wakeford, Tom	University of Edinburgh, School of Health in Social Science
23	Walker, Ian	New Democracy Foundation (Sydney)
24	Worthington, Richard	The Loka Institute (Claremont, California)

TABLE 12.2 Overview of scenarios on the future development of citizen panels

Scenario A: Market for deliberation services	Scenario B: Toolkit of democracy	Scenario C: Public reason machine
Business interests drive the development of citizen panels. Methods are shaped according to the logic of supply and demand. Stimulation of demand and strategic creation of new markets for participation services are a core activity of a specialized consultancy and services sector. Scientific support is mobilized to establish the urgency of a crisis of representation, the requirement to directly engage citizens for democratic legitimation, and to push particular standards of citizen panel design and conduct – also with a view to saving shares in a highly competitive market. Demand can be generated with governments, particularly if they do not rely on liberal-representative procedures for legitimation. Firms, international organizations and large research projects also contract these services.	Competing visions of political order drive the development of citizen panels. Development of methods is shaped in view of their performative effects and how they enact a particular reality of citizenship, democracy and political order. Activists and stakeholder groups explicate alternative political visions and struggle over their realization; they engage with the negotiation of participatory procedures within particular political situations. Design knowledge for citizen panels takes shape as a diverse, controversially discussed repertoire of principles, storylines, methodical components and practices, which is selectively drawn on in local situations.	Scientific efforts at theorizing and optimizing deliberation drive the development of citizen panels. Development of methods is shaped in laboratory experiments to theorize and set up arrangements that produce public reason, and to technologically replicate their function. Institutional approaches are combined with neuro-biology and artificial intelligence to fix configurations of enhanced human interaction that reliably determine rational public will. Proven evidence of superior performance helps an emerging high-tech industry to install them globally and replace more primitive democratic techniques such as elections, voting, wild debate and protest.

In the aftermath of the workshop, tape-recorded discussions were transcribed and analysed with regard to different views on key issues. This was further translated into a list of eight 'critical issues' for an 'extended innovation agenda' which highlights contentious questions and shows up the political implications of implicitly or explicitly deciding them (Mann et al. 2014a). For a list of the section headings that elaborate issues in more detail, see Box 12.1.

TABLE 12.3 Agenda of the Challenging Futures of Citizen Panels workshop, 6 April 2013, Berlin

Friday, 26 April 2013	Venue: Berlin-Brandenburg Academy of Sciences and Humanities (BBAW)	
Introduction		
9:00–9:10	Welcome and overview	• Introduction to workshop objectives and expected outcomes • Overview of the agenda
9:10–9:30	Why this workshop?	• 'Challenging futures' in relation to dynamics of the innovation journey of citizen panels
Session 1: Challenging Futures of Citizen Panels		
9:30–11:00	Opening plenary discussion with general statements	• Round table: what characterizes the present situation of citizen panel development? • Open plenary discussion
11:00–11:30	Coffee break	
Session 2: Identifying and articulating future issues for citizen panels		
11:30–13:00	Group work: discussion of future developments and identification of issues	• Identify specific issues that require further attention and/or debate in the future development of citizen panels; produce issue briefs
13:00–14:00	Lunch break	
Session 3: Compiling issues, discussing challenges		
14:00–14:30	Strolling the 'wall of issues'	• Participants read and discuss issue briefs produced by working groups
14:30–16:15	Discussion of selected issues and challenges in plenary session	• Moderators present two clusters of issues for discussion • Two discussion rounds in plenary session, one for each cluster
16:15–17:00	Concluding discussion in plenary session	• Wrap-up of discussion of issues • Identify open questions and missed points • Outlook on further procedure

BOX 12.1 HEADINGS OF ISSUE DESCRIPTIONS FOR AN EXTENDED INNOVATION AGENDA FOR CITIZEN PANELS

1. Functions of citizen panels: a matter of worldviews and philosophies?
2. Standardization: toward unified citizen panel practices?
3. Quality: how can we control the quality of citizen panels?
4. Impact: do citizen panels need closer links with political decision-making?
5. Representation: which is the public that citizen panels produce a view of?
6. Neutrality: can power asymmetries and biases be evaded?
7. Context: is the working of citizen panels dependent on situational contexts?
8. Social life: what drives and shapes the innovation of citizen panels in practice?

The agenda suggests wider public attention for those questions and a concern for the legitimacy of respective decisions in order to make the innovation of citizen panels more robust for social and political complexities of the world in which they are to work. 'Critical issues' draw attention to impacts beyond narrowly modelled functions; they emphasize the constructedness and immanent bias of participation methods, and highlight their political salience.

Wider circulation of the 'extended innovation agenda' offers entry points for non-experts to engage with the ongoing process and to make the design and deployment of citizen panels itself a public issue (cf. Marres 2005). The aim is to contribute to the politicization of the ongoing innovation process, not by demonizing hidden interests and mobilizing for or against a particular design, but by showing up different partial realities of politics as they become articulated and selected in the process of articulating methods of participation. The orientation is to cultivate a reflexive discourse and allow for open political debate of the political ordering that is at stake in 'knowing governance' in any particular way (Voß and Freeman forthcoming).

The actual effect of this intervention remains to be seen. The interactions at the workshop will have further effects in the practical engagement of participants with the innovation process. What specifically this effect will be, and how strong, is likely to be different for participants. For the moment it is interesting to consider the workshop as another example of reflexively engaging with technologies of participation. No matter what the further impact will be, the workshop, the scenarios and the report have become part of the innovation journey of citizen panels. The process created new linkages between people, topics, concerns and arguments.

In a certain respect this engagement was special. It did not aim to add or strengthen a particular perspective in the innovation process, but sought to orchestrate the collective movement from which the innovation emerges. It was an engagement with the governance of the innovation process. This by itself may seem preposterous or overly assertive. But there is an aspect to it that may even be more crucial with a view to the technoscientization of politics in the course of innovating participation methods: did we, with the design of our 'challenging futures' method, also pursue a technoscientific approach to participation? The intervention was designed in a confined research collective, on the basis of a specifically reduced model of the innovation process. Important design decisions were taken 'in seclusion' such as the decision to invite diverse practitioners and stakeholders of citizen panels rather than citizen participants, or the particular setup for the scenarios, the agenda, the style of moderation, etc. These decisions were not put up for wider discussion, neither with participants nor with a broader public. The workshop, finally, was set up as an experiment to probe adopted CTA methods in application to social technologies like citizen panels and other instruments of governance.[10] It may thus be seen as just a further turn of the screw: now we are moving towards technocracy of a third order which does not regard substantial policy issues, nor public participation procedures, but now the procedures for assessing procedural innovations. Even if we do not claim objectivity, neutrality or epistemic authority for our approach, we articulated it as an abstract, general procedure to be tested on different cases.

In this respect it is important to note that also here we find the technoscientific mode of ordering countered by a contestation of particular decisions of complexity reduction and the assumption of authority by us as experts. A workshop participant, for example, resisted and problematized the proposed procedure for the 'challenging futures' workshop. And we received challenging questions and comments when we presented the project at academic conferences. Even our own questions and discussions during the preparation of the whole process may count as a kind of self-reflexive engagement with the trajectory of our own project. So, there is reflexive engagement also with this third order of technocracy. To date, no protest groups have emerged in our meetings. But this may come. If the 'challenging futures' method comes to be established as a universal standard for constructively assessing social technologies and governance instruments, it will be challenged, and seemingly technical design features will be politicized – hopefully.

Reflexive governance of social innovation

In conclusion, I come back to the broader topic of technoscientific innovation in political order. By having a closer look at the innovation of citizen panels, we encounter a precarious balance between technoscientization and politicization. The dynamic of the innovation process emerges from the interplay of those different modes of engaging with methods of participation (Voß forthcoming). Technoscientization and reflexive engagement are intertwined in a spiralling, dialectical movement: initially, the technicization of substantial policy issues is met by contestations which draw out political dimensions of policy appraisal and stimulate broader participation. This prompted attempts to strengthen public participation, ultimately also by technicizing the design of participatory procedures. Again this was met by reflexive engagements emphasizing the political dimensions of procedures. The case of our constructive assessment exercise for citizen panels can be seen as part of a broader set of activities to open the design of participation methods for re-negotiating their diverse purposes in different situational contexts. Yet, I have discussed how the design of the assessment exercise itself may again be contested as a move of technoscientific closure, now with regard to the methods of method assessment.

In the case of citizen panels, scientific objectification and control are obviously part of the innovation process. Questions of political constitution building are here, at least partly, decided by professional and epistemic authority. With regard to increasingly centralized practices of secluded research, we may speak of an emerging technoscience of democracy. A special irony is that this second-order technocracy of political procedure emerges from a struggle against first-order technocracy of substantial policy decision.

But the case also shows that technoscientization of political procedure is immediately accompanied by reflexive engagement and debate. Critical academic discourse, alternative procedural designs, direct protest actions and dedicated assessment exercises have a formative impact on the realization of new forms of political order. They thematize the 'collateral realities' of apparently technical devices (Law 2012; Law and Ruppert 2013) and raise critical issues which show up their political dimensions. Reflexive engagement activities effectively contribute to explore robust pathways of innovation, even if they appear to work against innovation and slow it down.[11]

A more general point follows from understanding the critical questioning, opening and politicization of designs as constitutive elements of innovation – just as much as the advocating, closing and objectifying of specific solutions. The dynamic balancing of technoscientific closure with political opening appears as a practical way of coping with the inherent ambiguities of innovation. Any process of bringing new orders into existence is a non-exigent, partial reduction of complexity that excludes alternative realities. It implies a deliberate disengagement with the emergent flows of situational interaction. It increases selectively directed, internal productivity at the expense of potential agencies. As such it is inherently dilemmatic (see March, 1991, for a similar but more strategic articulation of the exploration/exploitation dilemma in organizational learning).

Even if there may be no way to resolve the dilemma in a rationally coherent way, it may well be that the dynamic agonism of technoscientific and political rationalities is a practical way of coping with it. What appears as a struggle between worlds (Latour 2002) may indeed work as interplay and balancing – similar perhaps to what party competition and institutional checks and balances do for the shaping of nation state politics. The point is that the contribution of reflexive engagements may be appreciated as a constitutive component of innovation and as a way of practically coping with inherent ambiguities of ordering.

This may also be of wider relevance for the emerging 'social innovation' discourse. In contrast to technological innovations, social innovations are widely described as inherently empowering, just, beneficial, socially embedded and legitimate (e.g., BEPA 2011 and www.socialinnovationeurope.eu). There is no recognition, so far, of the need of accompanying assessment activities comparable to what is known as 'technology assessment'. I tried to show that the innovation of public participation methods, as a particular kind of social innovation, exhibits the patterns and problems of technoscience between worlds (Latour 2002). The transformation of issues of collective ordering into technical design problems, together with their withdrawal from public debate, is accompanied by various forms of reflexive engagement which *de facto* take on the role of an emerging assessment regime. What makes an innovation prone to critical inquiry and debate is not the stuff that it is most visibly made of, but the mode by which it is assembled (Callon *et al.* 2009). As much as social innovation is technoscientized, it also throws up questions of 'responsible research and innovation' (Owen *et al.* 2012) and the need for

> a transparent, interactive process by which societal actors and innovators become mutually responsive to each other with a view to the (ethical) acceptability, sustainability and societal desirability of the innovation process and its marketable products (in order to allow a proper embedding of scientific and technological advances in our society).
>
> (Von Schomberg 2012, 48)

Notes

1 This is a basic understanding of technoscience that has been developed on the basis of Gaston Bachelard's pioneering of a performative theory of science and further developed in laboratory studies (Bachelard 1984 [1934]; Hacking 1983; Knorr-Cetina 1995; Latour 1999; for an application to policy studies see Voß 2014).
2 Social technology here refers to 'configurations that work' (Rip and Kemp 1998), which are composed of heterogeneous (material, human, semiotic) elements but where, in public perception, social rules and practices are foregrounded. The STS perspective shows that social elements are also constitutive of the practical working of configurations such as bicycles, cars or electricity systems (Callon 1987; Hughes 1987; Pinch and Bijker 1987). Likewise, it acknowledges that material elements are part of practically working participation methods and other social technologies.
3 Robert Goodin, in a discussion following his talk 'How Deliberative Democracy Can Make International Law', 26 September 2012, University of Tübingen, at the German Political Science Association's conference Promises of Democracy.
4 I build on research which sought to trace instances in which citizen panels became articulated both as abstract models and as implemented configurations of specific political situations. Such instances were drawn from academic literature, project documents, method manuals, policy reports and websites as well as 30 personal interviews and a group discussion with 25 actors who were practically engaged in the process. The overall pattern of the process was reconstructed in an iterative process of pattern matching and abduction for which we referred to a repertoire of concepts from science, technology and innovation studies (cf. Van de Ven *et al.* 1999; Yin 2003; Van de Ven 2007). I acknowledge funding by the German Federal Ministry of Education and Research (BMBF) through Grant No. 01UU0906 and thank Nina Amelung and Louisa Grabner, who did large parts of the empirical research, and Carsten Mann and Till Runge, who organized the constructive assessment workshop on which I report later on.
5 Occasionally other terms are used such as deliberative fora (Hendriks and Carson 2008), mini-publics (Goodin and Dryzek 2006).
6 Platforms such as www.participedia.net, http://participationcompass.org, www.partizipation.at, the *Journal of Public Deliberation* (www.publicdeliberation.net/jpd), the Deliberative Democracy Consortium (www.deliberative-democracy.net) and the International Association for Public Participation (www.iap2.org).

7 Most prominent is the engagement of the Commission of the European Union in commissioning expertise and collaborative projects for negotiating design standards, as well as implementing and evaluating experiments on European policy issues. Such projects include 'Participatory Approaches in Science and Technology' (PATH, 2004–6), 'Citizen Participation in Science and Technology' (CIPAST, 2005–8) and 'Meeting of Minds – European Citizens' Deliberation on Brain Science' (2006) (IFOK 2003; Goldschmidt and Renn 2006; Abels 2009). National governments also sought to develop a robust methodological basis for public participation – for example, 'Sciencewise' in the UK (www.sciencewise-erc.org.uk; cf. Chilvers 2010) or activities by the German Office of Technology Assessment (Hennen et al. 2004).
8 The Danish Board of Technology developed 'World Wide Views: A Methodology for Global Citizen Deliberation' (www.wwviews.org) with a first demonstration project related to negotiations under the United Nations Convention on Climate Change (World Wide Views on Global Warming 2009) and a second one on negotiations under the United Nations Convention on Biodiversity (World Wide Views of Biodiversity 2013).
9 The workshop was introduced by members of the research team and moderated by Arie Rip, a scholar and experienced practioner of constructive technology assessment, with occasional support by members of the research team.
10 A second experiment with an environmental market instrument (biodiversity offsetting and banking) was carried out a week later (Mann et al. 2014b).
11 This interplay actually implies a dilemma for radical technological critique which becomes part of a reality that it seeks to disrupt. Perhaps disregard and refusal to engage with a rejected discourse would be more effective in this case, but this is another question which will not be discussed here.

References

Abels, G. 2009. *Citizens' Deliberations and the EU Democratic Deficit: Is There a Model for Participatory Democracy? Tübinger Arbeitspapiere zur Integrationsforschung 1/2009*. Tübingen: Universität Tübingen.

Amelung, N. 2012. The emergence of citizen panels as a de facto standard, *Quaderni* 79: 13–28.

Andersen, I.-E. and Jäger, B. 1999. Scenario workshops and consensus conferences: Towards more democratic decision-making, *Science and Public Policy* 26(5): 331–340.

Arendt, H. 1979 [1969]. *On Violence*. Orlando, FL: Harvest Book, Harcourt.

Bachelard, G. 1984 [1934]. *The New Scientific Spirit*. Boston, MA: Beacon Press.

Barben, D., Fisher, E., Selin, C. and Guston, D. H. 2008. Anticipatory governance of nanotechnology: Foresight, engagement, and integration. In: E. J. Hackett, O. Amsterdamska, M. Lynch and J. Wajcman (eds) *The Handbook of Science and Technology Studies*, 3rd edition. Cambridge, MA: MIT Press, pp979–1000.

Barry, A. 2002. The anti-political economy. *Economy and Society* 31(2): 268–284.

BEPA 2011. *Empowering People, Driving Change: Social Innovation in the European Union*. Brussels: Bureau of Policy Advisors for the European Commission.

Bogner, A. 2012. The paradox of participation experiments, *Science, Technology & Human Values* 37(5): 506–527.

Braun, K. and Schultz, S. 2010. '… a certain amount of engineering involved': Constructing the public in participatory governance arrangements, *Public Understanding of Science* 19(4): 403–419.

Brown, M. B. 2006. Survey article: Citizen panels and the concept of representation, *The Journal of Political Philosophy* 14(2): 203–225.

Brown, M. B. 2009. *Science in Democracy: Expertise, Institutions, and Representation*. Cambridge, MA: MIT Press.

Callon, M. 1987. Society in the making: The study of technology as a tool for sociological analysis. In: W. E. Bijker, T. P. Hughes and T. J. Pinch (eds) *The Social Construction of Technological Systems*, Cambridge, MA: MIT Press, pp83–106.

Callon, M., Lascoumes, P. and Barthe, Y. 2009. *Acting in an Uncertain World: An Essay on Technical Democracy*. Cambridge, MA: MIT Press.

Camic, C., Gross, N. and Lamont, M. 2011. The study of social knowledge making. In: C. Camic, N. Gross and M. Lamont (eds) *Social Knowledge in the Making*. Chicago, IL: Chicago University Press, pp1–40.

Carpini, M., Delli, X., Lomax Cook, F. and Jacobs, L. R. 2004. Public deliberation, discursive participation, and citizen engagement: A review of the empirical literature, *Annual Review of Political Science* 7: 315–344.

Chilvers, J. 2008. Environmental risk, uncertainty, and participation: Mapping an emergent epistemic community, *Environment and Planning A* 40(12): 2990–3008.

Chilvers, J. 2010. *Sustainable Participation? Mapping Out and Reflecting on the Field of Public Dialogue on Science and Technology*, Harwell: Sciencewise Expert Resource Centre.

Chilvers, J. 2013. Reflexive engagement? Actors, learning, and reflexivity in public dialogue on science and technology, *Science Communication* 35(3): 283–310.

Chilvers, J. and Burgess, J. 2008. Power relations: The politics of risk and procedure in nuclear waste governance, *Environment and Planning A* 40(8): 1881.

Coote, A. and Lenaghan, J. 1997. *Citizens' Jury: Theory into Practice*. London: Institute for Public Policy Research.

Crosby, N. 1974. *The Educated Random Sample: A Pilot Study on a New Way to Get Citizen Input into the Policy-Making Process*. Minneapolis, MN: Center for New Democratic Processes.

Crosby, N. 1975. *In Search of the Competent Citizen: A Research Proposal*, unpublished manuscript.

Desrosières, A. 1998. *The Politics of Large Numbers: A History of Statistical Reasoning*. Cambridge, MA: Harvard University Press.

Deuten, J. J. 2003. *Cosmopolitanising Technologies: A Study of Four Emerging Technological Regimes*. Enschede: Twente University Press.

Dewey, J. 1971. Wie können die Bürger an Planungsprozessen beteiligt werden? Planwahl und Planungszelle als Beteiligungsverfahren, *Der Bürger im Staat* 21(3): 151–156.

Dewey, J. 2012 [1954]. *The Public and its Problems: An Essay in Political Inquiry*. University Park, PA: Penn State University Press.

Dienel, P. 1971. Wie können die Bürger an Planungsprozessen beteiligt werden? Planwahl and Planungszelle als Beteilungsverfahren, *Der Bürger im Staat* 21(3): 151–160.

Dienel, P. 1978. *Die Planungszelle. Eine Alternative zur Establishment-Demokratie*. Opladen: Westdeutscher Verlag.

Disch, L. 2012. Democratic representation and the constituency paradox, *Perspectives on Politics* 10(3): 599–616.

Disco, C., Rip, A. and van der Meulen, B. 1992. Technical innovation and the universities: Divisions of labour in cosmopolitan technical regimes, *Social Science Information* 31(3): 465–507.

Elliott, J., Heesterbeek, S., Lukensmeyer, C. J. and Slocum, N. 2005. *Participatory Methods Toolkit: A Practitioner's Manual*. Brussels: King Baudouin Foundation and the Flemish Institute for Science and Technology Assessment.

Ezrahi, Y. 1990. *The Descent of Icarus: Science and the Transformation of Contemporary Democracy*. Cambridge, MA: Harvard University Press.

Ezrahi, Y. 2012. *Imagined Democracies: Necessary Political Fictions*. Cambridge, MA: Cambridge University Press.

Felt, U. and Fochler, M. 2010. Machineries for making publics: Inscribing and de-scribing publics in public engagement. *Minerva* 48(3): 319–338.

Freire, P. 2000 [1970]. *Pedagogy of the Oppressed*. New York: Continuum International Publishing Group.

Fung, A. 2006. Varieties of participation in complex governance, *Public Administration Review* 66(s1): 66–75.

Gastil, J. and Keith, W. M. 2005. A nation that (sometimes) likes to talk: A brief history of public deliberation in the United States. In: J. Gastil and P. Levine (eds) *The Deliberative Democracy Handbook: Strategies for Effective Civic Engagement in the 21st Century*. San Francisco, CA: Jossey-Bass, pp3–19.

Geels, F. and Deuten, J. J. 2006. Local and global dynamics in technological development: A socio-cognitive perspective on knowledge flows and lessons from reinforced concrete, *Science and Public Policy* 33(4): 265–275.

Goldschmidt, R. and Renn, O. 2006. *Meeting of Minds – European Citizens' Deliberation on Brain Sciences. Final Report of the External Evaluation*. Stuttgart: DialogikGmbH.

Gomart, E. and Hajer, M. A. 2003. Is that politics? For an inquiry into forms in contemporary politics. In: B. Joerges and H. Nowotny (eds) *Social Studies of Science and Technology: Looking Back, Ahead*. Dordrecht: Kluwer, pp33–61.

Goodin, R. E. and Dryzek, J. S. 2006. Deliberative impacts: The macro-political uptake of mini-publics, *Politics & Society* 34(2): 219.

Grönlund, K., Bächtiger, A. and Setälä, M. (eds) 2014. *Deliberative mini-publics: Involving citizens in the democratic process*. Colchester: ECPR Press.

Greenpeace 2007. *Talking Nonsense – The 2007 Nuclear Consultation*, www.greenpeace.org.uk/files/pdfs/nuclear/2007-consultation-nuclear-dossier.pdf (accessed 22 September 2015).

Guston, D. 2014. Understanding 'anticipatory governance', *Social Studies of Science* 44(2): 218–242.

Guston, D. and Sarewitz, D. 2002. Real-time technology assessment, *Technology in Society* 24(1–2): 93–109.

Habermas, J. 1981. *Theorie des kommunikativen Handelns. Handungsrationalität und gesellschaftliche Rationalisierung*. Vol. 1. Frankfurt: Suhrkamp.

Habermas, J. 1993. *Justification and Application: Remarks on Discourse Ethics*. Cambridge, MA: MIT Press.

Hacking, I. 1983. *Representing and Intervening: Introductory Topics in the Philosophy of Natural Science*. Cambridge: Cambridge University Press.

Hendriks, C. M. 2005. Consensus conferences and planning cells: Lay citizen deliberations. In: J. Gastil and P. Levine (eds) *The Deliberative Democracy Handbook: Strategies for Effective Civic Engagement in the 21st Century*. San Francisco, CA: Jossey-Bass, pp80–110.

Hendriks, C. M. and Carson, L. 2008. Can the market help the forum? Negotiating the commercialization of deliberative democracy, *Policy Sciences* 41(4): 293–313.

Hennen, L., Petermann, T. and Scherz, C. 2004. *Partizipative Verfahren der Technikfolgenabschätzung und parlamentarische Politikberatung. Neue Formen der Kommunikation zwischen Wissenschaft, Politik und Öffentlichkeit*. Berlin: Büro für Technikfolgenabschätzung beim deutschen Bundestag (TAB).

Heurtin, J.-P. 2005. The circle of discussion and the semicircle of criticism. In: B. Latour and P. Weibel (eds) *Making Things Public: Atmospheres of Democracy*. Cambridge, MA: MIT Press, pp754–769.

Hörning, G. 1999. Citizens' panels as a form of deliberative technology assessment, *Science and Public Policy* 26(5): 351–359.

Horst, M. and Irwin, A. 2010. Nations at ease with radical knowledge: On consensus, consensusing and false consensusness, *Social Studies of Science* 40(1): 105–126.

Hughes, T. P. 1987. The evolution of large technical systems. In: W. E. Bijker, T. P. Hughes and T. J. Pinch (eds) *The Social Construction of Technological Systems: New Directions in the Sociology and History of Technology*. Cambridge, MA: MIT Press, pp51–82.

IFOK 2003. *Governance of the European Research Area: The Role of Civil Society*. Bensheim: Institut für Organisationskommunikation.

Irwin, A. 2001. Constructing the scientific citizen: Science and democracy in the biosciences. *Public Understanding of Science* 10(1): 1–18.

Irwin, A. 2008. STS perspectives on scientific governance. In: E. J. Hackett, O. Amsterdamska, M. Lynch and J. Wajcman (eds) *The Handbook of Science and Technology Studies*, 3rd edition. Cambridge, MA: MIT Press, pp583–608.

Joss, S. and Durant, J. 1995. *Public Participation in Science: The Role of Consensus Conferences in Europe*. London: Science Museum.

Karpowitz, C. F. and Mendelberg, T. 2011. An experimental approach to citizen deliberation. In: J. N. Druckman, D. P. Green, J. H. Kuklinski and A. Lupia (eds) *Cambridge Handbook of Experimental Political Science*. New York: Cambridge University Press, pp258–272.

Knorr-Cetina, K. 1995. Laboratory studies: The cultural approach to the study of science. In: S. Jasanoff, G. E. Markle, J. C. Petersen and T. Pinch (eds) *Handbook of Science and Technology Studies*. Thousand Oaks, CA: SAGE, pp140–166.

Laclau, E. and Mouffe, C. 2001. *Hegemony and Socialist Strategy: Towards a Radical Democratic Politics*. London: Verso.

Latour, B. 1999. *Pandora's Hope: Essays on the Reality of Science Studies*. Cambridge, MA: Harvard University Press.

Latour, B. 2003. What if we talked politics a little?, *Contemporary Political Theory* 2(2): 143–164.

Laurent, B. 2009. Replicating participatory devices: The consensus conference confronts nanotechnology. *CSI Working Paper Series*, 2009 (018).

Laurent, B. 2011a. *Democracies on Trial: Assembling Nanotechnology and its Problems*, PhD thesis. Paris: Mines Paris Tech, Centre de Sociologie de l'Innovation.

Laurent, B. 2011b. Technologies of democracy: Experiments and demonstrations, *Science and Engineering Ethics* 17(4): 649–666.

Law, J. 2012. Collateral realities. In: F. D. Rubio and P. Baer (eds) *The Politics of Knowledge*. London: Routledge, pp156–178.

Law, J. and Ruppert, E. 2013. The social life of methods: Devices, *Journal of Cultural Economy* 6(3): 229–240.

Levidow, L. 1998. Democratizing technology – or technologizing democracy? Regulating agricultural biotechnology in Europe, *Technology in Society* 20(2): 211–226.

Lezaun, J. 2007. A market of opinions: The political epistemology of focus groups. *The Sociological Review* 55: 130–151.

Lezaun, J. and Soneryd, L. 2007. Consulting citizens: Technologies of elicitation and the mobility of publics, *Public Understanding of Science* 16(3): 279–297.

Lövbrand, E., Pielke, R. and Beck, S. 2011. A democracy paradox in studies of science and technology, *Science, Technology and Human Values* 36(4): 474–496.

Maasen, S. and Merz, M. 2006. *TA Swiss Broadens its Perspective: Technology Assessment with Asocial and Cultural Sciences Orientation. Working Document of the Centre for Technology Assessment DT-37/2006*. Bern: TA-Swiss.

Maier, J. 2009. Nanotechnologie: Warten auf den Störfall, *Die Zeit*, 2009 (27).

Mann, C. and Voß, J.-P. forthcoming. Articulating future scenarios of policy instrument development: Enhancing sustainable innovation in governance by opening design processes for debate. In: M. Padnamabhan (ed.) *Transdisciplinarity for Sustainability*. London: Routledge.

Mann, C., Voß, J.-P., Amelung, N., Simons, A., Runge, T. and Grabner, L. 2013. *Preparatory Document for the Workshop 'Challenging Futures of Citizen Panels' on 26 April 2013*. Berlin: Technische Universität Berlin.

Mann, C., Voß, J.-P., Amelung, N., Simons, A., Runge, T. and Grabner, L. 2014a. *Challenging Futures of Citizen Panels: Critical Issues for Robust Forms of Public Participation. A Report Based on Interactive, Anticipatory Assessment of the Dynamics of Governance Instruments, 26 April 2013.* Berlin: Technische Universität Berlin.

Mann, C., Voß, J.-P., Simons, A., Amelung, N., Runge, T. and Schroth, F. 2014b. *Challenging Futures of Biodiversity Offsetting and Trading: Critical Issues for Robust Forms of Biodiversity Conservation. A Report Based on Interactive, Anticipatory Assessment of the Dynamics of Governance Instruments, 19 April 2013.* Berlin: Technische Universität Berlin.

March, J. G. 1991. Exploration and exploitation in organizational learning, *Organization Science* 2(1): 71–87.

Marres, N. 2005. *No Issue, No Public: Democratic Deficits after the Displacement of Politics*, PhD thesis. Amsterdam: University of Amsterdam.

OECD. 2001. *Citizens as Partners: OECD Handbook on Information, Consultation and Public Participation in Policy-Making.* Paris: OECD.

Osborne, T. and Rose, N. 1999. Do the social sciences create phenomena? The example of public opinion research, *The British Journal of Sociology* 50(3): 367–396.

Owen, R., Macnaghten, P. and Stilgoe, J. 2012. Responsible research and innovation: From science in society to science for society, with society, *Science and Public Policy* 39(6): 751–760.

Parkinson, J. 2004. Why deliberate? The encounter between deliberation and new public managers. *Public Administration* 82(2): 377–395.

Parkinson, J. 2006. *Deliberating in the Real World: Problems of Legitimacy in Deliberative Democracy*, Oxford: Oxford University Press.

Pinch, T. J. and Bijker, W. E. 1987. The social construction of facts and artifacts: Or how the sociology of science and the sociology of technology might benefit each other. In: W. E. Bijker, T. P. Hughes and T. J. Pinch (eds) *The Social Construction of Technological Systems.* Cambridge, MA: MIT Press, pp17–50.

Porter, T. M. 1996. *Trust in Numbers: The Pursuit of Objectivity in Science and Public Life.* Princeton, NJ: Princeton University Press.

Renn, O. and Schweizer, P.-J. 2009. Inclusive risk governance: Concepts and application to environmental policy making, *Environmental Policy and Governance* 19(3): 174–185.

Renn, O., Webler, T. and Wiedemann, P. 1995. *Fairness and Competence in Citizen Participation: Evaluating Models for Environmental Discourse.* Dordrecht: Kluwer.

Rip, A. 1987. Controversies as informal technology assessment, *Knowledge: Creation, Diffusion, Utilization* 8(2): 349–371.

Rip, A. and Kemp, R. 1998. Technological change. In: D. S. Rayner and E. L. Malone (eds) *Human Choice and Climate Change.* Columbus, OH: Battelle Press, pp327–399.

Rip, A. and Schot, J. W. 2002. Identifying loci for influencing the dynamics of technological development. In: K. H. Sorensen and R. Williams (eds) *Shaping Technology, Guiding Policy: Concepts, Spaces and Tools.* Cheltenham: Edward Elgar, pp155–172.

Rip, A., Misa, T. J. and Schot, J. P. 1995. *Managing Technology in Society: The Approach of Constructive Technology Assessment.* London: Pinter.

Rosanvallon, P. 2006. *Democracy Past and Future.* New York: Columbia University Press.

Rose, N. S. 1999. *Powers of Freedom: Reframing Political Thought.* Cambridge, MA: Cambridge University Press.

Rowe, G. and Frewer, L. J. 2000. Public participation methods: A framework for evaluation, *Science, Technology & Human Values* 25(1): 3–29.

Rowe, G. and Frewer, L. J. 2005. A typology of public engagement mechanisms, *Science, Technology & Human Values* 30(2): 251–290.

Schot, J. and Rip, A. 1997. The past and future of constructive technology assessment, *Technological Forecasting and Social Change* 54(23): 251–268.
Sloterdijk, P. and Mueller von der Haegen, G. 2005. Instant democracy: The pneumatic parliament®. In: B. Latour and P. Weibel (eds) *Making Things Public: Atmospheres of Democracy*. Cambridge, MA: MIT Press, pp952–955.
Smith, G. and Wales, C. 2002. Citizens' juries and deliberative democracy, *Political Studies* 48(1): 51–65.
Stewart, J., Kendall, E. and Coote, A. 1994. *Citizens' Juries*. London: Institute for Public Policy Research.
Stirling, A. 2008. Opening up and closing down power, participation, and pluralism in the social appraisal of technology, *Science, Technology & Human Values* 33(2): 262–294.
Sulkin, T. and Simon, A. F. 2001. Habermas in the lab: A study of deliberation in an experimental setting, *Political Psychology* 22(4): 809–826.
Ureta, S. 2015. A failed platform: The consensus conference travels to Chile, *Public Understanding of Science*, http://pus.sagepub.com/content/early/2015/01/07/0963662514561940.abstract (accessed 22 September 2015).
Van de Ven, A. H. 2007. *Engaged Scholarship: A Guide for Organizational and Social Research*. Oxford: Oxford University Press.
Van de Ven, A. H., Polley, D., Garud, R. and Venkatamaran, S. 1999. *The Innovation Journey*. Oxford: Oxford University Press.
Vergne, A. 2010. *Das Modell Planungszelle – Citizens Juries: Diffusion einer politischen Innovation*. Unpublished manuscript.
Von Schomberg, R. 2012. Prospects for technology assessment in a framework of responsible research and innovation. In: M. Dusseldorp and R. Beecroft (eds) *Technikfolgen abschätzen lehren. Bildungspotenziale transdisziplinärer Methoden*. Wiesbaden: VS Verlag, pp39–61.
Voß, J.-P. forthcoming. Governance-Innovationen: Epistemische und politische Reflexivität in der Herstellung von 'Bürgerpanelen' als neuer Form von Demokratie. In: W. Rammert, M. Hutter, H. Knobaluch and A. Windeler (eds) *Innovationsgesellschaft Heute*. Wiesbaden: VS Verlag.
Voß, J.-P. 2014. Performative policy studies: realizing 'transition management', *Innovation – The European Journal of Social Science Research* 27(4): 317–343.
Voß, J.-P. and Amelung, N. forthcoming. The innovation journey of 'citizen panels': Technoscientization and reflexive engagement in developing methods of public participation. Paper accepted for publication in *Social Studies of Science*.
Voß, J.-P. and Freeman, R. (eds) forthcoming. *Knowing Governance: The Epistemic Construction of Political Order*. Basingstoke: Palgrave-Macmillan.
Wakeford, T. 2003. *Teach Yourself Citizen Juries: A Handbook*, www.researchgate.net/profile/Tom_Wakeford/publication/275218861_Teach_Yourself_Citizens_Juries_%282nd_Edition%29/links/5535e80c0cf218056e92abea.pdf (accessed 22 September 2015).
Wakeford, T. and Singh, J. 2008. Towards empowered participation: Stories and reflections. *Participatory Learning and Action* 58(June): 6–10.
Wakeford, T., Singh, J., Murtuja, B., Bryant, P. and Pimbert, M. 2007. The jury is out: How far can participatory projects go towards reclaiming democracy? In: P. Reason and H. Bradbury (eds) *The SAGE Handbook of Action Research: Participative Inquiry and Practice*. London: SAGE, pp333–349.
Wynne, B. 2006. Public engagement as a means of restoring public trust in science: Hitting the notes, but missing the music?, *Public Health Genomics* 9(3): 211–220.
Wynne, B. 2007. Public participation in science and technology: Performing and obscuring a political-conceptual category mistake, *East Asian Science, Technology and Society* 1(1): 99–110.
Yin, R. K. 2003. *Case Study Research: Design and Methods*. Thousand Oaks, CA: SAGE.

13
REMAKING PARTICIPATION
Towards reflexive engagement

Jason Chilvers and Matthew Kearnes

Introduction

The title of this book, *Remaking Participation: Science, Environment and Emergent Publics*, might be read in at least three ways. First of all, through taking forward a co-productionist approach, the preceding chapters in this volume have sought to remake the reality of participation itself. Countering dominant 'realist' approaches that assume a pre-given or external reality of participation and 'the public', the authors have put forward an alternative reading of the *realities* of participation with science, technology and the environment as multiple, situated, co-produced and in the making. Based on this core perspective, two further important meanings flow from this alternative way of seeing and being with science and participation in late-modern society.

A second reading of the book's title is also evident in the ways the preceding chapters confirm that participation *is continually being remade* and offer snapshots of a set of ongoing reconfigurations and diverse ways in which democratic societies make collective choices about science, technology and the environment. As the chapters suggest, the ways in which participation is being remade are indicative of the shifting sensitivities of contemporary technological culture and society (Barry 2001; van Loon 2002). While scientific representation and the promises of technological innovation continue to be central preoccupations of contemporary political discourse, the relationship between science and democracy has never been more uncertain. The modes of political authority that scientific representation made available to – and were exploited by – democratic systems of governance no longer seem to enjoy an unassailable position of social and moral legitimacy. Perhaps the most poignant indication of these emerging fault lines is that the politics of technological risk, that characterized the latter half of the twentieth century, continues to challenge the constitutional consensus through which democratic nations both deploy and regulate scientific and technological innovation.

In this light, the chapters that comprise this volume paint a picture of the emergence of new modes of public legitimation – framed by notions of authenticity, responsibility and transparency – and the enactment of quasi-public spaces of deliberative decision-making and distributed innovation that characterize the everyday practices of contemporary technological politics. At the same time democratic politics itself has undergone an equally significant set of transformations. In an era characterized by popular withdrawal from formal political participation, the emergence of a reflexive and questioning attitude toward expertise and authority, and the more or less overt manipulation of electoral processes by economic and corporate elites (Crouch 2004; Wolin 2008), new forms of participation have emerged that disrupt the traditional temporalities and spatialities of contemporary political dialogue. Taken together, these dynamic processes have underpinned the proliferation of new sites for popular mobilization, participation and contestation, both above and below formalized processes of democratic representation. As documented in the preceding chapters, across a multitude of these sites – ranging from consultative committees, citizens' panels, national debates, stakeholder dialogues, scenario workshops, collective experiments, hackerspaces and 'mundane' engagements with technologies in everyday life – the lines between science, innovation and democracy are being redrawn and remade.

In addition to this epochal diagnosis, a third reading of the book's title is also possible. While participation is continually being remade, the volume poses a distinctly normative problem asking how participation *might be remade* in ways that are more cosmopolitan, reflexive, responsible and pluralist. In this concluding chapter our goal is to take these questions together by considering the ways in which the relational and co-productionist accounts that have characterized the contributions to this volume, provide a vantage point from which to reimagine and reconfigure participation. This is a matter of political and democratic theory and also of practical importance for the coordination, evaluation and transformation of participatory practices, institutions and constitutions. In turning to more obviously instrumental and normative concerns, in this final chapter we ask how participatory democracy might be reconfigured in light of analyses that demonstrate the performativity and contingency of science *and* its relations with publics. In turn, we explore how the insights developed in this book might inform the ways in which the diverse practices of the democratic engagement with science and the environment might be reconfigured and put back together again in more intentionally reflexive and responsible ways.

In the first main section of this concluding chapter we bring together some cross-cutting themes from the volume. In particular we articulate the alternative way of seeing, imagining and being with participation promoted by the relational and co-productionist perspectives brought forward in the chapters of this book. In order to scope out the difference our interpretive insights make in practice, we first consider their implications for reconceiving, renewing and in some senses resuscitating notions of *reflexivity* and *critique* which have been so central to moves to democratize science and the environment, but under existing arrangements have been insufficiently or inappropriately attended to.

This paves the way in the sections that follow to consider more programmatically what a relational and co-productionist perspective means for remaking participation in terms of reconfiguring participatory practices, institutions and constitutions. As noted in Chapter 1, co-productionist studies of participation have been quick to deconstruct and offer valuable interpretive insights, but have often shown a reluctance to engage with more normative and instrumental questions about reconfiguring participation. In the remainder of this concluding chapter we move beyond this stance to consider how insights developed in this book can offer more practical lessons for remaking democratic engagement with science and the environment in more intentionally reflexive ways. To this end we set out four interrelating and interdependent paths that are central to this constructively critical project, each of which forms the focus of the four remaining sections of the chapter as follows:

- The first move is the need to build more deliberately *reflexive participatory practices* which are based in ideas of collective experimentation but also actively attend to the inherent uncertainties, effects and experimental normativities of participation itself.
- Second, recognizing the impossibility of including all relevant actors in 'the collective experiment' – and the paradox of translating democratic norms of representation into methodological form via notions of representativeness – shifts the focus of intervention beyond the performance of discrete participatory collectives. Under such conditions practice should become just as interested in mapping and being responsive to diversities and ecologies of participatory collectives, their public articulations and productive effects. We frame this in terms of a commitment to *ecologize participation*.
- A third imperative attends to the spaces where technologies of participation have become closed down, standardized and 'black boxed'. Here we turn approaches for prompting reflection and anticipation (about purposes, social implications, ethical dimensions) in the responsible development of technoscience on to social innovations in participation. Reflexive engagement with technologies, institutions and expertises of participation in this way has the potential to prompt more *responsible democratic innovations*.
- Finally, we outline the challenge of *reconstituting participation* and nurturing relational reflexivities in a more thoroughly systemic and distributed sense, which are continually attentive to the stabilities and possible forms of emergence in constitutional relations between citizens, science and the state.

For each of these four paths we outline trajectories that take this up in practice, thus helping to attend to the politics and performativity of participatory collectives in terms of relational framing effects, closures, and uncertainties, and in doing so opening up alternatives and new possibilities of democratic engagement in science and the environment. We reject dismissive forms of critique and a wholesale replacement of existing participatory formations in favour of cultivating a critical standpoint for social science that attempts to be constructive, collaborative, additive and transformative in reconfiguring and reimagining participation from a relational and co-productionist perspective. We end with a brief coda that draws together key arguments made across the volume.

Reconfiguring participation, reflexivity and critique

Before outlining the practical possibilities for remaking participation, in this section we consider how a co-productionist standpoint offers a radically different way of seeing and imagining participation, and how it alters key notions of reflexivity, learning and critique that have become associated with participatory-democratic innovation and 'progress'. There is, of course, a tension between the interpretive and analytical emphasis of the volume – which focuses attention on the situated pragmatics of participation *as it is* in-the-making – and the more normative concerns with what participation *could and should be*. As we noted in the first two chapters of this book, realist accounts of participation have often adopted taken-for-granted normative commitments concerning the value of public participation in science and the environment, resulting in the development of specific techniques and methods that reflect these commitments in ways that are not openly reflected upon. Such realist perspectives are associated with particular renderings of reflexivity and critique. Set against a critique of technocracy (Habermas 1971), participatory interventions have often set out with a reflexive intent to engender learning and inclusion. Learning has been varyingly presented as a central design principle, evaluative metric and expected product of public engagement (e.g., Webler *et al.* 1995), with respect to participating publics, engaged experts and relevant institutions.

However, the meanings of reflexivity at play here are most often reduced to concerns of methodological revisionism and instrumental forms of learning (Felt and Wynne 2007; Petts 2007). Normative commitments have been conceived in instrumental ways and narrowly presented as a problem of methodological development. One effect has been the overwhelming focus, in much of the participation literature and practice, on evaluating institutionally orchestrated participatory initiatives against largely static normative criteria. This form of critique has come at the expense of an appreciation of the diverse ways in which publics are assembled around matters of collective concern (Latour 2004a, b; Mahony *et al.* 2010). Drives to normalize and professionalize participatory practices – particularly through the circulation of standardized methodologies and evaluative frameworks and the formation of a community of participation experts (Chilvers 2008a, b) – have also perpetuated such forms of instrumental learning. An emphasis on methodological revisionism together with the development of learning infrastructures that sustain the drive toward growing and 'scaling up' participation – while also prescribing and defining 'good' participatory practice – are indicative of the instrumentalization of particular forms of evaluation and methodological revision.[1]

A further relevant and revealing feature of realist perspectives on participation noted in Chapter 1 is a preoccupation with the 'impact' of participatory processes upon decision-making as a marker of successful public engagement and source of critique – whether they be decisions by policy-makers (see, for example, Rowe and Frewer 2000) or by citizens themselves (e.g., engagements that seek to overcome the 'value-action' gap in encouraging more sustainable behaviours; see Blake 1999). This relatively unitary notion of impact evokes a linear model of the relationship between knowledge and action. Echoing the implied 'social contract for science' – in which scientific authority is marshalled to 'speak truth to power' (Jasanoff and Wynne 1998) – much of the political and academic commentary concerning public participation tends to assume that the *prima facie* goal of public engagement initiatives is to speak *social* science and 'public truths to power'.

In contrast, more critical analyses of participation have often been reluctant to engage with normative and instrumental questions concerning the reconfiguring of participatory practices. While realist notions of public engagement have tended to assume that definitive representations of 'the public' and public issues can be unproblematically settled, various kinds of critical analyses have resisted the implicit methodological commitment of much of this literature. Indeed, much of this work has demonstrated the ways in which normative principles of democratization, accountability and engagement obscure unacknowledged political and institutional commitments and the asymmetric relations that persist between publics and science (see Chapter 5 in this volume; Welsh and Wynne 2013). This mode of analysis has helpfully demonstrated the ways in which realist accounts of participation tend to limit the scope of institutional and reflexive learning in ways that obscure in-built policy cultures, the (im)mobility of incumbent institutional commitments and other centres of power and calculation (Lezaun and Soneryd 2007).

However, as we have argued in the opening chapter of this volume, both realist and critical accounts of public participation have tended to present science and politics as two separate and distinct domains of human endeavour and collective social life. In both accounts science and politics remain clearly demarcated, expertise is seen as located in individuals or institutions, while publics are presented as possessing unique preferences or experience-based expertise. Indeed, it is the demarcation of science from politics that provides the basis for normative critique and the instrumentalization of claims concerning the prospects for the democratization of science. Though these alternative conceptions of participation have occasioned substantial commentary, the proposition that the normative assessment of participation is only possible from a position *outside* of participatory practice – marshalling democratic and political theory or instrumental evaluative criteria – mirrors older debates in science and technology studies (STS) and the sociology of scientific knowledge (SSK) concerning the relationship between situated ethnographic studies of scientific practice and the positionality of sociological observation and interpretive description (e.g., Pickering 1992). As Wynne (2013) has recently argued, 'the need for social sciences and humanities to question the meanings and material forms of technoscience as symbolic action with political-cultural dimensions, and how these are articulated with respect to publics, and *vice versa*, was' – and we argue continues to be – 'an important lacuna' (p66).

What then are the grounds for remaking participation? In contrast to both realist and critical analyses of participation, the chapters of this volume share a constructivist and additive notion of critique, with an emphasis on *reconfiguring* participatory practices, knowledges, imaginaries, institutions and constitutions that is held in tension with detailed analyses of the contingent enactment of participation (following Latour 2004b; Boltanski 2011). The chapters that comprise this volume vividly demonstrate the emergent qualities of public collectives (Chapters 3–5 and 10), the contingent technicalities of participation (Chapters 6, 7, 9 and 12) and the experimental methodologies entailed in enacting 'atmospheres of democracy' (Chapters 8, 10 and 11). In place of the kind of methodological revisionism that has tended to characterize debate and 'progress' in the participation field, these accounts refuse a position outside of participatory practice from which to mount normative analyses. The normative commitments that the chapters do adopt – what we have characterized, following Latour and Weibel (2005), as 'atmospheres of democracy' – are highly mediated and contingent. What matters is not simply the 'starting conditions' of participatory processes – formal commitments to openness, equality, representativeness and transparency (Dryzek 1990; Cohen and Arato 1992; Renn et al. 1995) – but a reflexive analysis of the inevitable openings, closures, framing conditions, ambivalences, imagined social orders and lacunae of *all* participatory forms, including participation in institutionally orchestrated settings and in counterpublic spaces (Warner 2002).

Put simply, viewed from a relational and co-productionist perspective, participation can no longer be viewed in the naïvely realist terms that have characterized much existing scholarship and practice. However, while problematizing the position from which to mount normative critique, the implications of the model of participation pursued in the volume is not antithetical to either instrumentalization or normative problematization. As Jasanoff (2011a) argues, 'there are three possible roles, analytically distinct yet practically interconnected, for STS scholars concerned with turning points in science–state–society relations: instrumental, interpretive, and normative' (p624). In this volume each of these roles has been held in tension. Indeed a co-productionist approach to participation gestures toward the interweaving of each of these roles where interpretive critique always projects normative dimensions, while interventions in practice in the 'field' will always have theories and political philosophies working through their performance. Likewise, all forms of collective participatory practice are not only seeking to *represent* public truths and mobilize societal action, but they are always simultaneously articulating how the world *ought* to be in terms of social and democratic orders, assumed relations between science and society, the imagined roles of publics, and so on. It is these inherent 'normativities of participation' that realist approaches routinely bracket out, deny or simply ignore.

In addition to reimagining patterns of critique, a co-productionist and relational conception of participation has far-reaching consequences for received notions of learning and reflexivity. Against a backdrop defined by the instrumentalization of prescriptive forms of learning, simplistic linear assumptions concerning the links between participation, learning and action that predominate participatory scholarship and practice are exposed as highly problematic and potentially damaging. The kinds of co-productionist conceptions of participation that characterize this volume lend themselves to more radical and relational definitions of reflexivity and reflexive learning. On the one hand, this approach is rooted in the 'process of identifying, and critically examining (thus rendering open to change), the basic, pre-analytic assumptions that frame knowledge-commitments' (Wynne 1993, 324). On the other, it is based in the recognition of a reflexive capacity embedded in the emergence and construction of public collectives.

Of course, relational analyses of science and debates around the notion of reflexivity – which have themselves provided an impetus for new forms of public engagement and democratization (Irwin 2006) – have a deep hinterland in STS debate and scholarship (Woolgar 1988; Lynch 2000). Indeed, much of this work might be characterized as simultaneously promoting a reflexive view of scientific knowledge while resisting overly instrumental versions of reflexivity as institutional learning. At the same time, in the interface between STS research and policy development, models of reflexivity have often been translated into procedural principles deployed instrumentally. Our contention in this volume is, however, that the 'new governance' of science and the development of participatory practices has remained largely insulated from the more radical understandings of relational and reflexive learning implied by a co-productionist conception of participation outlined in this book. This is partly an effect of the rather naïve view that public engagement practices are simply a way of building reflexivity *into* science and technology, rather than attending to the reflexivities *of* participatory practices and the partiality of public representations. To do the latter would be to openly question the reflexivity of participation arrangements themselves – the actors, practices, institutions, spaces, systems, cultures and constitutions implicated in their co-production. We suggest that having a more radical-relational perspective of *reflexive participation*, then, can provide the foundations and wellsprings for remaking participation.

Before setting out what this means for remaking participation in more practical terms, it is important to clarify how the meanings of reflexivity with respect to participation are transformed under the alternative relational and co-productionist approach developed in this book. Three aspects are particularly significant. First is the way in which it emphasizes what we would call the inherent indeterminacy and *uncertainties of participation*. No matter how hard actors (human or non-human) try to include 'all relevant actors' (human or non-human), a relational approach emphasizes that all forms of participation are partial, framed in particular ways, inherently uncertain and subject to exclusions, and thus produce 'overflows' (Callon 1998; Callon *et al.* 2009). Notions of reflexivity thus shift from concerns over methodological advancement in order to produce 'better', more accurate and complete representations of 'the public' and public issues under realist forms of participation. Reflexivity is turned around to at least involve being aware of, responsible for, and accounting for the ways in which participatory experiments frame and produce particular versions of the objects (issues), subjects (participants/

publics) and procedures (philosophies) of participation.[2] As we will see, there are many possible practical responses in attending to the contingencies, emergence and indeterminacies of participation on these three dimensions.

A second key aspect of our relational approach centres on the principle of *symmetry*. In place of the 'problem of extension' – defined either as the scaling-up of participatory experiments or demarcating the 'proper limits' of public input into policy problems (Collins and Evans 2002; Durant 2011) – a relational co-productive perspective redefines participation as a 'problem of relevance' (Marres 2012). Rather than attempting to prescribe the limits of participation in advance, this perspective aims to attend to and account for the multiple and diverse ecologies of collective public articulations and commitments that constitute the issue in question and shape trajectories of socio-material change. Reflexivity, in this guise, is not simply an external evaluative criterion, but an index of the reflexive work entailed by the construction and enactment of public collectives. A relational co-productionist perspective on participation therefore upholds the principle of symmetry. This demands a certain openness and responsiveness, an outward-looking disposition, where the emergent qualities of publics are seen as a positive force – a source of dynamic opportunity, creativity and transformation that should be encouraged rather than delimited and denied.

Third, one of the most profound implications of taking a relational and co-productionist perspective seriously is the way in which it moves beyond questions of institutional reflexivity to a more systemic and 'ecological' perspective on participation (as outlined in Chapter 2). This urges us to consider how reflexivity might be understood as a thoroughly *distributed, systemic and relational* quality or disposition, rather than as a policy prescription embodied by individual actors and procedures (Stirling 2014a,b). In place of a methodological revisionism, which is often defined by the attempt to perfect new participatory tools, an implication of the modes of analyses pursued in this volume is that all manner of participatory practices might be recast to embody more radically reflexive dispositions. It emphasizes the reflexivity of and between all actors and practices that make up a given system – including public collectivities themselves. On the one hand, reflexivity in this sense means attending to and being open about framing preconditions of constitutional stabilities tied up in durable relations between citizens, science and the state (Jasanoff 2011a) and systemic (ir)responsibilities which continually reproduce singular reductionist representations of plural publics – 'public ghosts' of the neoliberal machine and scientized policy cultures, as Wynne (see Chapter 5 in this volume) puts it. On the other hand, reflexivity is equally about being open to the possibilities of emergent participatory collectives that come with the inherent sense of distributed agency emphasized by a relational ecologies of participation perspective.

As we have outlined in this section, the co-productionist and relational perspectives evident in the chapters of this volume bring forward alternative ways of seeing, imagining and being with participation that transform notions of reflexivity and critique which have been central in moves to democratize science and the environment. In the remainder of this chapter we move from sketching out these broad contours to programmatically consider what this all means for remaking participation in more practical terms. If participation can never be the same again under a relational co-productionist perspective, what might it look like in practice? In the four sections that follow we develop the four suggested pathways in this constructively critical project that we introduced earlier, namely the need to forge *reflexive participatory practices*; *ecologize participation*; bring about *responsible democratic innovations*; and *reconstitute participation* as constitutive of, rather than separate from, science and democracy. Each of these trajectories should not be considered in isolation, however. They should all be seen as interdependent and interrelating efforts to remake participation in more intentionally reflexive ways.

Reflexive participatory practices

The most obvious and immediate point of departure in prompting reflexive participation is the notion of collective participatory experiments and practices. Reflexive practice in this context is defined by an openness concerning the forms of mediation through which participatory practices enact and perform publics and public knowledge. As Wynne (2006 and Chapter 5 in this volume) shows, even well-established social science methodologies such as focus groups can be made reflexive in this manner in ensuring an acute self-awareness of how the research intervention shapes the construction of publics (whether upland sheep farmers facing radioactive fallout from Chernobyl or publics getting to grips with the promises of nanoscience). In this sense, reflexive participation means attending to and being open about the framing preconditions of participatory processes, and the ways in which publics are enacted and performed. This conception of reflexivity aims to reveal rather than obscure the emergent quality of publics and the partiality of any representation of public meaning produced through collective participatory practices. This deeper model of reflexive practice – which avoids questions of methodological closure – provides the basis for analysis and enactment of diverse forms of participatory procedure, including both formalized and informal collectives of public participation.

This notion of reflexivity therefore focuses attention inward *and* outward – emphasizing the partiality of constructions of 'the public' while at the same time seeking to situate public participation processes in the constitutional transformations of relations between science and democracy. Collective participatory experiments can vary from localized processes to those that attempt to engage the global citizenry.[3] The key definitional point here is that specific events of public engagement across these different spatial registers represent a single collective of participation irrespective of their particular scalar focus.

Reflexive experimentation

This understanding of reflexive practice imbues participation with the kind of experimental ethos that has characterized many of the chapters of this volume.

Such thinking has brought forward a wave of work devising new forms of 'collective experimentation' that seek to expand the horizon of participatory practice and introduce an explicitly reflexive definition of participation. Waterton and Tsouvalis's case of the Loweswater Care Project (Chapter 10; see also Tsouvalis and Waterton 2012) provides a good example of moves toward an experimental approach to the composition of participatory practice. In their chapter Waterton and Tsouvalis document the ways in which the participatory practice itself functioned to reframe the nature of the problem, producing an experimental approach responsive to emerging natural and social complexities and contingencies. Mirroring Latour's (2004a) notion of a 'parliament of things' and Callon et al.'s (2009) conception of the emergence of 'hybrid fora' around issues of public contention (see also Gross 2010), Waterton and Tsouvalis's participatory approach extended to include a range of non-human actors enrolled in the collective.

In a similar fashion, Landström et al.'s (2011) 'competency group' experiment, centred on the issue of flood risk management in Pickering, UK (see also Lane et al. 2008), takes seriously the materiality of participation in an effort to move beyond deliberative discursive-linguistic models of participation and include material and non-human connections in more visceral ways as part of collective participatory practice (Carolan 2007; Whatmore 2009). Expanding the definition of participation and conceiving of participatory practices as both contingent and experimental introduces a reflexive openness to the diverse material forms of engagement, connection and experience with science, technology and the environment (see Chapters 4 and 11 in this volume). This opens up to the visceral connections between environment, science, art, narratives and so on (Yusoff and Gabrys 2011; Marres 2012; Michael 2012; Gabrys 2014). As Davies (Chapter 8) shows, a collective and experimental approach should also attend to the place of emotion and affect in public engagement, in ways largely neglected hitherto (see also Harvey 2009).

Opening up participatory practices

Another means to remake participatory practices is by constructively reconfiguring existing participatory formats and designs with the deliberate intention of imbuing them with a reflexive intent. This involves thinking through how such existing participatory practices, and the actors and institutions associated with them, can be rendered more reflexive. This requires an openness concerning the framing preconditions and the issue/object(s) of participation; the cognitive, normative, material and human inputs to such processes; and the possible outputs of participatory methods. Critical in achieving this reflexivity is an open expression of difference, uncertainty and contingency in the material and discursive commitments produced through participatory processes.

The case of reconfiguring focus group methodology (Chapter 5 in this volume) highlighted at the beginning of this section is an example of this approach, demonstrating a responsibility for and reflexivity about the ways in which research interventions shape the construction of publics. Two further models that have sought to open up participatory practices include multi-criteria mapping (MCM), a form of multi-criteria options appraisal involving experts and stakeholders (Stirling and Mayer 2001; Stirling 2007), and its sister methodology deliberative mapping (DM) (Burgess et al. 2007; Chilvers and Burgess 2008), which links the practice of MCM with open processes of public deliberation. A recent example of this approach is a deliberative mapping experiment relating to climate geoengineering (Bellamy et al. 2014) which sought to introduce a reflexive assessment of the prior framings of these emerging technologies through opening the process of technology assessment up to a wider deliberation on climate change more generally. Within this broader issue-space the DM process opened up to consider diverse courses of action for addressing the problem (e.g., placing geoengineering options such as stratospheric aerosol injection and biochar in the context of alternatives like low carbon technologies through to more sustainable ways of living). The criteria against which to judge these pathways were also opened up, beyond narrow metrics of cost and temperature reduction potential, to encompass a range of social, ethical and political implications and concerns. Finally, the DM experiment opened up participatory appraisal outputs in the form of maps that depict the differences, diversities and uncertainties in the positions and commitments of all participating citizens and specialists (in addition to areas of commonality) as opposed to producing consensual or definitive representations of public and expert knowledges.

What 'opening up' approaches such as deliberative mapping, MCM and related reflexive designs (e.g., the use of Q methodologies; Cairns and Stirling 2014) can do is serve as heuristics that indicate the potential for building reflexivity about the subjects, objects and normativities of participation into distributed participatory practices, including incumbent institutions and centres of power (what Stirling 2014a,b refers to as a 'Trojan horse' strategy for prompting more transformative democratic practice). All participatory practices have the potential to be opened up in such ways, to a greater or lesser extent.

Reflexive normativities

These reflexive and experimental practices require careful real-time scrutiny of mediating processes through which publics are enacted and performed. However, while these approaches have provided useful provocations for how participation may be remade in more reflexive ways, one dimension of participatory practice that has received relatively less attention is the normativities of participation and democracy evident in the design of participatory processes. As we argued in Chapter 2, this has been a particular blind spot in both participatory practice and STS scholarship alike (de Vries 2007). In contrast, one implication of the analyses pursued in this volume is an attention to the ways in which particular 'atmospheres of democracy' (Latour and Weibel 2005) and political ontologies are produced through the performance of collectives of public engagement (Marres 2013). One way of approaching this is by being continually attentive to the 'experimental normativities' of participation and democracy (Van Oudheusden and Laurent 2013), while also being open, transparent and actively communicating the partial and particular atmospheres of democracy they produce. A further response would be to experiment with participatory interventions that intentionally bring into being more than one collective of participation, each based on the cultivation of alternative atmospheres of democracy, to make normativities of democracy themselves a comparative experimental focus (e.g., mediating multiple dialogues on a particular issue that seek to cultivate more adversarial, agonistic, consensual, autonomous or other atmospheres of democracy which are then subject to comparative analysis).

Reflexive participatory practices thus demand continual 'reflection in action' concerning the subjects, objects and formats of participation. Michael (Chapter 4 in this volume) proposes ways to achieve this through an experimentation with more creative, speculative, design-oriented and open-ended forms of engagement that allow for and encourage multiple attachments, framings and purposes as a way of striving for reflexive engagement practice. He has evoked the notion of 'the idiot' as a productive way of actively exploring and becoming sensitive to these contingencies, in particular how actors associated with a particular participatory collective can act in unexpected ways which problematize taken-for-granted framings, procedural formats and matters of concern (Horst and Michael 2011; Michael 2012).

Taking this approach further, we argue that a key challenge for reflexive participation is not only to be continually responsive to contingencies and emergence during the performance of participatory practices, but to also actively acknowledge, publicize and value the inherent uncertainties of participation and the public with(in) the outputs and productions of participatory experiments. This could include 'making public' aspects of participatory practices such as the 'real' underlying purposes of a participatory collective; how the issue or object of participation has been framed and alternative framings closed out; how the identities of participating subjects have also been constructed in particular ways while other identities have been excluded; the particular normativity of democracy produced (to the exclusion of others); and so on. So one can envisage practical steps that attempt to communicate aspects of the performativity, contingency and partiality of participatory practices and social science methods along with the knowledge and artefacts produced by participatory processes. In turn, this approach helps to enable the reflexive interpretation of these participatory products by distant others and has the potential to work against systemic forces of closure around definitive singular, individualist or deficit representations of publics. However, as we note below in relation to the constitutional dimensions of the relationship between science and politics, this should not be done naïvely, but with an acute awareness of the systemic forces of closure involved.

Ecologizing participation

No matter how reflexive individual participatory practices strive to be, as we outlined in Chapter 2 it is not possible to include all relevant actors within a single socio-material collective of participation. Collective experiments are always partial and subject to overflows. Whether we define the collective experiment at the level of a situated participatory collective (Laurent 2011), a wider relational system of science and environmental governance (Felt and Wynne 2007) or planet Earth (Macnaghten and Szerszynski 2013; Stilgoe 2015), collective experiments will always be partial and exist in relation to, and be entangled with, other forms of networked arrangements. This wider context problematizes attempts to 'resolve' collective public issues and actions through discrete collective experiments and leads to a second move in reconfiguring participation – that of attending to, working with and accounting for diverse interrelating 'ecologies of participation'.

This commitment to ecologize participation that flows from a relational co-productionist perspective on public engagement has yet to be sufficiently worked through in contemporary social science scholarship and participatory practice. Its possible transformational implications for practice are quite clear, however. At its most immediate, ecologizing participation means identifying and mapping out the diverse socio-material collectives of participation and public involvement that make up a particular space of coherence – whether that be an issue-space, arena of technoscientific development, a controversy about participatory democracy itself, a wider political constitution, or a convergence of these. It also means attending to the interrelations between multiple collectives of participation and opens up possibilities for devising new ways of mapping across multiple collectives of public involvement as they already exist *in situ*.

One of the key challenges in ecologizing participation is to extend beyond the in-depth and situated understandings that characterize most studies of participation (including co-productionist ones), toward a more systemic understanding of the diversities of participatory practices that characterize particular issue-spaces. However, in practical terms it is simply not possible to exhaustively map the forms of participation that cohere around complex socio-technical systems or mark the diverse positions taken up in contemporary technological controversies. One response to this challenge is to conceive of mapping approaches that chart the ecologies of those collectives that have gained relevance in a controversial context or issue-space. One category of such methods are issue-mapping approaches (Rogers 2002; Marres and Rogers 2005; Marres 2012) such as *issuecrawler* and *googlescraper*.[4] Such tools enable the identification of diverse actors and collectives that have become publicized and make up an issue-space. What they do not capture is the specific constructions, productions and performative effects of each of these situated collectivities. In addition, meta-analyses (Macnaghten and Chilvers 2014), comparative case analysis (Jasanoff 2005; Chilvers and Longhurst 2015) or multi-sited ethnography (Marcus 1995) offer methods that might be deployed to yield additional insights into these dimensions.

A key touchstone for such analyses is to attend to *diversity* in gaining empirical insight into the co-productive dimensions of participatory collectives differentially situated in spaces of participatory coherence, such as an issue-space or arena of technological development. Such approaches depend on defining and mapping out the space of coherence in question as a starting point, which can be assisted by work in the digital humanities and issue mapping techniques, in addition to more traditional approaches such as literature searches, the collation of documentary evidence, ethnographic interviews and workshops with diverse system actors. This provides a basis to select diverse collectives of public engagement to undertake symmetrical and empirically detailed analyses of their respective co-productions in relation to perspectives on the issues in question, material commitments and constructions of publics and participation itself.

Chilvers and Longhurst (2015) offer one example of such a comparative multi-case approach using documentary analysis to map out the co-productive dimensions of diverse collectives of public engagement in UK energy transitions. The analysis mapped across forms of engagement not usually subject to *symmetrical* comparison – ranging from a government-sponsored public dialogue on energy futures, a domestic smart meter trial, public protest in the form of a climate camp, and distributed innovation in the case of a community solar energy initiative. It showed all the collectives of participation to be productive in producing different issue definitions and visions of the future energy transition, varying forms of commitment to bringing about change, and alternative normativities of participation in how the public engagements were organized. The co-productions of each collective on these dimensions were partial and subject to exclusions, but always in situated and differentiated ways. This illustrates one possible (albeit limited) way of accounting for diversities of situated public articulations and commitments around issues like energy transitions, which conventional individual approaches for knowing energy publics – such as formal government consultations, deliberative processes, or public opinion surveys – will always underplay.

In attempting to account for diverse ecologies of participation and their dynamic interactions, the mapping of an issue-space is itself a collective actor-network. The emphasis is not to bring new collectives of participation into being (i.e., the focus of approaches discussed under reflexive participatory practices, above) but to map across multiple already-existing collectives of participation. Such issue mappings themselves will always be partial and subject to overflows. As such, attempts to define and frame spaces of coherence need to be continually questioned in light of the ecology of issues that make up the wider political situations in which they are located (Barry 2012). At the same time, attempts to map ecologies of participation should also reflect on the extent to which any approach to mapping diverse public engagements accounts for endangered or de-publicized collectives of participation (Marres 2007). These will include processes of public participation that are written out of institutionalized versions of acceptable public engagement – activism, direct action, silences, refusals to participate and so on. The uncertainties of participation again reveal themselves, and would need to be acknowledged and made transparent when speaking for, representing or communicating the findings of issue mapping exercises and other attempts to ecologize participation.

We begin to appreciate how the paths for reconfiguring participation from a co-productionist perspective outlined in this concluding chapter are, and should be, closely intertwined. Moves to ecologize participation can help bring further sensitivity and awareness to participatory practices about the other collectives of participation relationally acting on them and being shaped by them. In turn, appreciating the indeterminacy of all forms of participation and public representation in reflexive participatory practices (as outlined in the previous section) alert us to the ways in which issue-maps can mask their own partialities. Furthermore, as we elaborate below, an ecologies of participation perspective holds important implications for how we govern science and the environment, where mapping the diversities of public articulations and actions disrupts the dominant focus on procedural issues of fairness, representativeness and competence through discrete collectives of participation (Renn et al. 1995). This opens up the possibility for forms of participatory intervention that address social inequalities and public asymmetries at more systemic levels, based on an analysis of patterns across multiple collectives of participation.

The mappings of participatory diversities produced through the approaches considered in this section are not only for powerful actors and institutions, but should serve all actors who make up the given political situation they find themselves in, especially those who are marginalized or decentred. Through making diversities of public engagement in matters of concern apparent in more systemic and transparent ways, ecologizing participation has the potential to disrupt institutional tendencies to deny and dismiss the legitimate existence of *other* alternative voices. Mapping diversities of participation may therefore have an effect on institutional reflexivities, while also serving to empower marginalized public collectivities and forms of collective action. Yet, ecologizing participation means much more than formal mappings. It should also be seen as inherent to the craft and experiential disposition of *all* distributed agencies. In this sense, ecologizing participation means promoting senses of otherness (including that of other emergent publics) and of alerting actors to their inherent interdependence, material connectedness with, and potential effects on other (distant) people and things.

Responsible democratic innovations

A key theme across the chapters of this book has been spaces of participatory prescription, in terms of technologies, expertise and institutions of participation, where particular configurations of participatory collectives become more or less stabilized and replicated at different sites in space and time. Any interest in reconfiguring and remaking participation needs therefore to reflexively engage with such technological spaces where and when distinct 'powers of participation' have become lodged. Indeed, despite the reflexive potential and intent of the above moves to build reflexive participatory practices and ecologize participation, these two important paths can in themselves bring forth standardized procedures which will in turn also require reflexive engagement of the sort outlined in this section.

In many respects the approach we advocate here mirrors work in STS designed to promote the reflexive consideration and anticipation of the underlying purposes, consequences, social assumptions and ethical dimensions of the technosciences and open them up to social shaping through technology assessment, anticipatory governance, responsible innovation and related approaches (Rip *et al.* 1995; Guston and Sarewitz 2002; Barben *et al.* 2008; Owen *et al.* 2012; Guston 2014). Given the technologization of participation documented in Chapters 2, 6, 7 and 12, there is a clear need for a new wave of *social* technology assessment directed specifically toward processes of democratic innovation (Chilvers 2010; see also Chapter 12). This includes asking the same sorts of 'upstream questions' that have been aimed at emerging technosciences in existing anticipatory governance arrangements: 'Why this technology [*of participation*]? Why not another? Who needs it? Who is controlling it? Who benefits from it? Can they be trusted?' (Wilsdon and Willis 2004, 28, emphasis added).

There is scope here for new interventions in interactive social science and reflexive engagement, as well as reflection in action, in order to reconfigure technologies, institutions and expertises of participation and render them 'technologies of humility' (Jasanoff 2003). This includes opening up and accounting for the inherent uncertainties, framing effects, politics, power relations, social assumptions and unintended consequences of emergent technologies of participation. Furthermore, these commitments to building *responsible democratic innovations* might be brought into productive conversation with developing notions of responsible research and innovation (Owen *et al.* 2013; Stilgoe *et al.* 2013). Such 'real time' constructive engagement with emergent technologies of participation could take on many forms. Indeed, reflexively opening up the politics of participatory expertise will depend on a diversity of social scientific and other interventions rather than one-off 'procedural fixes' (Chilvers 2013). Here we sketch out the contours of some possible ways forward, while acknowledging that others will be necessary.

First of all, a set of possibilities open up around forms of interactive social science where STS scholars and critical social scientists engage with mediators and experts of participation to prompt reflexive learning about the uncertainties, framing effects, politics and social implications of the technosciences *of* participation and democratic engagement. This might take a similar form to reflexive engagements between sociologists and laboratory scientists (Kearnes *et al.* 2006; Doubleday 2007; Barben *et al.* 2008), in this case extending out to the social laboratory of *in vitro* participatory experiments. There are also opportunities for reflexive learning through building more constructive relations at the critical social science–policy/ practice interface in work on democratic engagement, through communication, knowledge transfer, challenge and exchange (cf. Webster 2007; Wynne 2007; Chilvers 2008b; Burchell 2009; Chilvers 2010). This includes the development of platforms for collective reflection and experimentation between critical social scientists, scientists, participatory practitioners and policy-makers, such as the *Critical Public Engagement* series of workshops and seminars that provided the impetus for this book, as introduced in the Preface (Chilvers 2009).

Such forms of collective experimentation expose the blurred boundaries between scholarly reflection and participatory practice. STS scholars, other social scientists and practitioners of participation are not external to participatory systems but are continually intervening and being enrolled in the technicalization of democratic practice in diverse ways. What we are advocating here is just as much a case of studying and prompting 'ourselves and each other' (Irwin and Michael 2003; Chilvers 2008b) in order to take responsibility for our interventions in participation and their effects. This highlights the need for distributed and embodied means of fostering reflexive learning about technologies of participation from catalysing reflection in action and 'irony as practice' (Schön 1983; Felt and Wynne 2007) through to generating spaces of learning and reflection, including those that are more experiential, informal and outward looking, or so-called 'shadow spaces' (Pelling et al. 2008; Pallett and Chilvers 2015). We further reflect on this need for distributed reflexivities in the final section below.

A further set of possible reflexive engagements with technologies of participation might take the form of infiltrating existing infrastructures for learning, evaluation and governing participatory practices that have grown up around the professionalization and standardization of public engagement. As noted above and earlier in Chapter 2, existing learning infrastructures most often act as forces of closure through promoting best practices that fix particular models of participation and the public, rather than opening up these normativities of participation and of the powers that maintain them to wider scrutiny. In keeping with our constructively critical approach, rather than completely dismantling established infrastructures for participatory learning and evaluation, one possibility is to consider how they can be harnessed, sensitized and repurposed. One practical example is to explore how the burgeoning frameworks and methods for evaluating public engagement processes – a characteristic feature of contemporary audit cultures (Power 1997; Strathern 2000) and typically based on normatively defined models of 'good' democratic engagement (Rowe and Frewer 2000) – might be imbued with reflexive dispositions and intent.

A relatively modest measure with potential transformative effects would be to modulate and add new evaluation criteria that acknowledge the inescapably political and partial nature of participatory processes. As we note above with reference to communicating uncertainties of participation, such criteria could question things such as the 'real' underlying purposes of participation; why the process was framed in a particular way; the exclusions (of knowledges, visions and normativities of participation) that occur; and how the outcomes of the process compare to that of other public engagements on the same issues. Rather than the standard 'legitimatory' evaluation criteria that tend to accompany participatory exercises – defined by notions of inclusivity and impact upon decision-making – reflexively oriented criteria such as these would help to foster openness concerning the exclusions, uncertainties and the politics of closure in technologies of participation. The same could be said for good practice guidelines, training courses, capacity-building and so on. Rather than simply building technical capacities and scaling up narrow models of participation, more focus should be placed on cultivating sensibilities and capacities in reflexive participation. Again, a whole series of practical interventions can be taken to do this in an experimental fashion, the effects of which would need to be open to ongoing monitoring and reflection. Once the move is made to remake participation as *reflexive participation*, the possibilities multiply.

Moving beyond an engagement with mediators and infrastructures of participation, the *social* assessment of democratic innovations can also draw inspiration from established frameworks in STS for assessing and anticipating emerging areas of technoscience (Rip *et al.* 1995; Guston and Sarewitz 2002). These relatively established approaches might be turned around to focus on technologies of democracy and their future pathways, consequences, social implications and effects (Chilvers 2013). For example, Voß (see Chapter 12 in this volume) reports on one of the first examples of a systematic technology assessment and foresight process that focused on a particular technology of participation – the global development of citizen panels. This experiment in anticipatory governance of participation shows that the translation of systematic technology assessment procedures to social technologies is a productive move, which can play a valuable role in building reflexive consideration of the underlying purposes, politics and possible future consequences into the 'real-time' development of instruments and practices of public engagement. Other anticipatory approaches can provide valuable comparative insights here too, including horizon scanning (e.g., in mapping forthcoming issues that may demand public dialogue in the governance of science and technology; see Parker *et al.* 2014) or backcasting to provide hindsight from past reconfigurations in relations between science and society (Wilsdon 2014).

In advocating technology assessment of the social technologies of participation, it is important to recognize that we do not envisage assessment processes as a separate initiative. Rather, these assessments should also form a critical element of wider frameworks for anticipatory governance. This will allow the principles of inclusion, public engagement and public good – that have become touchstone definitions of responsible innovation – greater reflexive scrutiny than has been apparent hitherto. Such a perspective also alerts us to how frameworks and procedures for responsible innovation are themselves technologies of governance that need to be open to the same level of reflexive scrutiny as the authors have levelled at democratic innovations in this book (cf. Macnaghten and Chilvers 2014).

Finally, in keeping with the call for distributed-relational reflexivities in other parts of this book, we would emphasize the importance of seeing anticipatory governance of social technologies such as participation as informal, distributed and ongoing. There are now numerous examples of controversies that erupt around the object of participatory democratic procedures themselves – the UK government's troubled nuclear consultation in 2007 (Chilvers and Burgess 2008), the protest and counter-demonstrations relating to the 2006 Citizen Conference on nanotechnology in France (Joly and Kaufmann 2008; Laurent 2011) and controversies in Australia focused on consultation processes on the development of alternative water sources through desalination and recycled water (Kearnes *et al.* 2014). Through these episodes we see how controversies can act as an 'informal technology assessment' (Rip 1986) of technologies of participation and forms of democratic practice.

The rise of third-party critique – from social science, civil society organizations and process participants (essentially anyone or anything!) – relating to all things participatory (Irwin *et al.* 2013) is further evidence of this iterative and ongoing social appraisal of democratic innovations (see also Chapter 12). The trick is to be able to see criticism and controversy for what it is, as constructive and generative of our co-produced technological-democratic futures, rather than a definitive contest to determine which competing 'truth' of democracy most accurately represents an externalized public. Informal reflexive feedbacks over time will transform science and society, but this does not diminish the need for *deliberately* reflexive participation. Indeed, we argue that the alternative co-productionist way of seeing participation outlined in this book releases the potential to learn through controversies of democracy rather than see them as something to be dampened down, avoided and ignored.

Reconstituting participation

While it is important to focus on participation as an object of analysis and reflexive intervention in itself, the paths for remaking participation opened up in this book go beyond this focus on socio-material collectives of participation to consider ways to locate spaces of participatory coherence in wider constitutional formations that shape relations between citizens, science and the state (Jasanoff 2011a, 2011b). Thinking about participation in constitutional terms disrupts popular democratic imaginaries of citizen participation in terms of discrete discursive events linked to particular decision moments, to viewing participatory democracy more in terms of a continually emerging symphony of participatory collectivities and distributed agencies (Ezrahi 2012). Reconstituting (the imaginary of) democratic participation in this way opens up at least three implications for remaking participation.

First, reconstituting participation means acknowledging and actively attending to the ways in which emergent publics and diverse participatory practices are powerfully co-produced in relation to wider constitutional formations. This calls for a certain responsiveness and openness to the cultural-historical antecedences in contemporary modes of governance; socio-technical imaginaries and citizen–state relations that have evolved over time; political cultures in which certain participatory knowledge ways become more credible than others in particular settings; and driving forces of neoliberal globalization and science-led progress (Ong and Collier 2005; Jasanoff 2011b; Sunder Rajan 2012). These dynamics of political closure and modes of stabilisation (themselves co-productions of the past and present) cannot simply be wished away from insulated compartmentalized participatory experiments. They need to be openly acknowledged, exposed and *worked with* in becoming an integral part of reflective participatory practice and the reflexive horizons of the three moves in remaking participation we have outlined above. In this context, questions concerning systemic reflexivities and (ir)responsibilities thus become a key focus of concern. Sensing ourselves and each other as entangled in the wider constitutional relations between science, democracy and society should be an inherent part of the consciousness and wisdom of public participation and its governance in all its diverse forms.

Second, reconstituting participation in this way entails taking seriously a more systemic and ecological understanding of the emergence of public collectives and the ways in which public meanings are constructed and represented. The challenge of participatory democracy is not simply one of aggregating the unique preferences of autonomous and *individual* publics. Rather, it lies in attending to the multiply diverse socio-material collectives of participation through which publics make meaning, know and act in the world. As Michael (see Chapter 4) suggests, everyday acts of technoscientific consumption and engagement are intertwined with issues as varied as energy production, climate change and the production of petrochemicals through numerous distinct but connected social material collectives. 'Individual' beliefs, actions and choices – the focus of a range of 'behaviour change' initiatives – are, in this sense, always mediated through collective practices (cf. Shove 2010) that involve numerous non-human artefacts, devices and material settings, in material connection with distant others. As Irwin and Horst demonstrate (see Chapter 3), the figure of the engaged citizen of technical democracies is thus infinitely multiplied. The challenge becomes one of opening up to these diverse collective *constituencies* and ecologies of participation and their framings, knowings and doings with respect to the objects of governance and engagement. As noted above, this recasts the problem of participation from notions of extension, demarcation and control to become one of relevance, where the responsibilities for accounting for and responding to diverse ecologies of public participation become a central challenge for incumbent powers and institutions.

The more open and outward-looking disposition that this promotes has practical implications for the design of more adaptive systems of governance that are responsive to emergent publics and public issues (just as much as emergent natures). This is in keeping with Wynne's (1993) contention that the burden of reflexivity should disproportionately lie with the powerful and incumbent institutions (see also Chapter 5). As we have already alluded to, commitments towards reflexive governance and responsible (democratic) innovations have set out a range of tools and procedures, but it is important to recognize that the challenge of institutional reflexivity cannot be met through the design of new instruments and procedures alone. The stabilities and powers tied up in dominant political cultures and constitutional formations mean that a more reflexive outward-looking disposition that attends to the relevance of diverse collective public constituencies will always have to be consciously crafted (Pallett and Chilvers 2015). In place of simple procedural adjustment, this disposition will, in turn, be dependent on third-party critique and modest witnessing, and on the presence of 'tricksters' and mischievous actors that precipitate disruptions to existing orders (Haraway 1997; Rip 2006; Wynne 2007). A further way in which a relational–ecological perspective will inform the reconstitution of modes of participatory governance is through an appreciation of – and indeed strategies that nurture – the value of *diversities of participation*[5] across the socio-material collectives of distributed public concerns, knowledges and actions (Stirling 2011). The diversities of socio-material constituencies become, in this mode, a key source of transformative democracy. It relates to the capacities and potentials of institutions to respond to diverse public collectives. It becomes a way of understanding and intervening in the interests of justice and equity at more systemic levels, and is necessary for distributed reflexivities (in that the homogenization of participatory collectives is a sure sign of diminished reflexive potential).

Third, reconstituting participation in relational co-productionist terms goes beyond understandings of reflexivity as something held by particular actors or situated in particular institutional complexes. Rather, reflexivity is seen as a thoroughly relational and distributed quality. Taking the distributed agencies of science and democracy seriously entails a commitment to nurturing *relational-reflexivities* in a more thoroughly systemic and distributed sense. This decentres longstanding questions of institutional reflexivity and reframes them from the perspective of all actors that make up wider constitutional formations between science, politics and society. Normatively, this calls for 'the development of citizens, society and scientific cultures as nothing more and nothing less than a "modest witness", asking pertinent questions and inviting pertinent reflection' (Wynne 2006, 78). Taken to its logical conclusion, this ethos exercises a call on *all constituencies*, as socio-material collectives of participation in public affairs, to reflect on their own co-productions and those of others, in terms of their framings, knowledges, commitments and atmospheres of democracy. Perhaps this begins to form an unattainable ideal, but one that all citizens, actors and institutions can struggle to pursue, if only ever in partial and in seemingly contradictory ways.

Coda

In setting out these four broad interrelating paths for remaking participation, we are not suggesting a theory–practice dialectic where more practical and interventionist considerations are viewed as somehow separate from the more interpretive theoretical insights developed earlier in this book. Viewing participation from a relational and co-productionist perspective, one thing that has been apparent across this volume is that interpretive-analytical observation and questions of design are held in tension. To theorize and undertake interpretive analyses, no matter how 'pure' or 'abstract', is also to act and thereby intervene in the participatory objects of study, with normative commitments about how (participatory) democracy ought to be. The co-productionist idiom that undergirds much of the discussion here suggests that analysis is always, and inextricably, implicated in the 'cycles of world making' (Jasanoff 2004, 12) that characterize both technoscientific constructions and democratic orderings.

The ways in which the normative, material, social and cognitive are deeply intertwined and inseparable in *both* theory and practice highlights issues of responsibility and reflexivity that we have summarized in this chapter, in relation to the construction and constitution of democratic systems and practices. It also serves to release participatory practices from a negative spiral of destructive critiques and a reluctance, by some commentators, to engage with practice and instrumentalization due to concerns over institutional 'capture'. The programmatic sentiment that has characterized the final passages of this book deliberately diverges from many constructivist studies of participation.

Our aim has been to advance the provocative claim that after the deconstruction of participatory democracy and science and society relations, these relations might be put back together again and reconstituted in constructively critical ways. In short, remaking participation in the terms set out in this book depends on the interpretive, normative and instrumental research traditions in STS and social studies of participation working together and being held in tension.

So, in closing this volume, we return to a central question that animates *Remaking Participation*: what happens to the democratization of science once we recognize that democracy itself is neither external to science, nor singular or static? The legacies of the constructivist social studies of science are clear here. These bodies of analytical and empirical insight that have demonstrated the social shaping of technological artefacts and the inescapability of questions of social, moral and political judgment in the composition of publically authorized knowledge have brought forward a series of approaches designed to democratize science that have largely left democracy and normativities of participation intact or unquestioned. In contrast, central to the approaches that characterize this volume is the observation that one of the reasons why the democratization of science has not been as transformative as had been hoped is due to the underlying assumption of participatory democracy as a given external norm. In arguing for the democratization of science, constructivist accounts have not sufficiently extended their analytical tools and interventions to consider the always-partial and always-contingent composition of democractic systems of social and political order, public participation and public representation.

We have argued that the future vibrancy of this field – in both scholarship and practice – depends on attending to this lacuna; of remaking participation and public relations with science as co-productive, relational and emergent. In laying out some pathways towards this goal, we have sought to open up an interpretive theoretical terrain that is grounded in the co-productionist idiom that has characterized the sociological interpretation of scientific knowledge and its interface with systems of public ordering. The goal of this approach is to bring foward new avenues for the interpretive study of participation that include but extend beyond analyses of situated experiments and practices of participation in the making – to encompass analyses of the technologies and expertise of participation, spaces of controversy and wider constitutional formations in relations between science and democracy. This, in turn, opens up new paths for remaking participation. By attending to practices of democracy in the making, we can recognize that there are other forms of remaking and practising participatory democracy and science–society relations that come into view under a vision of reflexive participation. Our ambition is to create an opening for new possibilities of what participation is, was, and could be. This approach represents a conception of participation that is sensitive to its productions, circulations and effects and is applicable across a range of temporal, spatial and cultural contexts. In this light, we are acutely aware that while this book draws from international and transnational practice, the authors are predominantly (though not exclusively) European scholars, and we must be aware of the ways in which our perspectives and visions of participation will be contingent on the particular places, political cultures and institutions in which they are situated.

After considering the conceptual, empirical and practical concerns of this volume, we are left with the conclusion that *participation and democracy is always to come but already exists in powerful ways*. It is this dynamic interplay between stabilities and emergence – between the stabilization and standardization of the collectives, technologies, issues and cultures of participation, and the continual destabilization and emergence of new participatory collectivities – which has not formed a significant element of the imagination of participatory democracy. The chapters that comprise this volume have shown that the patterns of stability and destabilization that characterize the social life of science and the everyday practices of democratic engagement are always negotiated settlements. Attending to the practices of negotiation offers a vantage point for the hard work necessary in renegotiating and remaking the patternings of democratic engagement in an age of technoscience.

Notes

1. These developments in instrumental learning and 'scaling-up' are not unique to deliberative democratic spaces of participation, and are evident across diverse forms of public engagement including pro-environmental behaviour change (e.g., Jackson 2005, 2006) through to citizen-led social innovations and social movements (e.g., cases analyzed in Seyfang and Haxeltine 2012).
2. In the same way that scientists are often called on to express uncertainties when modelling natural and physical systems, such as global climate change (see Hulme 2009), perhaps it is time for those 'modelling' publics and their relations with 'science' and 'the environment' to be just as careful about their social representations?
3. For example, in the case of energy and climate change-related public engagements, this would range from collectives of localized engagement through everyday practices or behaviour change interventions in the home (e.g., Hargreaves *et al.* 2013) and grassroots community energy initiatives (e.g., Seyfang and Haxeltine 2012), forms of public protest around energy infrastructure and national public dialogues on low carbon energy futures (e.g., Pallett and Chilvers 2013; Chilvers and Longhurst 2015), through to global citizen engagements such as the Worldwide Views on Climate Change process in the lead up to the Copenhagen Climate Change Conference in 2009 (e.g., Ely *et al.* 2011).
4. For a useful overview of issue mapping methods and resources, see http://issuemapping.net.
5. In a similar way to which the value of diversity is appreciated in discourses on biodiversity or diversity in the workplace.

References

Barben, D., Fisher, E., Selin, C. and Guston, D. 2008. Anticipatory governance of nanotechnology: Foresight, engagement and integration. In: E. J. Hackett, O. Amsterdamska, M. Lynch and J. Wajcman (eds) *The Handbook of Science and Technology Studies*, 3rd edition. Cambridge, MA: MIT Press, pp979–1000.

Barry, A. 2001. *Political Machines: Governing a Technological Society*. London: Athlone Press.

Barry, A. 2012. Political situations: Knowledge controversies in transnational governance, *Critical Policy Studies* 6(3): 324–336.

Bellamy, R., Chilvers, J. and Vaughan, N. E. 2014. Deliberative mapping of options for tackling climate change: Citizens and specialists 'open up' appraisal of geoengineering. *Public Understanding of Science*, http://pus.sagepub.com/content/early/2014/09/12/0963662514548628.abstract (accessed 24 September 2015).

Blake, J. 1999. Overcoming the 'value–action gap' in environmental policy: Tensions between national policy and local experience, *Local Environment* 4(3): 257–278.

Boltanski, L. 2011. *On Critique: A Sociology of Emancipation*. Translated by Gregory Elliott. Cambridge, MA: Polity Press.

Burchell, K. 2009. A helping hand or a servant discipline? Interpreting non-academic perspectives on the roles of social science in participatory policy-making, *Science, Technology & Innovation Studies* 5(1): 49–61.

Burgess, J., Stirling, A., Clark, A., Davies, G., Eames, M., Staley, K. and Williamson, S. 2007. Deliberative mapping: A novel analytic-deliberative methodology to support contested science-policy decisions, *Public Understanding of Science* 16(3): 299–322.

Cairns, R. and Stirling, A. 2014. 'Maintaining planetary systems' or 'concentrating global power'? High stakes in contending framings of climate geoengineering, *Global Environmental Change* 28: 25–38.

Callon, M. 1998. *The Laws of the Markets*. Oxford: Blackwell.

Callon, M., Lascoumes, P. and Barthe, Y. 2009. *Acting in an Uncertain World: An Essay on Technical Democracy*. Cambridge, MA: MIT Press.

Carolan, M. S. 2007. Introducing the concept of tactile space: Creating lasting social and environmental commitments, *Geoforum* 38(6): 1264–1275.

Chilvers, J. 2008a. Deliberating competence: Theoretical and practitioner perspectives on effective participatory appraisal practice, *Science, Technology & Human Values* 33(2): 155–185.

Chilvers, J. 2008b. Environmental risk, uncertainty, and participation: Mapping an emergent epistemic community, *Environment and Planning A* 40(12): 2990–3008.

Chilvers, J. 2009. *Critical Studies of Public Engagement in Science and the Environment*. Workshop report. Norwich: University of East Anglia.

Chilvers, J. 2010. *Sustainable Participation? Mapping Out and Reflecting on the Field of Public Dialogue on Science and Technology*. Harwell: Sciencewise Expert Resource Centre.

Chilvers, J. 2013. Reflexive engagement? Actors, learning, and reflexivity in public dialogue on science and technology, *Science Communication* 35(2): 283–310.

Chilvers, J. and Burgess, J. 2008. Power relations: The politics of risk and procedure in nuclear waste governance, *Environment and Planning A* 40(8): 1881–1900.

Chilvers, J. and Longhurst, N. 2015 (forthcoming). Participation in transition(s): Reconceiving public engagements in energy transitions as co-produced, emergent and diverse, *Journal of Environmental Policy and Planning*.

Cohen, J. L. and Arato, A. 1992. *Civil Society and Political Theory*. Cambridge, MA: MIT Press.

Collins, H. and Evans, R. 2002. The third wave of science studies: Studies of expertise and experience, *Social Studies of Science* 32(2): 235–296.

Crouch, C. 2004. *Post-Democracy*. Cambridge, MA: Polity Press.

de Vries, G. 2007. What is political in sub-politics? How Aristotle might help STS, *Social Studies of Science* 37(5): 781–809.

Doubleday, R. 2007. The laboratory revisited: Academic science and the responsible development of nanotechnology, *Nanoethics* 1(2): 167–176.

Dryzek, J. S. 1990. *Discursive Democracy: Politics, Policy and Political Science*. New York: Cambridge University Press.

Durant, D. 2011. Models of democracy in social studies of science, *Social Studies of Science* 41(5): 691–714.

Ely, A., Van Zwanenberg, P. and Stirling, A. 2011. *New Models of Technology Assessment for Development*. STEPS Working Paper 45. Brighton: STEPS Centre.

Ezrahi, Y. 2012. *Imagined Democracies: Necessary Political Fictions*. Cambridge: Cambridge University Press.

Felt, U. and Wynne, B. 2007. *Science and Governance: Taking European Knowledge Society Seriously*. Report of the Expert Group on Science and Governance to the Science, Economy and Society Directorate, Directorate-General for Research. Brussels: European Commission.

Gabrys, J. 2014. Programming environments: Environmentality and citizen sensing in the smart city, *Environment and Planning D: Society and Space* 32(1): 30–48.

Gross, M. 2010. The public proceduralization of contingency: Bruno Latour and the formation of collective experiments, *Social Epistemology* 24(1): 63–74.

Guston, D. 2014. Understanding 'anticipatory governance', *Social Studies of Science* 44(2): 218–242.

Guston, D. and Sarewitz, D. 2002. Real-time technology assessment, *Technology in Society* 24: 93–109.

Habermas, J. 1971. *Towards a Rational Society: Student Protest, Science and Politics*. London: Heinemann.

Haraway, D. J. 1997. *Modest Witness@Second_Millennium.FemaleMan©_Meets_OncoMouse™: Feminism and Technoscience*. London: Routledge.

Hargreaves, T., Nye, M. and Burgess, J. 2013. Keeping energy visible? Exploring how householders interact with feedback from smart energy monitors in the longer term, *Energy Policy* 52: 126–134.

Harvey, M. 2009. Drama, talk and emotion: Omitted aspects of public participation, *Science, Technology & Human Values* 34(2): 139–161.

Horst, M. and Michael, M. 2011. On the shoulders of idiots: Re-thinking science communication as 'event', *Science as Culture* 20(3): 283–306.

Hulme, M. 2009. *Why We Disagree About Climate Change: Understanding Controversy, Inaction and Opportunity*. Cambridge: Cambridge University Press.

Irwin, A. 2006. The politics of talk: Coming to terms with the 'new' scientific governance, *Social Studies of Science* 36(2): 299–320.

Irwin, A. and Michael, M. 2003. *Science, Social Theory and Public Knowledge*. Maidenhead: Open University Press.

Irwin, A., Jensen, T. E. and Jones, K. E. 2013. The good, the bad and the perfect: Criticizing engagement practice, *Social Studies of Science* 43(1): 118–125.

Jackson, T. 2005. Motivating sustainable consumption: A review of evidence on consumer behaviour and behavioural change. A report to the Sustainable Development Research Network, January 2005, www.sustainablelifestyles.ac.uk/sites/default/files/motivating_sc_final.pdf (accessed 28 August 2015).

Jackson, T. 2006. *The Earthscan Reader on Sustainable Consumption*. London: Earthscan/Routledge.

Jasanoff, S. 2003. Technologies of humility: Citizen participation in governing science, *Minerva* 41(3): 223–244.

Jasanoff, S. 2004. The idiom of co-production. In: S. Jasanoff (ed.) *States of Knowledge: The Co-Production of Science and Social Order*. London: Routledge, pp1–12.

Jasanoff, S. 2005. *Designs on Nature: Science and Democracy in Europe and the United States*. Princeton, NJ: Princeton University Press.

Jasanoff, S. 2011a. Constitutional moments in governing science and technology, *Science and Engineering Ethics* 17: 621–638.

Jasanoff, S., ed. 2011b. *Reframing Rights: Bioconstitutionalism in the Genetic Age*. Cambridge, MA: MIT Press.

Jasanoff, S. and Wynne, B. 1998. Science and decisionmaking. In: S. Rayner and E. Malone (eds) *Human Choice and Climate Change. Volume 1: The Societal Framework*. Washington, DC: Battelle Press, pp1–87.

Joly, P. B. and Kaufmann, A. 2008. Lost in translation? The need for upstream engagement with nanotechnology on trial, *Science as Culture* 17(3): 225–247.

Kearnes, M., Macnaghten, P. and Wilsdon, J. 2006. *Governing at the Nanoscale: People, Policies and Emerging Technologies*. London: Demos.

Kearnes, M., Motion, J. and Beckett, J. 2014. *Australian Water Futures: Rethinking Community Engagement*. Report of the National Demonstration, Education and Engagement Program, University of New South Wales.

Landström, C., Whatmore, S. J., Lane, S. N., Odoni, N. A., Ward, N. and Bradley, S. 2011. Coproducing flood risk knowledge: Redistributing expertise in critical 'participatory modelling', *Environment and Planning A* 43(7): 1617–1633.

Lane, S. N., Bradley, S., Landstrom, C., Odoni, N., Ward, N. and Whatmore, S. 2008. Environmental competency groups: The case of flood risk modelling, *Geophysical Research Abstracts* 10.

Latour, B. 2004a. *Politics of Nature: How to Bring the Sciences into Democracy*. Cambridge, MA: Harvard University Press.

Latour, B. 2004b. Why has critique run out of steam? From matters of fact to matters of concern. *Critical Inquiry* 30(2): 225–248.

Latour, B. and Weibel, P. 2005. *Making Things Public: Atmospheres of Democracy*. Cambridge, MA: MIT Press.

Laurent, B. 2011. Technologies of democracy: Experiments and demonstrations, *Science and Engineering Ethics* 17(4): 649–666.

Lezaun, J. and Soneryd, L. 2007. Consulting citizens: Technologies of elicitation and the mobility of publics, *Public Understanding of Science* 16(3): 279–297.

Lynch, M. 2000. Against reflexivity as an academic virtue and source of privileged knowledge, *Theory, Culture & Society* 17(3): 26–54.

Macnaghten, P. and Chilvers, J. 2014. The future of science governance: Publics, policies, practices, *Environment and Planning C* 32(3): 530–548.

Macnaghten, P. and Szerszynski, B. 2013. Living the global social experiment: An analysis of public discourse on solar radiation management and its implications for governance, *Global Environmental Change*, 23(2): 465–474.

Mahony, N., Newman, J. and Barnett, C., eds. 2010. *Rethinking the Public: Innovations in Research, Theory and Politics*. Bristol: Policy Press.

Marcus, G. E. 1995. Ethnography in/of the world system: The emergence of multi-sited ethnography, *Annual Review of Anthropology* 24: 95–117.

Marres, N. 2007. The issues deserve more credit: Pragmatist contributions to the study of public involvement in controversy, *Social Studies of Science* 37(5): 759–780.

Marres, N. 2012. *Material Participation: Technology, the Environment and Everyday Publics*. Basingstoke: Palgrave.

Marres, N. 2013. Why political ontology must be experimentalized: On eco-show homes as devices of participation, *Social Studies of Science* 43(3): 417–443.

Marres, N. and Rogers, R. 2005. Recipe for tracing the fate of issues and their publics on the web. In: B. Latour and P. Weibel (eds) *Making Things Public: Atmospheres of Democracy*. Cambridge, MA: MIT Press, pp922–935.

Michael, M. 2012. 'What are we busy doing?': Engaging the idiot, *Science, Technology & Human Values* 37(5): 528–554.

Ong, A. and Collier, S., eds. 2005. *Global Assemblages: Technology, Politics and Ethics as Anthropological Problems*. Oxford: Blackwell.

Owen, R., Bessant, J. and Heitz, M., eds. 2013. *Responsible Innovation: Managing the Responsible Emergence of Science and Innovation in Society*. Chichester: Wiley.

Owen, R., Macnaghten, P. and Stilgoe, J. 2012. Responsible research and innovation: From science in society to science for society with society, *Science and Public Policy* 39(6): 751–760.

Pallett, H. and Chilvers, J. 2013. A decade of learning about publics, participation, and climate change: Institutionalising reflexivity? *Environment and Planning A* 45(5): 1162–1183.

Pallett, H. and Chilvers, J. 2015. Organizations in the making: Learning and intervening at the science–policy interface, *Progress in Human Geography* 39(2): 146–166.

Parker, M., Acland, A., Armstrong, A. J., Bellingham, J. R., Bland, J., Bodmer, H. C., Burrall, S., Castell, S., Chilvers, J., Cleevely, D. D., Cope, D., Costanzo, L., Dolan, J. A., Doubleday, R., Feng, W. Y., Godfray, H. C. J., Good, D. A., Grant, J., Green, N., Groen, A. J., Guilliams, T. T., Gupta, S., Hall, A. C., Heathfield, A., Hotopp, U., Kass, G., Leeder, T., Lickorish, F. A., Lueshi, L. M., Magee, C., Mata, T., McBride, T., McCarthy, N., Mercer, A., Neilson, R., Ouchikh, J., Oughton, E. J., Oxenham, D., Pallett, H., Palmer, J., Patmore, J., Petts, J., Pinkerton, J., Ploszek, R., Pratt, A., Rocks, S. A., Stansfield, N., Surkovic, E., Tyler, C. P., Watkinson, A. R., Wentworth, J., Willis, R., Wollner, P. K. A., Worts, K. and Sutherland, W. J. 2014. Identifying the science and technology dimensions of emerging public policy issues through horizon scanning, *PLoS ONE* 9(5): e96480.

Pelling, M., High, C., Dearing, J. and Smith, D. E. 2008. Shadow spaces for social learning: A relational understanding of adaptive capacity to climate change within organisations, *Environment and Planning A* 40(4): 867–884.

Petts, J. 2007. Learning about learning: Lessons from public engagement and deliberation on urban river restoration, *Geographical Journal* 173: 300–311.

Pickering, A., ed. 1992. *Science as Practice and Culture.* Chicago, IL: University of Chicago Press.

Power, M. 1997. *The Audit Society: Rituals of Verification.* Oxford: Oxford University Press.

Renn, O., Webler, T. and Wiedemann, P., eds. 1995. *Fairness and Competence in Citizen Participation: Evaluating Models for Environmental Discourse.* Dordrecht: Kluwer.

Rip, A. 1986. Controversies as informal technology assessment, *Knowledge: Creation, Diffusion, Utilization* 8(2): 349–371.

Rip, A. 2006. A co-evolutionary approach to reflexive governance - and its ironies. In: J.-P. Voß, D. Bauknecht and R. Kemp (eds) *Reflexive Governance for Sustainable Development.* Cheltenham: Edward Elgar, pp82–100.

Rip, A., Misa, T. and Schot, J. 1995. *Managing Technology in Society: The Approach of Constructive Technology Assessment.* London: Thomson.

Rogers, R. 2002. Operating issue networks on the web, *Science as Culture* 11(2): 191–213.

Rowe, G. and Frewer, L. 2000. Public participation methods: A framework for evaluation, *Science, Technology & Human Values* 25(1): 3–29.

Schön, D. A. 1983. *The Reflective Practitioner: How Professionals Think in Action.* New York: Basic Books.

Seyfang, G. and Haxeltine, A. 2012. Growing grassroots innovations: Exploring the role of community-based initiatives in governing sustainable energy transitions, *Environment and Planning C: Government and Policy* 30: 381–400.

Shove, E. 2010. Beyond the ABC: Climate change policy and theories of social change, *Environment and Planning A* 42(6): 1273–1285.

Stilgoe, J. 2015. *Experiment Earth: Responsible Innovation in Geoengineering.* Abingdon, Oxon: Routledge.

Stilgoe, J., Owen, R. and Macnaghten, P. 2013. Developing a framework for responsible innovation, *Research Policy* 42(9): 1568–1580.

Stirling, A. 2007. Risk, precaution and science: Towards a more constructive policy debate. *EMBO Reports* 8: 309–315.

Stirling, A. 2011. Pluralising progress: From integrative transitions to transformative diversity, *Environmental Innovation and Societal Transitions* 1(1): 82–88.

Stirling, A. 2014a. *Emancipating Transformations: From Controlling 'The Transition' to Culturing Plural Radical Progress*. STEPS Working Paper 64. Brighton: STEPS Centre.

Stirling, A. 2014b. Transforming power: Social science and the politics of energy choices, *Energy Research & Social Science* 1: 83–95.

Stirling, A. and Mayer, S. 2001. A novel approach to the appraisal of technological risk: A multicriteria mapping study of a genetically modified crop, *Environment and Planning C: Government and Policy* 19: 529–555.

Strathern, M., ed. 2000. *Audit Cultures: Anthropological Studies in Accountability, Ethics and the Academy*. London: Routledge.

Sunder Rajan, K., ed. 2012. *Lively Capital: Biotechnologies, Ethics and Governance in Global Markets*. Durham, NC: Duke University Press.

Tsouvalis, J. and Waterton, C. 2012. Building 'participation' upon critique: The Loweswater Care Project, Cumbria, UK. *Environmental Modelling & Software* 36: 111–121.

van Loon, J. 2002. *Risk and Technological Culture: Towards a Sociology of Virulence*. London: Routledge.

Van Oudheusden, M. and Laurent, B. 2013. Shifting and deepening engagements: Experimental normativity in public participation in science and technology, *Science, Technology & Innovation Studies* 9(1): 3–22.

Warner, M. 2002. Publics and counterpublics, *Public Culture* 14(1): 49–90.

Webler, T., Kastenholz, H. and Renn, O. 1995. Public participation in impact assessment: A social learning perspective, *Environmental Impact Assessment Review* 15: 443–463.

Webster, A. 2007. Crossing boundaries – Social science in the policy room, *Science, Technology & Human Values* 32: 458–478.

Welsh, I. and Wynne, B. 2013. Science, scientism and imaginaries of publics in the UK: Passive objects, incipient threats, *Science as Culture* 22(4): 540–566.

Whatmore, S. 2009. Mapping knowledge controversies: Science, democracy and the redistribution of expertise, *Progress in Human Geography* 33(5): 587–598.

Wilsdon, J. 2014. From foresight to hindsight: The promise of history in responsible innovation, *Journal of Responsible Innovation* 1(1): 109–112.

Wilsdon, J. and Willis, R. 2004. *See-Through Science: Why Public Engagement Needs to Move Upstream*. London: Demos.

Wolin, S. S. 2008. *Democracy Incorporated: Managed Democracy and the Specter of Inverted Totalitarianism*. Princeton, NJ: Princeton University Press.

Woolgar, S., ed. 1988. *Knowledge and Reflexivity: New Frontiers in the Sociology of Knowledge*. London: SAGE.

Wynne, B. 1993. Public uptake of science: A case for institutional reflexivity, *Public Understanding of Science* 2: 321–337.

Wynne, B. 2006. Afterword. In: M. B. Kearnes, P. Macnaghten and J. Wilsdon (eds) *Governing at the Nanoscale*. London: Demos, pp70–78.

Wynne, B. 2007. Dazzled by the mirage of influence? STS–SSK in multivalent registers of relevance, *Science, Technology & Human Values* 32: 491–503.

Wynne, B. 2013. Further disorientation in the hall of mirrors, *Public Understanding of Science* 23(1): 60–70.

Yusoff, K. and Gabrys, J. 2011. Climate change and the imagination, *Wires: Climate Change* 2(4): 516–534.

INDEX

actants 69, 85, 204–5, 224
activism 3, 38, 40, 46, 167, 168, 170, 274
activists 7, 8, 70, 126, 129, 131, 139, 247, 249
actor network theory (ANT) 13, 81, 85, 89, 147–50, 224
Adam, Barbara 178–9, 182, 187–8, 191–4
affect 36, 39, 164, 165, 173, 229, 270
affective 85, 95, 163, 172–3
agonistic 11, 37, 272
ambiguity 76–8, 93, 116, 210, 253
anticipatory governance 8, 181, 218–19, 233, 247, 278
anti-science 107, 112, 117
Appadurai, Arjun 184
Arendt, Hannah 112, 116
ARGONA (Arenas for Risk Governance) Project 149–57
Arnstein, Sherry xvi
assemblage 14, 32, 48, 68–9, 73–6, 83, 208–11
assembly 48, 205
Atelier Populaire xv–xvii
attachments 36, 68, 92
authority: epistemic 4, 239–40, 252–3; political 1, 4, 240; scientific 1–2, 7, 99–109, 116–18, 265

band-aid (plaster) 82, 92–5
Barad, Karen 201, 206–9, 215
Barry, Andrew 2, 5, 6, 10, 45, 49, 53, 102
Beck, Ulrich 65, 68
Bennett, Jane 49, 164

Bennett, Tony 89
best practice 45, 145, 157, 180, 277, 283
Bijker, Wiebe 2, 223, 231
Black, Julia 65
black-box 45, 218
blueprints 37, 144–5, 148–52, 156
blurring 40, 69, 70, 139
Bogner, Alexander 145
boundaries 40, 43, 68, 69, 85, 114, 127, 138, 148, 156, 221
Braidotti, Rosa 201–2, 214
Braun, Kathrin 147
Brown, Mark 48, 49, 144
Brown, Wendy 231
BSE (Bovine spongiform encephalopathy) 7, 54, 60, 66, 80

Callon, Michel 32, 38–9, 68–71, 76, 85, 124, 156, 193, 239, 246, 267, 270
CEA (Commissariat à l'Energie Atomique) 126, 131
CEH (Centre for Ecology and Hydrology) 203–4, 208–9
centredness 74–7
Chilvers, Jason 40, 146, 246
circulation 6, 12, 14, 32, 43–4, 57, 128, 146, 150, 241, 264, 282
citizen panels 34, 37, 43, 59, 149, 156, 241–54, 278
citizen science 3, 8, 46, 163
citizenship 147, 162–3, 169–71, 194
citizens' juries 242–4
citizen summit 71–3

Citizen Visions of Science Technology and Innovation (CIVISTI) Project 149, 151–2, 154, 156, 158
civic epistemology 53
civil society organizations (CSOs) 113, 127, 129, 131, 134
climate change 49, 64–75, 133–41, 271
closing down and closure 7, 16, 18, 19, 35, 41, 78, 238, 253, 263, 266, 272, 277
CNDP (National Comission for Public Debate, France) 127–32, 139–41
co-design 3, 91
collective experiments xiv–xv, 47, 48, 194, 213, 263, 270, 272–3, 276
collective participatory practices 13, 16, 34, 38, 46, 52–3, 266, 269, 270
collectives 13, 15, 34, 48, 205, 271–4
Collingridge, David 219
commercialization 7, 12, 55, 99, 109, 117
communities of practice 44
configurations 13, 15, 37, 42, 48, 130, 131, 136, 140, 150, 153, 157–8
consensual 10, 71, 167, 271, 272
consensus 53, 82, 154, 174, 213, 229, 241, 242
consensus conference 11, 37, 69–71, 84, 149, 154, 167, 241–4
constituencies 36, 49, 83, 280–1
constitution 13, 14, 47–8, 68, 126, 138–9, 205–7, 238, 281
constitutional 204, 269, 272, 279; arrangements xvi, xvii, 18, 55; changes 31; conditions 103, 108; formations 55, 279–82; moments 55; relations 15, 32, 279, 281, 282; stabilities 268
construction 13, 37, 40, 51, 69, 239, 268, 273; social 33, 223
constructive technology assessment (CTA) 8, 238, 246–7, 255
constructivist 13, 266, 282
consultants and consultancies 44, 133–4, 136–9, 148, 249
contingency 16, 37, 41, 46, 75, 116, 118, 231, 232, 245, 262, 268, 270, 272
control 35, 37, 65, 100–4, 111, 129, 171, 178, 182–3, 192, 209, 280
controversies 2, 7, 14, 18, 47–9, 54, 70, 81–8, 94, 273, 278–9, 282
Coordination Nationale (Democratic Republic of the Congo) 133–40
co-production 14, 17, 40–1, 46, 99, 107–9, 204, 213, 267, 274, 279, 281; instrumental-organisational perspective on 22; interpretive-philosophical perspective on 22

co-productionist: approach xvi, 3–5, 9, 13–22, 31–3, 35, 40–1, 51–52, 56, 262–8, 273–82; idiom 5, 13, 14, 281–2
credibility xiii, 18, 40, 44, 54
critical inquiry 129
critical social science xiv, 276
critique xiv–xv, 4–9, 17, 67, 131, 166–7, 204–13, 246, 262–8, 278–81
CSEC (Centre for the Study of Environmental Change) xviii, 105
cultures 11, 42, 43, 224, 261, 267; attestive 3; audit 277; deliberative 230, 237; institutional 111, 115; popular 92; scientific 99–106, 281; *see also* political culture
cyanobactera (blue-green algae) 201–15

Danish Board of Technology (DBT) 43, 69, 71–3, 149, 151, 154, 158, 255
de Beauvoir, Simone 76
de-centredness 65, 73–7
deficit model (of the public) 66, 68, 77, 82, 102, 107, 118
deforestation 133–9
deliberative, the deliberative society 168–9, 172
deliberative democracy xvi, 9, 50, 51, 57, 167, 169, 230, 244, 245
deliberative ecologies 51, 169
Deliberative Mapping (DM) 271
deliberative moments 169–72, 174
deliberative processes 36, 50, 51, 145, 156, 167, 169, 170, 175, 222
deliberative systems 50, 51, 169
democracy: atmospheres of 10, 38, 266, 271–2, 281; delegative 68, 69, 75; technological 9, 56, 57; technologies of 42, 278
democratic innovations 21, 45, 49, 264, 276, 278, 279
democratic practices 33, 41, 42, 169, 276, 278
democratization 4, 8, 12, 20, 67, 144
democratization of science xiv, 262, 265, 268, 282
demonstration 1, 53, 54, 56, 126, 130, 137, 225
denial 99–100, 104, 106, 108–9
de-politicization 8, 208
devices 6, 13–14, 33–4, 37, 123–4, 128–33, 153–5, 157, 206, 244, 271, 280
Dewey, John 35–7, 49, 106, 114, 117, 135
dialogue 67, 100–2, 106, 148–58; stakeholder 148, 151, 262

diversities 3, 15, 21, 36, 46, 49, 52, 130, 131, 135, 150, 271–6, 280, 283

ecologies of participation 48, 51–2, 54–6, 268, 273–4, 280
ecology of institutions 48
Edensor, Tim 188
emergence 8, 15, 32, 33, 40, 69, 82–4, 95, 105, 114, 135, 206, 262, 267–2, 280–3
emerging technologies 45, 54, 218–22, 233, 271
emotions 40, 66, 164, 166–7, 172, 270
energy 2, 90–1, 283; communities 91; demand reduction 91; saving 91; solar 227; transitions 2, 274
entanglements 178, 179, 184, 205, 231
Environment Agency (England) 202–3, 208, 213
Environment and Society Research Unit (ESRU) xviii
epistemic communities 44
ethno-epistemic assemblage (EEA) 40, 69
ethnography 57, 110, 273
European Commission (EC) 100, 118, 145–51, 158, 179
eutrophication 20, 202, 215
evaluation 4, 9, 12, 45, 57, 148, 157, 195, 244, 262, 264, 277
events 17, 32, 41, 48, 50, 81–6, 89–93, 163, 193, 195, 233; eventuation 14, 86, 88, 89, 92
everyday life 81–4, 88–94, 163, 188
everyday technologies 82, 86, 89
exclusions 33, 34, 45, 267, 272, 274, 277
experimental approach 37, 112, 270
experimental normativities 63, 288
Ezrahi, Yaron 3–4, 55, 101, 241

facilitation 39, 43, 147, 154, 157, 240
facilitators 39, 44, 152, 154
Felt, Ulrike 39, 147, 179, 188, 240
flooding 71, 72, 75
focus groups 34, 37, 42, 82, 163, 181, 183, 188, 244, 269
fragmentation 65, 73
framing(s) 16, 36, 43, 47, 67, 78, 111, 128, 147, 180–9, 201–12, 221, 272, 276, 280, 281
Fraser, Mariam 85, 86
Fries, Liv 150, 157
Futurescape City Tours 220, 225, 233

Gardiner, Michael 89
Gaver, Bill 92
geoengineering 45, 83, 222, 271, 284

geography xiv, 5, 9, 32
Gieryn, Thomas 221
globalization 6, 7, 18, 65, 107, 150
GMOs (genetically modified organisms) 112, 125, 186, 189
Greenpeace 113, 134, 135, 141, 245
Gregory, Jane 164

Habermas, Jurgen 9, 10, 36, 102, 167
hackerspaces 40, 164–5, 171, 262
Haraway, Donna 214
Hawkins, Gay 95
hearing 109, 111
Highmore, Ben 84, 89
Hommels, Anique 218–20, 223–5, 231, 233
Hughes, Thomas 224
humility 16
hybrid fora 38, 68, 69, 83, 156, 270
hybridity 65, 73

imaginaries 40, 47, 101, 104, 116, 117, 152, 181, 189, 193, 194, 279, 283
infrapublics 95
Ingold, Tim 192
innocent citizens 34, 43, 45, 54
innovation: distributed 3, 8, 40, 262, 274; journey 44, 241, 246; technological 3, 6, 12, 81, 146, 189, 242, 253, 261; trajectory 104, 108, 184, 185, 186
inscription 147, 151, 152, 158
instruments 4, 11, 37, 42, 43, 46, 182, 206, 278
International Association for Public Participation (IAP2) 44, 254
inventive problems 86, 88, 92, 94–5
invited participation 34–9, 50, 54, 104–9, 118, 167–70, 173–4
irony 6, 12, 241, 253, 277
Irwin, Alan 8, 39, 69, 146
issue space 49, 271, 273, 274

Jasanoff, Sheila 1, 7, 14, 31, 53, 55, 103, 266, 276, 281

Kirkman, Robert 222, 231
knowledge: co-production 16, 204, 208, 212; diversities of 2, 280; embodied 172; expert 42, 83, 271; fragmentation 65; global 244; knowledge-ways 15, 32, 47, 53; lay xiii, xiv; linear relation to power 48, 66, 265; local 242; public 16, 53, 82, 269; regimes 242; society 39, 184; transfer 185, 276
Koestler, Arthur 118

Laclau, Ernesto 10, 101, 113, 114
Latour, Bruno 36, 38–9, 45, 48, 56–7, 103, 114, 152, 192, 201–8, 224, 266
Laurent, Brice 42, 43, 246
Law, John 44, 85, 241
legitimacy 15, 32, 44, 47, 51, 53, 132, 143, 169, 176, 188, 239, 245, 251, 259
Lezaun, Javier 36, 37, 42, 84
Lindblom, Charles 74
Lippmann, Walter 70, 75
listening 100, 102
Loweswater Care Project (LCP) 201–15, 270

Mansbridge, Jane 50, 51, 169–74
Marres, Noortje 36, 39, 49, 52, 70, 75, 114, 117, 144, 193, 268
matters of concern 4, 36, 178, 183, 239, 272, 275
mediators 39–40, 276–8
methodological revisionism xiv, 9, 264, 266, 268
Michael, Mike 14, 39, 69, 85, 87
mini-publics 13, 34, 46, 47, 169
models of participation 10, 11
modest 74, 106, 115, 233, 277, 280
Mol, Annemarie 192
Mouffe, Chantal 10
Multi-criteria Mapping (MCM) 270–1
multiplicity 2, 15, 20, 36, 41, 51, 52, 73, 77–8, 86, 140, 168, 182
mundane 36, 41, 81–95; artefacts 82, 88, 94; events 84; technologies 81, 86–9, 94

nanotechnology 19, 57, 124, 126–32, 139–41, 168, 171, 181, 190, 219–30, 278
nation states 6, 54, 55, 187, 194
neoliberalization 7, 18, 26, 39, 88, 101, 104, 107, 117, 268, 279
neutrality of expertise 44, 132, 251, 252
new institutional theory 146–7
new technologies 124, 185, 194, 220–2, 227, 233, 241
NGOs (non-governmental organizations) 102, 111–12, 134–5, 137, 140
non-humans 69, 88, 204–5, 212, 214
normativities, experimental 41, 263, 271
nuclear power 111, 112, 114, 118, 153
nuclear waste management 145, 148–57

obduracy 218–33
opening up 8, 41, 45, 70, 103–4, 165, 168, 190, 214, 263, 270, 271, 276–7, 280

openness 6, 9, 16, 67, 77, 84, 86, 266, 268, 269, 277, 279
opinion polls 3, 36, 42, 240
organizational carriers 145, 148, 151, 155–8
organizers (of participatory processes) 128–9, 152, 154–8, 181, 186
Osborne, Thomas 39
overflows 8, 15, 17, 21, 36, 49, 68, 70, 73, 164, 168, 267, 272, 274

Parkinson, John 50, 169
participants 39, 40, 73, 82–6, 91–5, 128, 135–6, 147, 152–6, 166–72, 181–95, 203–14, 220–30
participation: collectives of xvi, 14–16, 33–4, 38–42, 47–55, 269, 272–5, 279–81; co-productionist conception of 13, 17, 40, 267; co-production of 32, 221; definition of 15, 51, 270; devices of 46; diversities of 51, 263, 275, 280; ecologizing 272–5; imaginary of 48, 50; lab-based 145; models of 34, 38, 57, 266; normativities of 38, 41, 266, 271, 277, 282; objects of 13, 34–8, 40, 54, 272; powers of 263, 275; procedures 240, 252; realities of xv, 13, 261; reconfiguring 262; reconstituting 279–81; uncertainties of 267, 272, 274, 277; uninvited 167, 170, 173, 174
participation expertise 41–2, 44, 46, 56, 263, 276, 282
participation experts 44, 54, 57, 128, 264
participatory: democracy 11, 14, 169, 170, 262, 273, 279–83; events 82, 145, 182, 183, 190, 193; experiments 32–49, 145, 147, 194, 201, 268, 272; methods xiv, 4, 9, 37, 238–45, 251–4; politics xvi, 3, 18, 46, 82, 83, 123; practice(s) 11–15, 33–43, 45–50, 145, 178–80, 185, 190, 208, 243, 267–74
participatory rural appraisal (PRA) 43
performance of participation 13, 15, 21, 34, 35, 36, 51, 53, 263, 266, 271
planning cells 241–4
pleasure 84, 88, 162–6, 173–4
pluralist 11, 53, 54, 55, 262
plurality 50, 65, 67, 84, 105, 132
PMO (Pièces et Main d'Oeuvre, protest group in France) 126, 129, 131, 246
policy 65, 74, 76–7, 101–6, 111, 169, 179, 184, 185, 276
policy cultures 55, 102, 265
policy-makers xiv–xv, 17, 65–78, 179, 183–4, 188, 190–2

political culture 32, 47, 48, 52–5, 74, 146, 241, 279, 282
political economy 14, 100, 104
political ontologies 13, 34, 35, 52, 271
political order 3, 13, 15, 60, 238, 239, 240, 249, 252, 253, 282
political participation 50, 238, 240, 245
political philosophies 3, 14, 37, 266
political science 9, 38, 53, 135, 239
political situations 2, 16, 17, 49, 54, 55, 249, 275
political theory 5, 9, 32, 35, 37, 169, 172, 265
politicians 65, 73, 106, 108, 109, 224
post-normal science 8, 38
post-political 26, 109
power 14, 34, 45, 48, 115, 147, 178, 182–3, 205, 277–80; centres of 7, 265; citizen xvi; political xv, 6, 31, 57, 62, 123, 140; relational xvi, 279; of science and expertise 109
practice-based perspective xv, 33
pragmatist 4, 31, 36, 37, 49, 53
probes 89–96; cultural 90, 92, 96
procedural formats (of participation) 15–19, 32, 35, 37–40, 42–6, 145–9, 151–3, 169, 179–82, 226, 272
professionalization 12, 44, 55, 244, 246, 277
progress, science-led 7, 279
protest 3, 19, 38, 39, 46, 57, 88, 108, 114, 167, 170, 172, 249, 274, 278, 283
public: accountability 102, 106, 116; alienation 50, 104, 109; attitudes 22, 107, 111, 117, 118; identities 14, 33; legitimation 5, 7, 262; mistrust 107, 115; opinion 3, 11, 15, 22, 33, 70, 112, 125; opposition 107, 118; space 68, 233; trust 54, 63, 66; truths xiv, 12, 266; values 11, 15, 113
public debate 127, 128, 129, 130, 131, 132, 141, 155, 173, 180, 254
public deliberation 7, 36, 41, 145–58, 271
public dialogue 45, 57, 102, 246, 278
public engagement: assemblage 73–6; citizen-led 46; event 14, 17, 82–6, 193; exercises 92, 146–53, 157, 181, 183, 188, 191, 220, 222, 230, 233; formalized 146; invited 108; methods 113, 226, 230, 233; models of 70, 225; practices 41, 56, 152, 218, 221, 225, 233, 267, 278; upstream 8, 127
public engagement with science (PES) 18, 65, 81, 83, 85
public interest 2, 11, 18, 101

public participation professionals 158
public reason 6, 47, 53, 241, 249
publics: construction of 269, 270, 274; emergent 268, 269; national 128, 135
public understanding of science (PUS) 66, 67, 71, 81, 83, 101–3, 163–4

realist understandings of participation 4, 8, 11, 12, 16, 33, 41, 51, 261, 264, 265, 266, 267
REDD+ (Reducing Emissions from Deforestation and Forest Degradation) 133–40
reflexive 18, 45, 77, 100, 105–6, 109, 115, 157, 262, 267–1; engagement 238, 245–6, 252–4, 261–3, 275–7; learning 246, 265, 267, 276–7; participation xv, 21, 267–9, 272, 277, 279, 282; participatory practices 269–72
reflexivity 16, 37, 106, 111, 115, 157, 262, 264, 266–9, 280–1; distributed 277, 281; institutional 102, 103, 106, 116, 267, 268, 275, 280, 281; of organizations 146, 148
regulation 55, 65, 127, 134, 138, 182
relational 47, 113, 267, 272; conditions 115; disposition 268; understanding of participation xvi, 5, 10, 12, 16, 35, 41, 51, 266, 268–9, 273
relationality 81, 105, 114, 117
remaking participation xv–xvi, 17–21, 263–9, 275, 279, 281–2
replication 37, 43, 128, 149, 239
representation 41, 48, 56, 75, 90, 128, 205, 206, 213, 214, 249, 251, 263; correspondence theory of 11; democratic 6, 138, 282; of nature xiv, 205; political 2, 6, 50, 134, 239, 240; of publics xiv, 11, 15, 16, 18, 32, 44, 101, 105, 108, 267, 268, 269, 275, 283; scientific 208, 261
representative democracy 19, 123, 162, 170, 175, 241
representativeness 9, 12, 75, 263, 266, 275
resistance xvi, 2, 6, 33, 34, 38, 43, 46, 66, 88, 245
responsibility 7, 72, 88, 105–7, 130–2, 180, 184–8, 192–4, 202, 214, 262, 268, 277–81
responsible democratic innovations 21, 263, 269, 275–6
responsible innovation 8, 186, 276, 278
Rifkin, Jeremy 182
Rip, Arie 254, 278

Index

risk 71, 83, 87–8, 99, 104, 106, 107, 108, 109, 117, 127, 225
risk assessment 7, 104, 115–16, 118
rolling luggage 18, 82, 84, 86, 87, 88, 89
Rose, Nikolas 42, 43, 46
Royal Society, London xiv, xvii, 78, 165

scale 47, 48, 53, 114, 115, 181, 239
scaling up participation 12, 47, 48, 277, 283
scenarios 72, 78, 88, 168, 183, 186, 247–52
Schaffer, Simon 3–4
science and democracy 2–5, 8, 12–16, 22, 34, 38, 47–8, 51, 55–6, 99, 104, 261, 269, 281, 282
science communication 66, 163–74
Sciencewise Expert Resource Centre (UK) xiv, 57, 255
scientific citizenship 20, 163–4, 169–74
scientific governance 69, 74, 77, 146, 152, 163
scientific institutions 7, 14, 67, 82, 102
scientific knowledge 3, 6, 32, 66, 74, 82, 83, 101, 103, 116, 118, 163, 166, 173, 265, 267, 282
scientism 103–4
scientists 20, 39, 100–8, 111, 115, 167, 183, 185, 188, 202, 212, 226, 230, 276
scientization 6, 158
Scott, James 101, 111
scripts 39, 89, 146, 147, 148, 155, 156
secluded research 238–9, 246, 253
Shapin, Steven 3–4
Shove, Elizabeth 84
social contract for science 22, 265
social innovation 186, 238, 252, 253, 254, 283
social science methods 12, 13, 37, 46, 54, 272
social scientists 18, 77, 84, 105, 144, 171
social studies of participation 42, 56, 282
social technologies 238, 240, 246, 252, 254, 278
sociotechnical imaginaries 189, 191, 194
sound science 7
spaces xv, 6, 14, 53–6, 74, 108–9, 174, 187–8, 194, 273–5, 279–82
spaces of coherence 32, 48, 49, 273
spaces of government 5, 6, 102, 104
spaces of negotiation 14, 17, 32, 47, 49, 52, 55–6
standardization xv, 11, 14, 17, 32, 37, 38, 42, 43, 44, 56, 57, 97, 128, 180, 243, 251, 277, 283

standards 9, 45, 144, 150, 156, 202, 244, 249
state: experiments 124–6, 132, 138–40; power 126, 139–40; the state 31, 55, 66, 104, 107, 114, 123–41, 279
Stengers, Isabelle 84, 85, 86, 212, 213
Stirling, Andy 103, 185, 271
stories 150, 151, 194, 207, 220
Strathern, Marilyn 2, 155
STS (Science and Technology Studies) xiv, 5, 13, 32, 33, 67, 125, 144, 163, 201, 221, 240, 265, 278
STS scholars 74, 103, 125, 145–9, 157, 173, 204, 231, 266, 276
subjects, of participation 13, 34–9, 43, 240–1, 267, 271–2
sub-politics 65
sustainability 218, 225, 228, 230, 234, 237, 254, 258
symmetrical analysis 41, 107, 274
symmetry 38, 268
synchronicity 190, 191
synthetic biology 57, 83, 118, 164, 172, 222
systems of participation 50, 52

Taylor, Charles 6, 162
technocracy 139, 241, 252, 253, 264
technocratic 78, 232
technological determinism 219, 231
technological zone 45, 54, 58
technologies of elicitation 42
technologies of humility 276
technologies of participation 37, 42–7, 49, 52, 53, 72, 145–7, 150, 186, 238, 241, 245–7, 251, 276–8
technology assessment 218–33, 242, 253, 255, 275
technosociality 87, 88, 89
temporalities 178–95, 219, 221, 226, 229, 234
Thorpe, Charles 164
time 178, 179, 181, 182, 183, 184, 187, 189, 193–5; clock time 20, 182–4, 187; multiplicities of 190; time-spaces of participation 13, 41, 42
tools 4, 11, 42, 43, 67, 86, 154, 168, 229, 232, 240, 273, 280
topological spaces of participation 17, 33, 52, 54
trajectorism 182, 184–7, 195
transformation 1, 5, 6, 16, 31, 71, 85, 115, 194, 254, 262, 268, 269

translation 32, 38, 45, 147, 149, 278
transparency 6, 7, 45, 145, 148, 150, 154, 262, 266

uncertainties 16, 21, 46, 68, 76, 77, 106, 116, 130, 190, 205, 208, 212, 263, 270, 271, 276, 277, 283, 284
uninvited participation xvi, 8, 78, 108, 118, 167–70, 173–4
Urry, John 64, 75
users 88, 89, 91, 94, 147, 150, 151, 174

Velcro 86–8
visions xvi, xvii, 36, 69, 145, 146, 147, 152, 154, 156, 158, 179, 185, 191, 222, 228, 232, 274, 277, 282
Von Schomberg, Rene 254

walking tour 220–30
Warren, Mark 169
Watts, Laura 187
Weibel, Peter 38, 266, 271
Welsh, Ian 100, 108, 152, 153
Whitehead, Alfred North 81, 84–5, 89
Wiek, Arnim 220
wild, in the 32, 35, 36, 239, 266
Wilsdon, James 8, 230, 276
Winner, Langdon 144, 221, 232
World Bank 133–7
World Trade Organization (WTO) 107, 148
Wynne, Brian 99, 100, 103, 109, 152–3, 157, 167, 265, 267, 269, 280, 281

Young, Marion Iris 48, 167

eBooks
from Taylor & Francis
Helping you to choose the right eBooks for your Library

Add to your library's digital collection today with Taylor & Francis eBooks. We have over 50,000 eBooks in the Humanities, Social Sciences, Behavioural Sciences, Built Environment and Law, from leading imprints, including Routledge, Focal Press and Psychology Press.

Choose from a range of subject packages or create your own!

Benefits for you
- Free MARC records
- COUNTER-compliant usage statistics
- Flexible purchase and pricing options
- 70% approx of our eBooks are now DRM-free.

Benefits for your user
- Off-site, anytime access via Athens or referring URL
- Print or copy pages or chapters
- Full content search
- Bookmark, highlight and annotate text
- Access to thousands of pages of quality research at the click of a button.

ORDER YOUR FREE INSTITUTIONAL TRIAL TODAY

Free Trials Available

We offer free trials to qualifying academic, corporate and government customers.

eCollections
Choose from 20 different subject eCollections, including:
- Asian Studies
- Economics
- Health Studies
- Law
- Middle East Studies

eFocus
We have 16 cutting-edge interdisciplinary collections, including:
- Development Studies
- The Environment
- Islam
- Korea
- Urban Studies

For more information, pricing enquiries or to order a free trial, please contact your local sales team:

UK/Rest of World: **online.sales@tandf.co.uk**
USA/Canada/Latin America: **e-reference@taylorandfrancis.com**
East/Southeast Asia: **martin.jack@tandf.com.sg**
India: **journalsales@tandfindia.com**

www.tandfebooks.com